Over the last fifty years plant breeders have achieved impressive improvements in yield, quality and disease resistance. These gains suggest that many more modifications might be introduced if appropriate genes can be identified. Current DNA techniques allow the construction of transgenic plants and this important new book reviews the current state of knowledge.

A team of leading researchers provide in-depth reviews at the cutting edge of technology for laboratory techniques for transformation of important soil microorganisms and recalcitrant plants of economic value. The book is divided into three sections: soil microorganisms; cereal crops; and industrially important plants. The most effective methods used to date are compared and their merits and limitations are discussed. Some chapters emphasize case studies and applications. In cases where obstacles remain to be overcome, an overview of progress to date is given.

The book will serve as a general guide and reference tool for those working on transformation in microbiology and plant science.

PLANT AND MICROBIAL BIOTECHNOLOGY RESEARCH SERIES: 3
Series Editor: James Lynch

Transformation of Plants
and Soil Microorganisms

PLANT AND MICROBIAL BIOTECHNOLOGY RESEARCH SERIES
Series Editor: James Lynch

Titles in the series

1. Plant Protein Engineering
   Edited by P. R. Shewry and S. Gutteridge

2. Release of Genetically Engineered and Other Microorganisms
   Edited by J. C. Fry and M. Day

3. Transformation of Plants and Soil Microorganisms
   Edited by K. Wang, A. Herrera-Estrella and M. Van Montagu

Transformation of Plants and Soil Microorganisms

Edited by

Kan Wang
ICI Seeds, Iowa, USA

Alfredo Herrera-Estrella
Centro de Investigación y Estudios Avanzados, Mexico

and

Marc Van Montagu
Universiteit Gent, Belgium

CAMBRIDGE
UNIVERSITY PRESS

PUBLISHED BY THE PRESS SYNDICATE OF THE UNIVERSITY OF CAMBRIDGE
The Pitt Building, Trumpington Street, Cambridge, United Kingdom

CAMBRIDGE UNIVERSITY PRESS
The Edinburgh Building, Cambridge CB2 2RU, UK
40 West 20th Street, New York NY 10011–4211, USA
477 Williamstown Road, Port Melbourne, VIC 3207, Australia
Ruiz de Alarcón 13, 28014 Madrid, Spain
Dock House, The Waterfront, Cape Town 8001, South Africa

http://www.cambridge.org

© Cambridge University Press 1995

First published 1995
First paperback edition 2004

A catalogue record for this book is available from the British Library

Library of Congress cataloguing in publication data
Transformation of plants and soil microorganisms / edited by Kan Wang.
Alfredo Herrera-Estrella, and Marc Van Montagu.
    p.   cm. – (Plant and microbial biotechnology research series ; 3)
Includes bibliographical references and index.
ISBN 0 521 45089 6 hardback
1. Plant genetic transformation.   2. Crops – genetic engineering.
3. Genetic transformation.   4. Soil microbiology.   I. Wang, Kan,
1959– .   II. Herrera-Estrella, Alfredo.   III. Van Montagu, Marc.
IV. Series.
SB123.57.T7   1995
631.5′23 – dc20   94-11609 CIP

ISBN 0 521 45089 6 hardback
ISBN 0 521 54820 9 paperback

And he gave it for his opinion, that whoever could make two ears of corn or two blades of grass to grow upon a spot of ground where only one grew before, would deserve better of mankind, and do more essential service to his country than the whole race of politicians put together.

Jonathan Swift (1667–1745)
Gulliver's Travels (1726), Chapter 7, 'Voyage to Brobdingnag'

This book is sponsored by ICI Seeds,
a Business Unit of ZENECA INC.

Contents

List of Contributors   xi
Series Preface   xiii
Preface   xv
Acknowledgements   xvi
Abbreviations and Terms   xvii

Part I Transformation of Soil Microorganisms

1 *Pseudomonas*   3
 Gareth Warren, Joyce Loper, Dallice Mills and Linda Thomashow
2 **Nocardioform and Coryneform Bacteria**   10
 Jan Desomer and Marc Van Montagu
3 *Agrobacterium, Rhizobium,* **and Other Gram-Negative Soil Bacteria**   23
 Alan G. Atherly
4 **Filamentous Fungi**   34
 Gustavo H. Goldman, Marc Van Montagu and Alfredo Herrera-Estrella

Part II Transformation of Cereal Crops

5 **Rice Transformation: Methods and Applications**   53
 Junko Kyozuka and Ko Shimamoto
6 **Maize**   65
 H. Martin Wilson, W. Paul Bullock, Jim M. Dunwell, J. Ray Ellis, Bronwyn Frame,
 James Register III, and John A. Thompson
7 **Barley, Wheat, Oat and Other Small-Grain Cereal Crops**   81
 Ralf R. Mendel and Teemu H. Teeri

Part III Transformation of Industrially Important Crops

8 **Leguminous Plants**   101
 Jack M. Widholm
9 **Spring and Winter Rapeseed Varieties**   125
 Philippe Guerche and Catherine Primard

10 Sunflower   137
Günther Hahne
11 Forest Trees   150
Ronald R. Sederoff

Index   164

Contributors

Alan G. Atherly
Department of Zoology and Genetics
Room 2216, Molecular Biology Building
Iowa State University
Ames, IA 50011
USA

W. Paul Bullock
ICI Seeds
Research Department
2369 330th Street, Box 500
Slater, IA 50244
USA

Jan Desomer
Laboratoire Central
Solvay, S.A.
Rue de Ransbeek 310
B-1120 Brussels
Belgium

Jim M. Dunwell
ZENECA Seeds
Jealott's Hill Research Station
Bracknell, Berkshire RG12 6EY
England

J. Ray Ellis (deceased)
ZENECA Seeds
Jealott's Hill Research Station
Bracknell, Berkshire RG12 6EY
England

Bronwyn Frame
ICI Seeds
Research Department
2369 330th Street, Box 500
Slater, IA 50244
USA

Gustavo Goldman
Universidade de São Paulo
Faculdade de Ciencias Farmaceuticas de Ribeirão
   Preto
Via do Café S/N
14040-903 Ribeirão Preto, SP
Brazil

Philippe Guerche
Laboratoire de Biologie Cellulaire
Institut National de la Recherche Agronomique
78026 Versailles Cedex
France

Günther Hahne
Institut de Biologie Moléculaire des Plantes
Centre National de la Recherche Scientifique
Université Louis Pasteur
12, rue du Général Zimmer
67084 Strasbourg Cedex
France

Alfredo Herrera-Estrella
Centro de Investigación y Estudios Avanzados
Unidad Irapuato
Departamento de Ingeniería Genética
Km. 9.6 del Libramiento Norte Carretera Irapuato-
   León
Apartado Postal 629
26500 Irapuato
Gto., Mexico

Junko Kyozuka
CSIRO Division of Plant Industry
GPO Box 1600
Canberra ACT 2601
Australia

Joyce Loper
USDA-ARS Horticultural Crops Research
Laboratory
3420 N.W. Orchard Avenue
Corvallis, OR 97330
USA

Ralf R. Mendel
Institute of Botany
Technical University of Braunschweig
Humboldtstr. 1
3300 Braunschweig
Germany

Dallice Mills
Department of Botany and Plant Pathology
Oregon State University
Corvallis, OR 97331
USA

Catherine Primard
Laboratoire de Biologie Cellulaire
Institut National de la Recherche Agronomique
78026 Versailles Cedex
France

James Register III
ICI Seeds
Research Department
2369 330th Street, Box 500
Slater, IA 50244
USA

Ronald R. Sederoff
Department of Forestry, Genetics and Biochemistry
North Carolina State University
Raleigh, NC 27695
USA

Ko Shimamoto
Laboratory of Plant Molecular Genetics
Nara Institute of Science and Technology
8916–5 Takayama
Ikoma
Nara 630–01
Japan

Teemu Teeri
Institute of Biotechnology
University of Helsinki
Karvaamokuja 3
FIN-00380 Helsinki
Finland

Linda Thomashow
USDA-ARS Root Disease and Biological Control
Research Unit
367 Johnson Hall
Washington State University
Pullman, WA 99164
USA

John A. Thompson
ZENECA Seeds
Jealott's Hill Research Station
Bracknell, Berkshire RG12 6EY
England

Marc Van Montagu
Laboratorium voor Genetica
Universiteit Gent
Ledeganckstraat 35
B-9000 Gent
Belgium

Kan Wang
ICI Seeds
Research Department
2369 330th Street, Box 500
Slater, IA 50244
USA

Gareth Warren
DNA Plant Technology Corporation
6701 San Pablo Avenue
Oakland, CA 94608-1239
USA

Jack M. Widholm
Department of Agronomy
University of Illinois at Urbana-Champaign
W-203 Turner Hall
1102 South Goodwin Avenue
Urbana, IL 61801-4798
USA

H. Martin Wilson
ICI Seeds
Research Department
2369 330th Street, Box 500
Slater, IA 50244
USA

Series Preface
Plant and Microbial Biotechnology

The primary concept of this series of books is to produce volumes covering the integration of plant and microbial biology in modern biotechnological science. Illustrations abound, for example the development of plant molecular biology has been heavily dependent on the use of microbial vectors, and the growth of plant cells in culture has drawn largely on microbial fermentation technology. In both of these cases the understanding of microbial processes is now benefiting from the enormous investments made in plant biotechnology. It is interesting to note that many educational institutions are also beginning to see things this way and integrating departments previously separated by artificial boundaries.

Having set the scope of the series, the next objective was to produce books on subjects that had not already been covered in the existing literature and, it was hoped, to set some new trends.

One of the most commonly used techniques to genetically engineer both plants and microorganisms is transformation. However, it seemed to me that, whereas transformation was of course covered in all molecular biology textbooks, a substantive research monograph that would cover this exciting and expanding field was not available.

The Genetics Laboratory at the University of Gent, under the direction of Marc Van Montagu, has been an earlier 'player' and is now a world leader in transformation. Particularly, their involvement in the characterization of the Ti (tumor inducing) plasmid from a soil bacterium (*Agrobacterium tumefaciens*) and its use in the transformation of plants set one of the first scenes for plant molecular biology to emerge. Marc Van Montagu was, therefore, an obvious choice as a volume editor. One of the great strengths of the Gent Laboratory has been its international flavor. As I attempted to delve into the transformation of *Trichoderma*, I found that the Gent team was already there and I was delighted to meet Alfredo Herrera-Estrella, who was active in the field in Gent. He was an obvious choice to join the editorial team on his return to Mexico. Then to complete the team, Marc and Alfredo suggested that Kan Wang, who had completed her Ph.D. in Gent and now leads a crop transformation team in ZENECA/ICI SEEDS in the United States of America, should take the editorial lead.

This editorial team has persuaded an outstanding international group to produce an excellent collection of accounts of transformation in plants and soil microorganisms. It should provide a good stimulus to accelerate the pace of development of agricultural and environmental biotechnology.

Jim Lynch

Preface

Advances in biology continue to be made at a striking and ever increasing rate, especially since the powerful technique of gene manipulation became available. For plant biologists, however, the possibility of engineering plants was not considered until the mid 1970s, when gene transfer mediated by the soil bacterium *Agrobacterium tumefaciens* was discovered. From this moment on, enormous developments have occurred concerning not only practical aspects such as crop improvement but also fundamental aspects relating to the understanding of the biology of plants in their interactions with the environment. One of the most important subjects is perhaps that of plant–microbe interactions.

For either applied or academic research, it is essential to establish efficient transformation systems. Although the basic transformation techniques have been well established for model systems such as *Escherichia coli*, yeast and tobacco, many important species have proven more difficult to manipulate. During the past several years, many research groups took up the challenge of transformation and have made significant progress. To keep abreast with this rapid development, we have invited in this volume several leading groups to share their experiences with prospective and practicing researchers in the microbe/plant field. Together, we have endeavored to give readers details of one particular problem of major importance – transformation of soil microorganisms and recalcitrant crops.

Part I consists of four chapters on the transformation of soil microorganisms. Each chapter describes most, if not all, techniques used for transformation in the laboratory of a specific microbe. In some cases conjugation, a natural DNA transfer process, is widely used for transformation; in other cases either a physical treatment such as electroporation or a physical–chemical treatment such as polyethylene glycol-CaCl$_2$ is used as an effective method. A comparison of these techniques and key points of each procedures are provided.

Many plant tissue culture systems, unlike their animal counterparts, allow for the regeneration of whole organisms. However, a number of plants, especially most of the economically valuable crops and trees cannot be easily manipulated in culture. Part II covers transformation of cereal crops; three chapters deal with transformation of the most important monocotyledonous plants – rice, maize, barley, wheat, etc. Great progress has been made over the past several years in these areas and transformation of these crops is now becoming routine in a number of laboratories. Progress in the transformation of recalcitrant dicotyledonous plants and woody species has also been significant. Four chapters in Part III provide up to date information on some industrially important plants, such as soybean, rapeseed, sunflower, and forest trees. All the chapters review current transformation techniques. The most effective methods are emphasized, technical problems are highlighted and potential applications are discussed.

We hope that this book encourages all of us to keep putting forward our best efforts and that it inspires new people to enter this field of biotechnology. We should not forget that the great impact of gene manipulation in more recent developments in agriculture has yet to be seen by the general public.

K. Wang, A. Herrera-Estrella and M. Van Montagu

Acknowledgements

We thank G. Angenon, A. Caplan, M. De Block, R. Deblaere, W. Dillen, J. Desomer, K. D'Halluin, G. Gheysen, E. Göbel, M. Holsters and D. Inzé for critical reading of the manuscripts, and M. De Cock for typing tables. Special thanks are due to A. Uytterhaegen for her efficiency and patience during the preparation of this book. Finally, we are grateful to R. Harington of Cambridge University Press for his constant support in the editing process.

Abbreviations and Terms

A	haploid genome of cabbage	B5	B5 medium
ABA	abscisic acid	BA	6-benzyladenine
AB^R	antibiotic resistance	*bar*	phosphinothricin acetyl
Ac	maize autonomous transposable element activator		transferase gene from *Streptomyces hygroscopicus*
act	actin gene	BC1	result of the first backcross
*act*1	a rice gene for actin	BCG	bacille Calmette–Guérin
achenes	sunflower kernels	*ben*A3	β-tubulin A3 benomyl resistance
ADH	alcohol dehydrogenase		gene from *Neurospora crassa*
*adh*1	a gene for alcohol dehydrogenase	*bgl*1	β-glucosidase gene from *Trichoderma reesei*
ADP	adenosine diphosphate	*ble*	transposon 5 bleomycin
agroinfection	transfer of viral DNA via *Agrobacterium*		resistance gene
		bml	β-tubulin benomyl resistance
ALS	promoter from the mutated acetolactate synthase gene from *Arabidopsis thaliana*		gene from *Aspergillus nidulans*
		*bml*R	β-tubulin benomyl resistance gene
als	acetolactate synthase gene	*bml*R3	β-tubulin R3 benomyl resistance
*amd*S	acetamidase gene from *Aspergillus nidulans*		gene from *Colletotrichum graminicola*
Ap	ampicillin	BMS	Black Mexican Sweet (maize
aph	aminoglycoside phosphotransferase gene		cultivar)
		bp	base-pair(s)
*arg*B	ornithine carbamoyl transferase gene from *Aspergillus nidulans*	*bz*1	bronze 1 gene
		B73	maize inbred line
aroA	5-enolpyruvylshikimate synthase gene; enzyme is active in the synthesis of aromatic amino acids	C	haploid genome of turnip
		C	capacitance
		cab	chlorophyll *a/b* binding protein gene
ARS	autonomously replicating sequences	CaMV	cauliflower mosaic virus
		CaMV 35S	cauliflower mosaic virus 35S RNA
*att*P	phage attachment site		
AVG	aminoethoxyvinyl glycine	CaMV 19S	cauliflower mosaic virus 19S RNA
A188	maize inbred line		
A188 × B73	F1 hybrid made between inbred lines A188 and B73	CAT	chloramphenicol acetyl transferase

cat	chloramphenicol acetyl transferase gene	*hsp*60	60 kilodalton heat shock protein gene
*cat*1	catalase 1 gene	Hup	hydrogen uptake-related hydrogenase
Cb	carbenicillin		
*cbh*1	cellobiohydrase I gene from *Trichoderma reesei*	Hyg	hygromycin
		IBA	indolebutyric acid
cDNA	complementary DNA	Ignite	herbicide containing phosphinothricin as active ingredient
c.f.u.	colony-forming units		
Cm	chloramphenicol		
cms	cytoplasmic male sterile	*ina*	ice nucleation gene
ColE1	colicin E1	Ina$^+$	ice nucleation active
cos	bacteriophage λ cohesive ends	Ina$^-$	ice nucleation inactive
cosduction	transduction of plasmids which carry cos sequences	*Inc*P, Q, W	incompatibility group P, Q and W
CP	coat protein	INH	isonicotinic acid hydrizide
CPMV	cowpea mosaic virus	*int*	integrase gene
CS	chlorsulfuron	Kan	kanamycin
cv.	cultivar	Kan 15	kanamycin (15 mg/l)
cvs.	cultivars	kb	1000 base-pairs
2,4-D	2,4-dichlorophenoxyacetic acid	Km	kanamycin
DGT	direct gene transfer	Km^R	kanamycin resistance
dhfr	dihydrofolate reductase gene	*lacZ*	*Escherichia coli* gene for β-galactosidase
dicot(s)	dicotyledonous plant(s)		
DNA	deoxyribonucleic acid	LB	Luria–Bertani medium
Ds	non-autonomous dissociation element	*leu*2	β-isopropyl malate dehydrogenase gene from *Saccharomyces cerevisiae*
EDTA	ethylenediaminetetra-acetic acid		
electroduction	electric pulse-mediated transfer of plasmids directly between bacterial strains	*luc*	luciferase gene from firefly
		lysA	*m*-diaminopimelate decarboxylase gene
Em	early methionine gene	MIJ	microinjection
EP	electroporation	*mob*	mobilization genes
EPSP	enolpyruvylshikimate-3-phosphate	*mob*::Tn5	transposon 5 carrying plasmid mobilization functions
EPT	efficient plasmid transformation	monocot(s)	monocotyledonous plant(s)
		mRNA	messenger ribonucleic acid
Fi	fasciation-inducing	MS	Murashige and Skoog medium
G418	synthetic aminoglycoside antibiotic, geneticin	Mt	millions of tons
		Mtx^R	methotrexate resistance
GA$_3$	gibberellic acid	4-MU	4-methylumbelliferone
glaA	glucoamylase gene	*Mu* 1	a maize transposable element
gltA	citrate reductase gene	N6	plant medium
goxA	glucose oxidase gene	NAA	naphthaleneacetic acid
GUS	β-glucuronidase	*niaD*	nitrate reductase gene
gusA	alternative name for *uidA*	*nif*	nitrogen fixation gene
Hfr	high frequency of recombination	*nod*	nodulation gene
Hm^R	hygromycin resistance	NOS	nopaline synthase promoter
H-NMR	high resolution nuclear magnetic resonance	*nos*	nopaline synthase gene
		NPT II	neomycin phosphotransferase II
hph	hygromycin phosphotransferase gene from *Escherichia coli*	*npt*II	transposon 5 neomycin phosphotransferase gene
HPT	hygromycin phosphotransferase	NT1	cell line of *Nicotiana tabacum*

ocs   *Agrobacterium* gene for octopine synthase on a Ti-plasmid

*oli*C^R oligomycin resistance gene from *Aspergillus niger*

ori   replication origin

*ori*T   transfer origin

OSC   osmotically sensitive cell

P-1   incompatibility group

p15A   incompatibility group

par   partitioning function genes

Paro   paromomycin

PAT   phosphinothricin acetyltransferase

pat   phosphinothricin acetyltransferase gene

*pcb*C   isopenicillin synthetase gene from *Cephalosporium acremonium*

PCR   polymerase chain reaction

PCV   packed cell volume

PDA   pisatin-demethylating ability

pda   pisatin demethylase gene from *Nectria haematococca*

PEG   polyethylene glycol

PenG   penicillin G

p.f.u.   plaque-forming units

*pga*II   pre-progalacturonidase gene

Phleo   phleomycin

pin2   proteinase inhibitor II gene from potato

polA   DNA polymerase A

PPT, Basta   phosphinothricin

proj.   microprojectile bombardment

proto.   protoplast direct DNA uptake

*psb*A   soybean atrazine resistance gene

p.s.i.   pounds per square inch

PTR   T-DNA 1′ and 2′ gene promoters from *Agrobacterium tumefaciens*

pyr4   orotidine 5′-monophosphate decarboxylase gene from *Trichoderma reesei* and *Neurospora crassa*

*pyr*F   orotidine 5′-monophosphate decarboxylase gene

*pyr*G   orotidine 5′-monophosphate decarboxylase gene from *Aspergillus*

Q4   transducing bacteriophage Q4

qa-2   catabolic dehydroquinase gene from *Neurospora crassa*

R   resistance

R0   primary transformants

R1   progeny of primary transformants

*rab*16A   a rice abscisic acid responsive gene

RAPD   rapid amplified polymorphic DNA

*rbc*S   ribulose-1,5-bisphosphate carboxylase oxygenase small subunit gene

RFLP   restriction fragment length polymorphism

Ri   root inducing

RIP   repeat-induced point mutation

*rol*C   a pathogenesis-related gene of the TL-DNA of the *Agrobacterium rhizogenes* Ri plasmid

rRNA   ribosomal ribonucleic acid

RSV   rice stripe virus

Rubisco   ribulose-1,5-bisphosphate carboxylase oxygenase

S   spring varieties

sh1   shrunken 1 gene

sh1   a maize sucrose synthase gene

Sm   streptomycin

Sp   spectinomycin

*spo*C1   sporulation-specific gene cluster

*spo*C1C   sporulation-specific C gene

SV40   simian virus 40

*sul*I   enterobacteria sulfonamide resistance gene

τ   time constant

Tc   tetracycline

T-DNA   transferred DNA

Ti   tumor inducing

TL   left region of the T-DNA

TMV   tobacco mosaic virus

Tn5   bacterial transposon 5

Tn5*seq*1   deletion in transposon 5 streptomycin resistance gene

Tn903   transposon 903

TR   right region of the T-DNA

tra   DNA transfer genes

*trp*C   anthranilate synthase gene from *Aspergillus nidulans*

tub2^R benomyl-resistant β-tubulin gene 2 from *Trichoderma viride*

ubi   a maize gene for ubiquitin

*uid*A   *gus*A or β-glucuronidase gene

ura3   orotidine 5′-monophosphate decarboxylase gene

*ura*5	orotate phosphoribosyl transferase gene	WT	wild-type
vir	virulence function	X-Gluc	5-bromo-4-chloro-3-indolylglucuronide, substrate for β-glucuronidase
vir	virulence-related operon on the Ti plasmid of *Agrobacterium tumefaciens*	*y*A	*Aspergillus nidulans* yellow spore mutation
*vir*A, B, C, E, F, G	virulence operons A, B, C, E, F, G from *Agrobacterium tumefaciens*	2 μ circle	*Saccharomyces cerevisiae* minichromosome
*vir*D4	virulence gene D4 from *Agrobacterium tumefaciens*	70	CaMV 35S RNA promoter with duplicated enhancer
W	winter variety		

PART I TRANSFORMATION OF SOIL MICROORGANISMS

1

Pseudomonas

Gareth Warren, Joyce Loper, Dallice Mills and Linda Thomashow

Introduction

Some pseudomonads exist in close association with plants. Best characterized are the phytopathogenic strains that are epiphytic colonists. Some of these are specific in their ability to infect only one plant species or are limited to particular cultivars within a species, as is true for some *Pseudomonas syringae* pathovars. Other pseudomonads colonize the rhizosphere. Best studied among the rhizosphere colonizers are strains that benefit plant health by their antagonism of pathogenic potential colonists (Défago & Haas, 1990). However, not all rhizosphere pseudomonads are of this sort: certain strains are neutral or even deleterious to the host plant.

Genetic manipulation is indispensible for the study of the interactions between plant and pseudomonad (Lindow, Panopoulos & McFarland, 1989). Introducing DNA into the pseudomonad is sometimes problematic: it is the subject of this chapter. The techniques developed for plant-related pseudomonads, based on prior expertise, have been those of conjugal transfer and direct transformation.

Transformation of *Pseudomonas*

Direct transformation of *Pseudomonas syringae* and *P. fluorescens* has been employed successfully (Mukhopadhyay, Mukhopadhyay & Mills, 1990), but optimal protocols may differ considerably from strain to strain. The effects of divalent cations used in inducing competence for transformation are highly strain and species dependent.

Electroporation is now an alternative to treatment with divalent cations for inducing DNA uptake; optimal protocols are probably less strain specific and the technique is potentially more efficient. In the future it may become the preferred way for introducing DNA into *Pseudomonas*. Its practicability has been demonstrated with *Pseudomonas aeruginosa* (Smith & Iglewski, 1989; Smith et al., 1990), *P. chlororaphis* and *P. oxalaticus* (Wirth, Friesenegger & Fiedler, 1989), and *P. putida* (Fiedler & Wirth, 1988; Trevors & Starodub, 1990).

Conjugal transfer to *Pseudomonas*

Conjugation has usually been the method of choice for introducing DNA into pseudomonads. Conjugal receptor function appears to be universal among Gram-negative bacteria, and conjugation techniques can often be adapted with relative ease to a new strain. It is thought that the single-stranded mode by which DNA enters the conjugal recipient reduces the probability of recognition by the recipient's restriction system, in comparison to the double-stranded mode of DNA entry that is usual in the transduction or transformation of Gram-negative bacteria.

The donor in conjugations with *Pseudomonas* is almost always *Escherichia coli*, because the latter is amenable to efficient direct transformation and is a reliable host for the plasmid vectors commonly utilized. When DNA appears in the conjugal recipient cell and causes a heritable change in its phenotype, that recipient is termed a **transconjugant**.

3

Some plasmids are able to replicate in a variety of bacterial hosts, including *E. coli* and *Pseudomonas*. They are said to possess a **broad host range.** When conjugation involves a broad host range plasmid, most transconjugants are recipients in which the plasmid is now established as a new replicon.

Other plasmids have a **narrow host range**; for example, ColE1 and pBR322 are limited to enteric bacteria (e.g. *E. coli*) by their inability to replicate in other types. The ability to be transferred during conjugation, however, does not depend on the ability to replicate in the recipient. Thus, narrow host range plasmids can be utilized to carry DNA into *Pseudomonas*, and since they do not persist as independent replicons, selection for the transferred DNA selects transconjugants in which the transferred DNA has recombined with the recipient genome. In this use, narrow host range plasmids are termed **suicide vectors.** Two narrow host range plasmids (one limited to *E. coli*, the other to *Pseudomonas*) can be joined to form a plasmid able to replicate in both (Van den Eede *et al.*, 1992): vectors constructed on this principle are sometimes known as **shuttle vectors.**

Conjugative ('self-transmissible') plasmids such as RK2 (also known as RP1, RP4, and R68; Palombo *et al.*, 1989) cause their host bacterium to conjugate. They have evolved this ability to permit transfer of their own DNA; however, it should be remembered that conjugation and DNA transfer (mobilization) are separate functions that require distinct sets of genes. The products of the mobilization genes recognize the transfer origin (*ori*T) from which the physical transfer of the plasmid DNA begins. Conjugative plasmids are necessary for conjugation but are inconvenient as vectors, because the conjugation genes occupy relatively large regions of DNA. **Transmissible** ('mobilizable') plasmids do not cause conjugation but can be transferred when conjugation occurs. They are generally used as vectors in combination with a coresident conjugative plasmid. In nature, transmissible plasmids such as RSF1010 (Derbyshire, Hatfull & Willetts, 1987) carry their own distinct *ori*T together with a set of mobilization genes whose products recognize that origin specifically. However, transmissible vectors without mobilization genes can be constructed by making use of the *ori*T from a conjugative plasmid. This limits them to mobilization by just one type of conjugative plasmid (a circumstance not favored by evolution, but adequate for experimental purposes). DNA mobilization and specificity were addressed in an excellent review by Willetts & Wilkins (1984).

Many of the vectors in common use are transmissible, broad host range plasmids that are deleted derivatives of the incompatibility group P (*Inc*P) plasmid RK2: they retain *ori*T and the tetracycline resistance gene of RK2. Plasmid pRK290 is a prototype of such *Inc*P vectors (Ditta *et al.*, 1985). They are most commonly mobilized by pRK2013, a conjugative plasmid that is also derived from RK2 but lacks tetracycline resistance and possesses a narrow host range (Figurski & Helinski, 1979). Other broad host range vectors have been derived from the transmissible *Inc*Q plasmid RSF1010 (Bagdasarian *et al.*, 1981) or from conjugative *Inc*W plasmids (Leemans *et al.*, 1982).

Many suicide vectors, including the pSUP series (Simon, Priefer & Pühler, 1983), employ the *ori*T of RK2 and utilize replicons of the ColE1 or p15A incompatibility groups. These are mobilized by pRK2013 or a 'mobilizer' strain in which part of a conjugative plasmid is integrated into the bacterial chromosome (Simon *et al.*, 1986). Plasmid pBR322, which contains an *ori*T, has been used as a suicide vector by supplying ColE1-type mobilization functions *in trans* (Van Haute *et al.*, 1983).

When a suicide plasmid is transferred, selection requires recombination in order to form the transconjugant. Since such recombination is a rare event, success requires a higher efficiency of conjugal DNA transfer than is necessary for selection of recipients of broad host range plasmids.

Efficiency of transfer varies considerably, depending on the identity of the recipient strain, the ages of the donor and recipient cultures, and how long the cultures are incubated together. In general, stationary-phase recipient cultures have proven more satisfactory than log-phase cultures but log-phase cultures are preferred for donor strains (including the helper if the mating is triparental). In a typical procedure, cells are washed by centrifugation, mixed on a nitrocellulose filter at a donor:recipient ratio of 1:1 or 1:5 and incubated for intervals ranging from 5 h (Hamdan, Weller & Thomashow, 1991) to 48 h (Thomashow & Weller, 1988) on Luria Bertani (LB) agar at 28 °C. Cultures are then washed from the filters into sterile liquid and plated directly on to selective media. Alternatively, cells from individual donor colonies have been transferred with a metal replicator to plates of LB agar seeded with approximately 10^8 recipient cells (and if the cross is triparental, seeded also with 10^7 cells of the helper); after 48 h, transconjugants are replicated to selective media.

Case studies

Pseudomonas syringae pv. *syringae* J900

Strain J900 is an epiphytic pathogen of beans. DNA introduction into J900 has been used to investigate its pathogenic interaction with the plant. A novel strategy for bringing in DNA was needed because *Inc*P vectors are unstable during growth of J900 *in planta*, so that after a period of 1 week, 75% of the cells may not contain the plasmid. Moreover, strains containing *Inc*P plasmids sometimes exhibit retarded growth and attentuated virulence.

The first element of the strategy was the construction of a shuttle vector capable of stable replication in the target strain, utilizing an origin of replication from a narrow host range plasmid indigenous to *P. syringae*. The replication region (*ori*) from the cryptic plasmid pOSU900 (Mukhopadhyay *et al.*, 1990), harbored by J900, was selected because it was expected to function in various pathovars of *P. syringae*. *ori* was recognized, among various fragments of pOSU900 cloned in pBR322, by its ability to replicate in, and thereby confer pBR322-derived antibiotic resistance on, a plasmidless *Pseudomonas* strain.

Direct transformation of *Pseudomonas* was attempted using a $CaCl_2$/heat-shock protocol developed for *E. coli* (Maniatis, Fritsch & Sambrook, 1982). Vectors containing the *Bam*HI fragment on which *ori* was originally cloned yielded approximately 2000 transformants per microgram of DNA. (An *Inc*P vector also gave transformants at approximately the same frequency: D. Mills & Y. Zhao, unpublished data). However, deletions made within the *ori*-containing fragment resulted in plasmids that gave increased transformation frequencies, with maximum levels approaching 10^6 transformants per microgram of DNA. The deletion derivatives had copy numbers in *Pseudomonas* up to six-fold higher than the original plasmid, and the derivatives most efficient in transformation were those with the higher copy

numbers. It is not obvious why this should be so.

The original pBR322-*ori* plasmid was stably maintained in *P. syringae*, with 98% of cells retaining the plasmid after a 2-week period of growth *in planta*. However, the stability of the deletion derivatives in *P. syringae* was extremely variable, and ranged from approximately 1% to 100% during a similar growth period *in planta*. Those derivatives that were stable in *P. syringae* could also transform and be maintained without antibiotic selection in *P. fluorescens* B10 (Kloepper, Schroth & Miller, 1980) and Pf-5 (Howell & Stipanovic, 1979) at frequencies ranging from 10^3 to 10^5 per microgram of DNA.

The inability to recover any transconjugants of *Agrobacterium tumefaciens*, *Rhizobium meliloti* or of a *polA* mutant of *E. coli* indicates that the *ori* cloned from pOSU900 has a narrow host range. This may be useful for limiting unwanted dissemination of the vector during studies in the ecosystem.

Pseudomonas fluorescens strain MS1650

Strain MS1650 is an ice nucleation active (Ina$^+$) epiphytic colonist (Warren *et al.*, 1987). It was considered desirable to introduce a defective *ina* gene into MS1650 for the purpose of generating Ina$^-$ derivatives for experiments in the biological control of frost damage by competitive exclusion.

Experiments with broad host range plasmids showed that MS1650 was a relatively poor recipient for conjugal transfer from *E. coli* donors. Suicide vectors utilizing the transfer origin of pBR322 did not yield detectable numbers of transconjugants. Therefore the defective *ina* gene was cloned into a vector that had been observed to give more efficient transfer than pBR322 into a few other strains of *Pseudomonas* (Warren, Corotto & Green, 1985). The new vector, pLVC18, contained the mobilization system of the broad host range *Inc*Q plasmid RSF1010 but lacked its replicative origin so that suicide mutagenesis was still possible. Conjugative functions were provided by an *Inc*P helper plasmid. It became possible to isolate very small numbers of transconjugants with this system – enough to proceed with the marker exchange strategy.

The availability of an alternative mobilization system with greater efficiency allowed selection of the desired recombination events. One conjugation/mobilization system may not be universally more efficient than all others, so it is necessary to test various systems empirically.

Pseudomonas fluorescens strain Pf-5

Strain Pf-5 is a biological control agent of plant diseases caused by the soil-borne fungi *Pythium ultimum* (Howell & Stipanovic, 1980) and *Rhizoctonia solani* (Howell & Stipanovic, 1979). Strain Pf-5 produces the antibiotics pyoluteorin, pyrrolnitrin, and 2,4-diacetylphloroglucinol. It has been desirable to introduce DNA into Pf-5 to analyze the role of antifungal compounds in the suppression of plant disease by this pseudomonad.

Initial attempts to introduce plasmids via conjugation with *E. coli* donors were unsuccessful. Although *E. coli* is relatively resistant to pyrrolnitrin and 2,4-diacetylphloroglucinol, it is sensitive to pyoluteorin (Takeda, 1958). It could be demonstrated that the *E. coli* donors were killed on the mating plates. Since Pf-5 is viable at 37 °C but does not produce pyoluteorin at this temperature, coculture of Pf-5 with *E. coli* was attempted at 37 °C. The *E. coli* donors survived and adequate numbers of transconjugants were recovered. Using this procedure, the suicide transposon vector pLG221 (Boulnois *et al.*, 1985) was used to generate Tn5 mutants of Pf-5 that were deficient in pyoluteorin production (Kraus & Loper, 1992). Matings at 37 °C have also been used to increase the recovery of transconjugants of *P. aeruginosa* LEC1, a soil isolate that suppresses *Septoria tritici* blotch of wheat (Flaishman *et al.*, 1990).

The crucial step in the development of a working protocol for conjugal transfer of DNA into Pf-5 was the recognition of a potential cause of failure with matings conducted under the usual conditions – in this case, the toxicological incompatibility of donor and recipient during coculture. In other cases, cultural incompatibility has been ameliorated by other means, for example manipulation of the media formulation or selection of a resistant donor strain.

Pseudomonas fluorescens strain HV37a

HV37a is a root-colonizing pseudomonad that inhibits the growth of several fungal pathogens (Gutterson *et al.*, 1988). It was desirable to introduce DNA for marker exchange to manipulate production of antifungal antibiotics, and see how this influenced the interaction of the bacteria with the plant and its fungal pathogens.

Transconjugants arose when broad host range *Inc*P plasmids were mobilized into HV37a by standard conjugation protocols. However, the low frequencies with which they were obtained made it unsurprising that suicide plasmids did not yield detectable numbers of transconjugants. An alternative method of obtaining marker exchange makes use of the instability of some broad host range *Inc*P vectors in HV37a (Jones & Gutterson, 1987). When selection for these vectors is relaxed, plasmidless strains arise in a few generations. If the unstable vector carries a chromosomal gene with a distinguishable allele, then after plasmid loss a detectable proportion of the resulting strains will have substituted the plasmid-borne allele for the original chromosomal allele. In some transconjugant cells, homologous recombination between plasmid and chromosome will occur and cause their cointegration. Subsequent excisive recombination can now leave an originally plasmid-borne allele in the chromosome. Why does an unstable vector favor the recovery of marker exchanges? Presumably, the instability provides an internal selection for the cointegration event.

This strategy illustrates an alternative means of introducing specific alterations into the *Pseudomonas* genome. It is probably more time consuming than the use of suicide plasmids, where the latter is feasible. The plasmids that are unstable in HV37a may not be sufficiently unstable in other pseudomonads, but it is likely that equivalent ones could be constructed. Instability or conditional stability can be generated in most plasmids by making appropriate deletions or by selecting temperature-sensitive mutations.

Pseudomonas fluorescens strain CHA0

Strain CHA0 is a rhizobacterium that suppresses tobacco black root rot caused by *Thielaviopsis basicola*. It excretes several metabolites with antifungal properties, including hydrogen cyanide (Voisard *et al.*, 1989). Introduction of DNA into this strain has been desirable for genetic analysis, in particular for transposon mutagenesis to identify the genes for antifungal traits and test their importance in disease suppression.

Strain CHA0 has proved refractory to conjugal reception of the broad host range *Inc*P plasmid RK2, and also to reception of other plasmids, both broad host range and suicide, when mobilized by RK2 or pRK2013. A large deletion in RK2, including its primase gene, increased the efficiency of recovering CHA0 transconjugants by two to three orders of magnitude (Voisard, Rella & Haas, 1988). The deletion derivative also mobilizes other $oriT_{RK2}$ vectors at concomitantly higher efficiencies, enabling transposon mutagenesis with suicide vectors.

The enhancement effect has not been definitively ascribed to the removal of the RK2 primase gene (and it should be noted that the primase gene is helpful in some types of interspecies conjugation). However, some alternative explanations can be eliminated. The effect cannot be due to restriction, since it extends to the mobilization of other plasmids, and also to transfer of RK2 between strains of CHA0. Likewise, the effect cannot be due to changes in plasmid maintenance, because RK2 replicates stably in CHA0 once introduced. The effect is likely to be exerted on the conjugal donor, either by affecting the structure of mating aggregates or by changing some attribute of the DNA to be transferred (for example, altering the make-up of the leader proteins).

The RK2 deletion that enhances transfer to CHA0 also provides a lesser enhancement for transfer into *P. fluorescens* S9, and would certainly bear testing with other recalcitrant pseudomonads. Other types of plasmid mutation could enhance transfer to other strains. Mutagenesis of the vectors used for transfer, or of the conjugative plasmids used to mobilize them, would appear to be a reasonable strategy for other cases. The discovery of this means of improving transfer efficiency to CHA0 adds to the diversity of empirical approaches brought to bear on the problem.

Pseudomonas fluorescens strain 2-79 and *P. aureofaciens* strain 30-84

Strains 2-79 and 30-84 colonize the roots of wheat and provide protection against take-all, an important soil-borne disease caused by the fungal pathogen *Gaeumannomyces graminis* var. *tritici*. Both strains produce the antibiotic phenazine-1-carboxylic acid, and strain 30-84 also produces two hydroxylated phenazine derivatives. Manipulation of the genes involved in antibiotic biosynthesis has been desirable to evaluate the role of the antibiotics both in disease suppression and as a factor that may contribute to competitiveness in rhizosphere colonization.

The phenazine-producing strains have proven recalcitrant as genetic recipients in conjugation largely because phenazines are toxic to *E. coli*. Viability was reduced in donor populations by up to three orders of magnitude within the first hour after mixing with a culture of *Pseudomonas fluorescens* 2-79 (L. Thomashow, unpublished data). Conjugal efficiency was increased by selecting spontaneous mutants of a donor strain that could grow in the presence of phenazine-1-carboxylic acid (Thomashow & Weller, 1988) and by briefly exposing recipient strains to 10 mM Tris (pH 8) buffer containing 1 mM ethylenediaminetetra-acetic acid (EDTA). Both Tris and EDTA, but especially the latter, are disruptive to the outer membrane of *Pseudomonas*, and prolonged exposure causes substantial cell lysis, as indicated by increased viscosity of the culture. It is not known whether the Tris/EDTA treatment facilitates conjugation by removing a physical barrier associated with the recipient cell envelope or if metabolic processes, including phenazine synthesis, are affected.

It was not possible to select donor strains resistant to the mixture of three phenazines produced by *P. aureofaciens* 30-84. However, supplementation of the media used for conjugation with either 100 μM ferric ammonium citrate or 10 mM *p*-aminobenzoic acid suppressed phenazine biosynthesis and enabled the recovery of transconjugants (Pierson & Thomashow, 1992).

Conclusions

The majority of the cases considered above involve conjugal transfer as the means of introducing DNA into a plant-associated pseudomonad. The efficiency of conjugal transfer depends on many factors: restriction systems in the recipient, anti-*E. coli* toxins produced by the recipient, and conjugal receptivity (which may vary considerably for different types of conjugative plasmid in the donor). It is not currently possible to predict conjugal receptivity; experimentation with various conjugation and mobilization systems is recommended. However, the presence of restriction systems and the production of toxins are amenable to experimental investigation and manipulation; knowledge of these characteristics can be used to improve transconjugant recovery frequencies.

Direct transformation has an appealing simplicity as a means of introducing DNA into *Pseudomonas*. When successful, it is also faster than the use of conjugal transfer. Electroporation is likely to be used increasingly as a means of direct transformation because its applicability is probably quite general.

Acknowledgement

We thank Jennifer Kraus for excellent editorial suggestions.

References

Bagdasarian, M., Lurz, R., Rukert, B., Franklin, F. C. H., Bagdasarian, M. M., Frey, J. & Timmis, K. N. (1981). Specific-purpose plasmid cloning vectors. II. Broad host range, high copy number, RSF1010-derived vectors, and a host-vector system for gene cloning in *Pseudomonas. Gene*, **16**, 237–247.

Boulnois, G. J., Varley, J. M., Sharpe, G. S. & Franklin, F. C. H. (1985). Transposon donor plasmids, based on ColIb-P9, for use in *Pseudomonas putida* and a variety of other Gram negative bacteria. *Molecular and General Genetics*, **200**, 65–67.

Défago, G. & Haas, D. (1990). Pseudomonads as antagonists of soilborne plant pathogens: modes of action and genetic analysis. In *Soil Biochemistry*, Vol. 6, ed. J. M. Bollag & G. Stotzky, pp. 249–291. Marcel Dekker: New York & Basel.

Derbyshire, K. M., Hatfull, G. & Willetts, N. S. (1987). Mobilisation of the non-conjugative plasmid RSF1010: a genetic and DNA sequence analysis of the mobilisation region. *Molecular and General Genetics*, **206**, 161–168.

Ditta, G., Schmidhauser, T., Yakobson, E., Lu, P., Liang, X. W., Finlay, D., Guiney, D. & Helinski, D. R. (1985). Plasmids related to the broad host range vector, pRK290, useful for gene cloning and for monitoring gene expression. *Plasmid*, **13**, 149–153.

Fiedler, S. & Wirth, R. (1988). Transformation of bacteria with plasmid DNA by electroporation. *Analytical Biochemistry*, **170**, 38–44.

Figurski, D. H. & Helinski, D. R. (1979). Replication of an origin-containing derivative of plasmid RK2 dependent on a plasmid function provided *in trans. Proceedings of the National Academy of Sciences, USA*, **76**, 1648–1652.

Flaishman, M., Eyal, Z., Voisard, C. & Haas, D. (1990). Suppression of *Septoria tritici* by phenazine- or siderophore-deficient mutants of *Pseudomonas. Current Microbiology*, **20**, 121–124.

Gutterson, N., Ziegle, J. S., Warren, G. J. & Layton, T. J. (1988). Genetic determinants for catabolite induction of antibiotic biosynthesis in *Pseudomonas fluorescens* HV37a. *Journal of Bacteriology*, **170**, 380–385.

Hamdan, H., Weller, D. M. & Thomashow, L. S. (1991). Relative importance of fluorescent siderophores and other factors in biological control of *Gaeumannomyces graminis* var. *tritici* by *Pseudomonas fluorescens* strains 2-79 and M4-80R. *Applied and Environmental Microbiology*, **57**, 3270–3277.

Howell, C. R. & Stipanovic, R. D. (1979). Control of *Rhizoctonia solani* on cotton seedlings with *Pseudomonas fluorescens* and an antibiotic produced by the bacterium. *Phytopathology*, **69**, 480–482.

Howell, C. R. & Stipanovic, R. D. (1980). Suppression of *Pythium ultimum*-induced damping-off of cotton seedlings by *Pseudomonas fluorescens* and its antibiotic, pyoluteorin. *Phytopathology*, **70**, 712–715.

Jones, J. D. G. & Gutterson, N. (1987). An efficient mobilizable cosmid vector, pRK2013, and its use in a rapid method for marker exchange in *Pseudomonas fluorescens* strain HV37a. *Gene*, **61**, 299–306.

Kloepper, J. W., Schroth, M. N. & Miller, T. D. (1980). Effects of rhizosphere colonization by plant growth-promoting rhizobacteria on potato plant development and yield. *Phytopathology*, **70**, 1078–1082.

Kraus, J. & Loper, J. E. (1992). Lack of evidence for a role of antifungal metabolite production by *Pseudomonas fluorescens* strain Pf-5 in biological control of *Pythium* damping-off of cucumber. *Phytopathology*, **82**, 264–271.

Leemans, J., Langenakens, J., De Greve, H., Deblaere, R., Van Montagu, M. & Schell, J. (1982). Broad host range cloning vectors derived from the W-plasmid Sa. *Gene*, **19**, 361–364.

Lindow, S. E., Panopoulos, N. J. & McFarland, B. L. (1989). Genetic engineering of bacteria from managed and natural habitats. *Science*, **244**, 1300–1307.

Maniatis, T., Fritsch, E. F. & Sambrook, J. (1982). *Molecular Cloning: A Laboratory Manual.* Cold Spring Harbor Laboratory Press, Cold Spring Harbor, NY.

Mukhopadhyay, P., Mukhopadhyay, M. & Mills, D. (1990). Construction of a stable shuttle vector for high-frequency transformation in *Pseudomonas syringae* pv. *syringae. Journal of Bacteriology*, **172**, 477–480.

Palombo, E. A., Yusoff, K., Stanisich, V. A., Krishnapillai, V. & Willetts, N. S. (1989). Cloning and genetic analysis of *tra* cistrons of the Tra2/Tra3 region of plasmid RP1. *Plasmid*, **22**, 59–69.

Pierson, L. S. & Thomashow, L. S. (1992). Cloning

and heterologous expression of the phenazine
biosynthetic locus from *Pseudomonas aureofaciens*
30–84. *Molecular Plant-Microbe Interactions*, 5,
330–339.

Simon, R., O'Connell, M., Labes, M. & Pühler, A.
(1986). Plasmid vectors for the genetic analysis and
manipulation of rhizobia and other gram-negative
bacteria. *Methods in Enzymology*, 118, 640–659.

Simon, R., Priefer, U. & Pühler, A. (1983). A broad
host range mobilization system for in vivo genetic
engineering: transposon mutagenesis in gram-
negative bacteria. *Bio/Technology*, 1, 784–790.

Smith, A. W. & Iglewski, B. H. (1989).
Transformation of *Pseudomonas aeruginosa* by
electroporation. *Nucleic Acids Research*, 17, 105–109.

Smith, M., Jessee, J., Landers, T. & Jordan, J. (1990).
High efficiency bacterial electroporation: 1×10^{10} *E.
coli* transformants/µg. *Focus*, 12, 38–40.

Takeda, R. (1958). Pseudomonas pigments. I.
Pyoluteorin, a new chlorine-containing pigment
produced by *Pseudomonas aeruginosa*. *Hako Kogaku
Zasshi*, 36, 281–290.

Thomashow, L. S. & Weller, D. M. (1988). Role of a
phenazine antibiotic from *Pseudomonas fluorescens* in
biological control of *Gaeumannomyces graminis* var.
tritici. *Journal of Bacteriology*, 170, 3499–3508.

Trevors, J. T. & Starodub, M. E. (1990).
Electroporation of pKK1 silver-resistance plasmid
from *P. stutzeri* AG259 into *P. putida* CYM318.
Current Microbiology, 21, 103–108.

Van den Eede, G., Deblaere, R., Goethals, K., Van
Montagu, M. & Holsters, M. (1992). Broad host
range and promoter selection vectors for bacteria
that interact with plants. *Molecular Plant–Microbe
Interactions*, 5, 228–234.

Van Haute, E., Joos, H., Maes, M., Warren, G., Van
Montagu, M. & Schell, J. (1983). Intergeneric
transfer and exchange recombination of restriction
fragments cloned in pBR322: a novel strategy for the
reversed genetics of the Ti plasmids of *Agrobacterium
tumefaciens*. *EMBO Journal*, 2, 411–417.

Voisard, C., Keel, C., Haas, D. & Défago, G. (1989).
Cyanide production by *Pseudomonas fluorescens* helps
suppress black root rot of tobacco under gnotobiotic
conditions. *EMBO Journal*, 8, 351–358.

Voisard, C., Rella, M. & Haas, D. (1988). Conjugative
transfer of plasmid RP1 to soil isolates of
Pseudomonas fluorescens is facilitated by certain
large RP1 deletions. *FEMS Microbiology Letters*, 55,
9–14.

Warren, G. J., Corotto, L. V. & Green, R. L. (1985).
Conjugal transmission of integrating plasmids into
Pseudomonas syringae and *Pseudomonas fluorescens*. In
*Advances in the Molecular Genetics of the
Bacteria–Plant Interaction*, ed. A. A. Szalay & R. P.
Legocki, pp. 212–214. Cornell University
Publishers: Ithaca, NY.

Warren, G. J., Lindemann, J., Suslow, T. V. & Green,
R. L. (1987). Ice nucleation-deficient bacteria as
frost protection agents. In *Biotechnology in
Agricultural Chemistry*, ACS Symposium Series no.
334, ed. H. M. Le Baron, R. O. Mumma, R. C.
Honeycutt & J. H. Duesing, pp. 215–227. American
Chemical Society.

Willetts, N. & Wilkins, B. (1984). Processing of
plasmid DNA during bacterial conjugation.
Microbiological Reviews, 48, 24–41.

Wirth, R., Friesenegger, A. & Fiedler, S. (1989).
Transformation of various species of Gram-negative
bacteria belonging to 11 different genera by
electroporation. *Molecular and General Genetics*, 216,
175–177.

2

Nocardioform and Coryneform Bacteria

Jan Desomer and Marc Van Montagu

Introduction

Nocardioform and coryneform bacteria are Gram-positive, soil-dwelling microorganisms with a high G + C content genome and include a large number of species of medical, agricultural, or industrial interest. Many mycobacterial diseases such as tuberculosis (caused by *Mycobacterium tuberculosis*) and leprosy (*M. leprae*) have been under investigation for a long time, whereas, more recently, opportunistic infections with mycobacteria (*M. avium*) have been described in patients treated with immunosuppressive drugs and in individuals with acquired immunodeficiency syndrome. *Mycobacterium bovis* bacille Calmette–Guérin (BCG), an avirulent strain of *M. bovis*, has been used as a vaccine for the prevention of tuberculosis and has nonspecific immuno-stimulating properties. Mycobacterial components are frequently added as adjuvants to stimulate the immune response to foreign antigens.

Nocardioform bacteria of the genus *Rhodococcus* are mostly saprophytic soil organisms but include human and animal pathogens (e.g. *Rhodococcus bronchialis* isolated from sputum of patients with pulmonary diseases; *R. equi*, which causes a purulent bronchopneumonia in foals, cattle, swine and occasionally humans) (Goodfellow & Minnikin, 1981; Goodfellow, 1986). *Rhodococcus fascians* is a pathogen on a large range of plants, causing fasciation, a disease characterized by the loss of apical dominance and the development of adventitious shoots. In severe infections, these stunted shoots have the appearance of a 'leafy gall' (Tilford, 1936; Lacey, 1939). In addition, rhodococci exhibit a wide range of notable metabolic activities including biotransformation of steroids (Ferreira *et al.*, 1984), lignin degradation (Rast *et al.*, 1980), degradation of xenobiotic compounds (Appel, Raabe & Lingens, 1984; Hasegawa *et al.*, 1985; Cook & Hütter, 1986), production of biosurfactants and flocculants (Cooper, Akit & Kosaric, 1982; Kurane *et al.*, 1986), and degradation of alkanes (Murai *et al.*, 1980).

The coryneform bacteria are the most important bacterial group involved in commercial production of L-amino acids, which are used as taste enhancers in human food and in supplementing animal feed with essential amino acids. At the same time, this group includes important human and animal pathogens (e.g. *Corynebacterium diphtheriae*, the etiologic agent of diphtheria, and *Corynebacterium xerosis*, a skin pathogen), whereas phytopathogenic species cause devastating wilting diseases of economically important crops such as tomato, potato, maize and ornamentals (Keddie & Jones, 1981).

Despite the medical, agricultural and industrial interest in coryneform and nocardioform bacteria, their study has been hampered by the lack of methods for the introduction of genetic material. During the last decade, cloning systems for these bacteria have been accumulating and are increasingly applied to complement the strain improvement of amino acid producers by 'classical' mutagenesis (Batt *et al.*, 1985), to study pathobiology by creating specific mutants (Crespi *et al.*, 1992), and for the development of live vaccines (Stover *et al.*, 1991).

Introduction of DNA into coryneform and nocardioform bacteria

Introduction of genetic material into bacteria can generally be achieved in any of the following ways.

First, transformation is the most convenient method to introduce exogenously added DNA into the bacteria of interest. Bacterial species that do not have an identified competent phase in their life cycle, or in which such a phase cannot be induced by starvation or addition of metals, can be transformed by polyethylene glycol (PEG)-mediated DNA uptake in protoplasts/spheroplasts or made transiently porous for DNA entry by high voltage electroporation of intact cells. Second, transduction of bacterial DNA by phage particles (including 'cosduction' of plasmids containing the cohesive ends of the phage genome) requires the existence of a generalized transducing phage that can at least recognize the receptor required for attachment to the bacterial surface. This latter prerequisite is not always available or easily identified, but, in cases where generalized transducing phages are described, they constitute a convenient tool for fine mapping of genes by cotransduction. Third, exchange of genetic material between bacterial strains can also be achieved by conjugation of autonomous, conjugative elements or by mobilization of nonautonomously transmittable plasmids. As the recent publications on plasmid exchange between distantly related genera point out, exciting new ways for introduction of DNA into recalcitrant bacteria are appearing as more is learned about the properties of transmissible plasmids. Fourth, protoplast fusion has been applied to a lesser extent to exchange genetic material between coryneform or nocardioform bacteria (Furukawa *et al.*, 1988).

Transformation

Polyethylene glycol-mediated DNA uptake in protoplasts/spheroplasts

PEG-mediated DNA uptake in protoplasts or spheroplasts is a process that proceeds in three distinct steps: (i) partial or complete removal of the cell wall, (ii) PEG-assisted introduction of the naked DNA, and (iii) regeneration of an intact cell wall followed by screening or selecting for transformants. Regeneration seems to be the bottleneck, given the strain specificity of the composition of regeneration media. Transfection by naked phage DNA (where no regeneration is required because viable phage particles can be assembled in protoplasts) has been used successfully to optimize the first two steps of the protocols (Brownell *et al.*, 1982; Ozaki *et al.*, 1984; Yeh, Oreglia & Sicard, 1985; Sánchez *et al.*, 1986), allowing a quicker development of a protocol for the third, apparently limiting, step of protoplast transformation.

The exact conditions to generate protoplasts or spheroplasts competent for transformation/transfection are as diverse as the transformed species, or the research groups working on these problems (Table 2.1), but some underlying common principles can be distilled from the different reports. Most nocardioform and coryneform bacteria are less sensitive to lysozyme action than are other Gram-positive bacteria, although lysozyme has been demonstrated to hydrolyze the glycosidic bonds between *N*-acetylglucosamine and *N*-acetylmuramic acid in these species. This recalcitrance has been attributed to the presence of a characteristic layer of mycolic acids (Yoshihama *et al.*, 1985). Most commonly used ways to increase the sensitivity to lysozyme action are the addition of cell wall synthesis inhibitors such as glycine (0.5% –2.5%), or antibiotics (0.3–0.5 U penicillin G/ml; 0.3 –200 µg ampicillin/ml) to the cultured cells prior to lysozyme treatment (Table 2.1). Penicillin G inhibits cross-linking of the glycans, whereas glycine is incorporated into cell walls instead of D-alanine; however, due to its less efficient incorporation and cross-linking abilities, this often results in growth inhibition (up to 40%; Best & Britz, 1986) and in the formation of a looser cell wall. Excessively high glycine concentrations result in exaggeratedly misshapen cells that are insensitive to lysozyme action (Serwold-Davis, Groman & Rabin, 1987). Periods required for the pretreatment vary from 1 to 14 h (Table 2.1). Pretreatment of cells with isonicotinic acid hydrazide (INH) or cerulin (inhibitors of mycolic acid synthesis (concentrations not mentioned)) did not result in significantly higher spheroplast production (Yoshihama *et al.*, 1985). Less common methods to increase protoplasts/spheroplasts production are the limited addition of surface-active compounds such as sodium dodecylsulfate (SDS) (Martín *et al.*, 1987), the less generally applicable use of lysozyme-sensitive mutants (Smith *et al.*, 1986), or morphologically distinct auxotrophic

Table 2.1. Polyethylene glycol-mediated DNA uptake in protoplasts/spheroplasts in nocardioform and coryneform bacteria

Organism	Buffer (osmotic stabilizer)[a]	Enzyme	Pretreatment	PEG addition	Regeneration (osmotic stabilizer)	Transformation frequency	Reference
Coryneform bacteria							
Corynebacterium glutamicum Corynebacterium herculis Brevibacterium flavum Microbacterium ammoniaphilum	13.5% Na-succinate	1 mg lysozyme/ml, 37°C for 3 h	0.5 U PenG/ml	20% PEG 6000	13.5% Na-succinate	2.4×10^2–3.5×10^5 4.1×10^3 3.2×10^2 2.6×10^2	Katsumata et al. (1984)
Corynebacterium lilium	10% Na-succinate	1 mg lysozyme/ml, 0.5 µg Ap/ml for 20h	4 µg Ap/ml in hypertonic medium at 34°C for 1 h	30% PEG 6000	10% Na-succinate	3.4×10^4	Yeh et al. (1985)
Corynebacterium glutamicum	0.41 M sucrose	1 mg lysozyme/ml, 30°C overnight	0.3 U PenG/ml	20% PEG 6000	0.41 M sucrose	$> 10^{10}$ (transfection)	Ozaki et al. (1984)
Corynebacterium glutamicum Brevibacterium lactofermentum	10% Na-succinate	2 mg lysozyme/ml, 2.5% glycine for 20h	2.5% glycine, 0.3 µg Ap/ml	30% PEG	10% Na-succinate	1.5×10^5	Yeh et al. (1986)
Corynebacterium glutamicum	0.5 M sorbitol	2.5 mg lysozyme/ml for 1.5 h	2% glycine	50% PEG 1000	0.5 M sorbitol	10^4	Yoshihama et al. (1985)
Corynebacterium diphtheria Corynebacterium ulcerans Corynebacterium glutamicum	13.5% Na-succinate	0.4 mg lysozyme/ml for 1 h (static)	2% glycine	20% PEG 8000	13.5% Na-succinate	3.8×10^3 (transfection) 0.2–150 (transformation)	Serwold-Davies et al. (1987)
Brevibacterium lactofermentum	0.41 M sucrose	300 µg lysozyme/ml at 35°C for 4 h	0.3 U PenG/ml	25% PEG 6000	0.25 M sucrose, 0.25 M Na-succinate	10^6/µg	Santamaria et al. (1984, 1985)
Clavibacter michiganense	10.3% sucrose	2 mg lysozyme/ml for 2 h	0.5% glycine	20% PEG 6000	0.5 M sorbitol	3×10^3 (transfection) 2×10 (transformation)	Meletzus & Eichenlaub (1991)
Nocardioform bacteria							
Rhodococcus erythropolis	0.5 M sucrose	2 mg lysozyme/ml, 37°C for 45 min		165–210 µl 20% PEG 4000	0.5 M sucrose	2–9×10^2	Dabbs & Sole (1988)
Rhodococcus erythropolis	7.33% D-mannitol	1 mg lysozyme/ml, 30°C for 10–15 min	3% glycine	20% PEG 1000	7.33% D-mannitol	10^5 (transfection)	Brownell et al. (1982)
Rhodococcus sp (H13-A) Rhodococcus globerulus Rhodococcus equi	10.3% sucrose	10 mg lysozyme/ml, 35°C for 2 h	200 µg Ap/ml for 2 h	25% PEG 8000	10.3% sucrose	6.4×10^5 2.4×10^5 3.4×10^6	Vogt Singer & Finnerty (1988)
Mycobacterium smegmatis	10.3% sucrose	2 mg lysozyme/ml, 37°C for 2 h	1% glycine	25% PEG 1000	0.5 M sucrose	10^3–10^4 (transfection)	Jacobs et al. (1987)

Notes:
PEG, polyethylene glycol; Ap, ampicillin; PenG, penicillin G.
[a] All percentages are w/v.

mutants (particularly isoleucine), that result in facilitated protoplast formation but are not more sensitive to lysozyme action than is the parent strain (Best & Britz, 1986).

Enzymatic treatments with lysozyme (1 mg/ml to 10 mg/ml) are performed in osmotically stabilized buffers (Table 2.1). No uniformity can be found in the required periods for enzymatic treatment that range from very short (10 min) to extremely long (24 h), although the tendency is to reduce exposure times to lysozyme, as prolonged incubation results in reduced regeneration and transformation efficiencies. Most protocols recommend shaking during lysozyme digestion, except for those described by Yoshihama *et al.* (1985) and Serwold-Davis *et al.* (1987).

Similarly, no consensus can be found in the effectiveness of enzyme treatments. Although Yoshihama *et al.* (1985) reported that the lysozyme treatment results only in osmotically sensitive but morphologically unaltered cells, several other groups report that up to 99% of the cells are transformed to round protoplasts as observed by light microscopy. The use of other hydrolytic enzymes (mutanolysin or lysostaphin) in combination with lysozyme did not result in increased spheroplast production (Yoshihama *et al.*, 1985).

Uptake of DNA by the spheroplasts/protoplasts is facilitated by the addition of PEG. Generally, PEG with an average molecular weight of 6000 is used, in concentrations that vary between 20% and 50% (Table 2.1). Coryneform and nocardioform bacteria are less sensitive to the presence of PEG than are other Gram-positive bacteria such as *Bacillus subtilis* (Chang & Cohen, 1979) and dilution with hypertonic media is sufficient to allow regeneration on osmotically stabilized media. These latter also differ greatly in their exact composition. Transformation efficiencies depend on several variables, e.g. the nature of the DNA used (phage versus plasmid), the host organism used to prepare the transforming DNA and the recipient organism. Under optimal conditions the values can vary from as low as less than one transformant/µg DNA to up to 10^6 transformants/µg DNA. Transfection efficiencies are higher than transformation efficiencies of identical organisms, reflecting possibly the regeneration frequency of the cell wall required for transformation.

Transformation efficiencies are linearly correlated with the concentration of transforming DNA, but saturation thresholds have been reported (Santamaría, Gil & Martín, 1985). An additional problem that might arise during PEG-mediated transformation of protoplasts/spheroplasts is the loss of endogenous plasmids during the protoplasting/regeneration process. Evidence for 'curing' following formation and regeneration of protoplasts/spheroplasts was provided for *Staphylococcus aureus* (Novick *et al.*, 1980), *Streptomyces coelicolor* (Hopwood, 1981), and *Rhodococcus fascians* (Desomer, Dhaese & Van Montagu, 1988).

High voltage electroporation of intact cells
Introduction of DNA into bacterial cells via brief high voltage electric discharges has become the preferred method for transformation of recalcitrant species during the last few years, because it is simple, fast and reproducible. The exact mechanism by which electrotransformation occurs, however, remains obscure. When cell membranes are placed in an electric field of a critical strength, a reversible local disorganization and transient breakdown occurs, allowing both molecular influx and efflux (Chassy, Mercenier & Flickinger, 1988). However, when the imposed electric field becomes too high, irreversible damage to the membranes may result in cell death. During 'pore' formation, transforming DNA can diffuse into the bacterial cells. Several protocols for electroporation and transformation efficiencies are summarized in Table 2.2. In general, no pretreatment of the cells is required, except for extensive washes of the harvested cells with the electroporation medium. Nevertheless, Hermans, Boschloo & de Bont (1990) and Haynes & Britz (1990) recommend the addition of 4 to 8 mg of INH, or INH in combination with 2.5% (w/v) of glycine to the growing cell cultures in order to obtain higher transformation efficiencies for *M. aureum* and *C. glutamicum*, respectively. However, Duncan & Shivnan (1989) found no significant improvement by pretreating the cultures with glycine, although the transformation efficiencies obtained by these authors were already higher than those reported by others. These high transformation efficiencies could be correlated with the low conductivity of the electroporation medium they used. Electroporation can be achieved in media as simple as water (Meletzus & Eichenlaub, 1991), supplemented with PEG (Desomer *et al.*, 1990), or glycerol (Duncan & Shivnan, 1989), which both

Table 2.2. *High voltage electroporation of intact bacterial cells*

Organism	Electroporation medium	Electric settings field (kV/cm)	R (Ω)	C (µF)	Time constant (ms)	Transformation efficiency	Remarks	Reference
Nocardioform bacteria								
Mycobacterium aureum	10% sucrose, 7 mM Hepes, pH 7.0, 1 mM MgCl$_2$	12.5	200	25	3.5–4.5	2.34×10^4	Pretreatment with 4 µg INH/ml	Hermans et al. (1990)
Mycobacterium smegmatis	7 mM Na-phosphate, pH 7.2	6.25	—	25	—	5×10^3	Phasmids	Snapper et al. (1988)
Mycobacterium bovis BCG Mycobacterium smegmatis	272 mM sucrose 10 mM Hepes, pH 7.0, 10% glycerol	6.25	—	25	—	10–500	Homologous integration	Husson et al. (1990)
Rhodococcus fascians	30% PEG 1000	12.5	400	25	3.5–4.5	10^5–10^7		Desomer et al. (1990)
Mycobacterium smegmatis (EPT)	10% glycerol	12.5	—	25	—	10^4 10^5	Transfection Shuttle plasmids	Snapper et al. (1990)
Coryneform bacteria								
Clavibacter michiganense	H$_2$O	12.5	—	25	13.5	2×10^3		Meletzus & Eichenlaub (1991)
Corynebacterium glutamicum	10% glycerol	12.5	200	25	4.5–5	1.5×10^3–1×10^7		Meletzus & Eichenlaub (1991)
Corynebacterium calluvae Brevibacter lactofermentum	10% glycerol 10% glycerol					10^2–10^3 10^3–10^5		Dunican & Shivnan (1989)
Brevibacter ammoniagenes	10% glycerol					10^3		Dunican & Shivnan (1989)
Corynebacterium glutamicum	0.5 M sucrose, 1 mM MgCl$_2$	12.5	200	25	4.5–5	5.2×10^5	Pretreatment 2.5% glycine, 4–8 mg INH/ml	Haynes & Britz (1989, 1990)

Notes:
INH, isonicotinic acid hydrizide.

enhance efficiency and allow freezing, in portions of the 'electrocompetent' cells.

The electric parameters are very similar in all cases. An exponentially decreasing pulse with an initial field strength of 12.5 kV/cm is delivered from a 25 μF capacitance. The total resistance (R) is a combination of internal resistance (determined by the conductivity of the electroporation medium, the volume of the cells to be electroporated, etc., all parameters that are less amenable to manipulation) and external resistances, which can be changed simply by switching the resistors of the electroporation apparatus. Both resistances determine the time constant ($\tau = RC$) that defines the time needed for the field strength to reach about one-third of the initial field strength, and for practical reasons can be regarded as the effective duration of the pulse. With the exception of *Clavibacter michiganense*, most time constants reported for efficient electroporation of nocardioform and coryneform bacteria are in the range 3.5–5 ms (Table 2.2). A lower time constant (< 3 ms) results, at least for *R. fascians*, in a drastic decrease in transformation efficiency (J. Desomer, unpublished data). Considering the time needed for transformants to form colonies, τ has proven to be a good time-saving evaluation parameter.

Other critical values affect transformation efficiency by electroporation. A uniform physiological state of the cells at the time of harvest is generally ensured by collecting cells from late exponential or early stationary phase cultures ($A_{600} = 0.5$ to 0.6). The density of electroporated samples fluctuates around 10^9 colony forming units (c.f.u.)/ml.

A saturation at higher DNA concentrations, which could be explained by the existence of an electrocompetent subpopulation, was observed for several coryneform bacteria (Dunican & Shivnan, 1989) and for *R. fascians* (Desomer *et al.*, 1990). For the latter species, electrotransformation of a 160 kDa plasmid, albeit with low transformation efficiency, has been reported (Desomer *et al.*, 1990), indicating that plasmid size need not limit electrotransformation. Interestingly, *M. smegmatis* mutants could be obtained that exhibited an enhanced (four to five orders of magnitude) plasmid transformation efficiency (EPT) phenotype, but were unaffected in efficiency of transfection by phage D29 DNA, or in the efficiency of transformation by integrating vectors (Snapper *et al.*, 1990). As it was shown that no

inactivation of a restriction/modification system was involved, it was assumed that EPT mutants are affected in some aspects of plasmid replication and maintenance (Snapper *et al.*, 1990).

Conjugation

Self-transmissible extrachromosomal elements have been described both in nocardioform and coryneform bacteria. In *Clavibacter flaccumfaciens*, a 69 kb circular plasmid that encodes arsenite, arsenate and antimony resistance genes (pDG101) could be conjugated between several subspecies with transfer frequencies ranging from 5×10^{-9} to 1.3×10^{-4}. Crosses between *C. flaccumfaciens* and *C. michiganense* were unsuccessful (Hendrick, Haskins & Vidaver, 1984). Also in the nocardioform bacteria, conjugative endogenous metal resistance plasmids were found. In *Rhodococcus fascians*, transfer of cadmium resistance plasmids (120–160 kb) was observed between wild-type and cured strains with frequencies of 10^{-4} to 10^{-2} (Desomer *et al.*, 1988). Lithoautotrophic, thallium-resistant, *Nocardia opaca* strains can transfer both the thallium resistance (frequency 10^{-1} to 10^{-2} per donor cell) and the ability to grow lithoautotrophically (frequency 10^{-4} to 10^{-5}) to cured *N. opaca* or *Rhodococcus erythropolis* strains (Sensfuss, Reh & Schlegel, 1986). Whereas the correlation between transfer of thallium resistance and conjugal transfer of 110 kb plasmids was obvious (Sensfuss *et al.*, 1986), it has only recently been discovered that the genetic information for hydrogen-autotrophic growth was located on large, conjugative, linear extrachromosomal elements (270–280 kb) (Kalkus, Reh & Schlegel, 1990). Similarly sized linear plasmids were discovered in the phytopathogen *Rhodococcus fascians*. In this case too, conjugal transfer (10^{-4}/acceptor) could be detected after appropriate tagging of these plasmids, but here the linear plasmids encode essential fasciation-inducing genes (Crespi *et al.*, 1992). No thorough genetic analysis of the conjugation mechanism in these species has been published.

Although this intraspecies conjugation of self-transmissible extrachromosomal elements has proven useful in the allocation of plasmid-residing genes, the technique has been of limited use in genetically engineering this group of bacteria. The recent description of interspecies mobilization of appropriate shuttle plasmids from *Escherichia coli* to both coryneform and nocardioform bacteria, therefore, opens up exciting new possibilities (Lazraq *et al.*, 1990; Schäfer *et al.*, 1990; Gormley & Davies, 1991; Mazodier & Davies, 1991).

In the first applications of this system, shuttle vectors were constructed that contained an origin of replication for *Mycobacterium* species (Lazraq *et al.*, 1990) or coryneform bacteria (Schäfer *et al.*, 1990), an origin of replication for *E. coli*, an *Inc*P-type origin of transfer and a kanamycin resistance gene known to be expressed in both *E. coli* and coryneform bacteria or mycobacteria. *Escherichia coli* helper strains that provided *Inc*P transfer function *in trans* could mobilize the resulting plasmids to *M. smegmatis* (pMY10: transfer frequency between 2.2×10^{-5} and 1.2×10^{-7}; Lazraq *et al.*, 1990) or different coryneform bacteria (transfer frequencies ranging from 10^{-7} to 10^{-2}, depending on the species used as recipient; Schäfer *et al.*, 1990). Interestingly, transfer frequencies to *C. glutamicum* were high enough to allow rescue of nonreplicating vectors by homologous recombination (Schwarzer & Pühler, 1991). Using the *Inc*Q plasmid pRSF1010 or derivatives known to replicate in both *E. coli* and mycobacteria (Hermans *et al.*, 1991), Gormley & Davies (1991) were able to establish plasmid transfer by mobilization from an *E. coli* helper strain to *M. smegmatis* (frequency 10^{-2} to 10^{-3}). Due to its relative independence of host-encoded functions, pRSF1010 can replicate stably in mycobacteria and overcome the necessity for multiple replicons to suit the different hosts. This property could turn pRSF1010 into a powerful tool for genetic manipulation of these industrially and medically important bacteria.

Transduction

Generalized transduction has not frequently been employed for the introduction of genetic material in nocardioform or coryneform bacteria. In 1976, Momose, Miyashiro & Oba reported isolating a bacteriophage that mediated generalized transduction in *Brevibacterium flavum* at a frequency of 10^{-6}/plaque-forming unit (p.f.u.). More recently, a generalized transducing bacteriophage for *Rhodococcus erythropolis* has been described (Q4; Dabbs, 1987). Under optimal conditions, transduction to prototrophy of auxotrophic markers or transduction of antibiotic resistance markers was found at a frequency of 10^{-8}/p.f.u.

More sophisticated are the phasmids for use in mycobacteria that consist of *E. coli* plasmid and the (pseudo)temperate phages of mycobacteria. These plasmids can be propagated in *E. coli* as plasmids, transfected to fast-growing mycobacteria, and packaged into phage particles that can then efficiently infect slow-growing mycobacteria (e.g. BCG; Jacobs, Tuckman & Bloom, 1987; Snapper *et al.*, 1990).

A cosmid-type vector (containing the cohesive ends of *Brevibacterium* phage f1a) for coryneform bacteria has been described that was packaged *in vivo* upon infection of the harboring bacteria with f1a and transduced with a frequency of 10^{-4}/p.f.u. (Miwa *et al.*, 1985).

Fate of transforming DNA

Plasmids and marker genes

Like other microorganisms, coryneform and nocardioform bacteria contain a multitude of plasmids. An early survey of phytopathogenic *Corynebacterium* species resulted in the detection of plasmids, ranging from 35 to 78 kb (Gross, Vidaver & Keralis, 1979), some of which are conjugative (Hendrick *et al.*, 1984). Because of their larger sizes, they had little value for development as cloning vectors. Therefore, several research groups sought and found smaller plasmids in coryneforms, most of which were cryptic (for reviews, see Sandoval *et al.*, 1985; Martín *et al.*, 1987). For the nocardioform genera *Rhodococcus* and *Mycobacterium*, fewer plasmids have been described. Large native plasmids that code for resistance to chloramphenicol and heavy metals (cadmium, antimony, and arsenate) have been described in *R. fascians* (Desomer *et al.*, 1988) and an unidentified *Rhodococcus* isolate (Dabbs & Sole, 1988). The former plasmids are also conjugative (Desomer *et al.*, 1988). In addition, large, linear plasmids (>200 kb) conferring the ability for hydrogen-autotrophic growth (Kalkus *et al.*, 1990) or phytopathogenicity (Crespi *et al.*, 1992) were detected in rhodococci. An intermediate-sized plasmid (80 kb) is probably correlated with the production of a virulence-associated protein in *Rhodococcus equi* (Tkachuk-Saad & Prescott, 1991), whereas smaller cryptic plasmids (2.6–19.5 kb) were found in *Rhodococcus* species (Vogt Singer & Finnerty, 1988; Hashimoto *et al.*, 1992). In mycobacteria, the best characterized, but cryptic, plasmid is pAL5000 (Labidi *et al.*, 1984; Rauzier, Moniz-Pereira & Gicquel-Sanzey, 1988; Ranes *et al.*, 1990).

Cloning vectors have been developed based upon these small cryptic plasmids (Hashimoto *et al.*, 1992) or by deletion of the larger plasmids (Desomer *et al.*, 1990). As marker genes, either heterologous genes from *E. coli* (aminoglycoside-phosphotransferase (*aph*) or chloramphenicol acetyltransferase (*cat*); Martín *et al.*, 1987) or *Streptomyces* species frequently been used (thiostrepton resistance; Vogt Singer & Finnerty, 1988) as well as endogenous marker genes (spectinomycin/streptomycin resistance (Katsumata *et al.*, 1984); chloramphenicol resistance (Desomer *et al.*, 1990, 1992); arsenate resistance (Dabbs, Gowan & Andersen, 1990)). Examples of such cloning vectors for corynebacteria, rhodococci, and mycobacteria are listed in Table 2.3.

Integration of nonreplicative vectors

Vectors that cannot replicate in coryneform or nocardioform bacteria, but can be selected for, due to the presence of a marker gene, can be rescued upon introduction into these bacteria by integration with an existing replicon. This integration can occur via either homologous or illegitimate recombination.

Integration by homologous recombination
Integration of transformed DNA by homologous recombination has been studied in coryneform and nocardioform bacteria, mainly for applied reasons (stable integration, removal of undesired genes) and is confined to the analysis of recombination between vector-borne DNA fragments and their chromosomal homologous counterparts. In mycobacteria, introduction of a pUC19-based vector, unable to replicate in *M. smegmatis*, and containing the *aph* gene of Tn*903* inserted into the *pyr*F gene of *M. smegmatis*, resulted in 10–500 transformants/µg of electroporated plasmid DNA. Southern analysis revealed that 60% of these were integrated by single homologous recombination through the *pyr*F DNA sequences flanking the *aph* gene. The remaining 40% of the transformants had undergone a double homologous recombination, resulting in the replacement of the intact chromosomal *pyr*F gene copy by the interrupted *pyr*F gene, and consequently resulting in uracil auxotrophy. Looping out of the vector sequences

by a second homologous recombination event in isolates where the plasmid had already been integrated by a single homologous recombination occurred at frequencies lower than 10^{-3} (Husson, James & Young, 1990).

Similar observations were obtained for integration by homologous recombination of nonreplicative vectors containing an interrupted *lys*A gene in *Corynebacterium glutamicum*. Since the used vectors contain the *mob* functions of RP4 and, therefore, are mobilizable in conjugations between *E. coli* S17-1 and *C. glutamicum*, single homologous recombinants were obtained with higher efficiencies (5×10^2 to 5×10^3 integrants/mating). Only 2% of these isolates had undergone a double recombination event, resulting in lysine auxotrophy. Loss of vector sequences by a second recombination event in mutant strains with an integrated vector was obtained only after 200 generations, at low frequency (Schwarzer & Pühler, 1991).

Reyes *et al.* (1991) constructed plasmids from which a repliconless cartridge, called 'integron', can be isolated (it contains a selectable marker and the *glt*A region of *C. melassecola*). These plasmids can be obtained from the coryneform host to be transformed, and thus provide a source of host-modified 'integron' DNA, that in turn can be used for ligation to the gene of interest and transformation of the host. In this way, stable integrants of corynebacteria were obtained without restriction modification barriers (Reyes *et al.*, 1991).

Integration after electrotransformation of non-replicating vectors by single recombination through homologous vector-borne DNA sequences and chromosomal counterparts has also been reported for rhodococci, with an efficiency of 10^2/µg of electroporated plasmid DNA (Desomer *et al.*, 1990). Single- and double-homologous recombination between a DNA fragment interrupted by the phleomycin resistance (Phleo^R) gene, and the wild-type counterpart located on a conjugative plasmid was forced by conjugation to a plasmidless acceptor strain. Cm^{S}Phleo^R isolates (double recombinants) were obtained with a frequency of only 5×10^{-3} of that of the Phleo^R mutants (M. Crespi & J. Desomer, unpublished data).

Integration by illegitimate recombination

In several nocardioform bacteria, integration of nonreplicative plasmid DNA in the genome by

Table 2.3. *Examples of cloning vectors*

	Size (kb)	AB[R] markers[a]	Comments	Reference
Coryneform bacteria				
pULRS6	6.1	Hyg Kan		Martín *et al.* (1990)
pULRS8	5.8	Cat Kan		
pULMJ51	11.3	Cat	Contains promoterless kanamycin gene for shotgun cloning of promoters	Cadenas, Martín & Gil (1991)
Mycobacterium				
pMGV261	4.5	Kan	Expression vector (*hsp*60) promoter	Stover *et al.* (1991)
pMGV361	4.5	Kan	Integrative expression vector	Stover *et al.* (1991)
Rhodococcus				
pRK4	5.3	Kan	Based on cryptic pRC3 plasmid	Hashimoto *et al.* (1992)
pRF30	13.2	Cm	Deletion derivative of pRF2 (160 kb)	Desomer *et al.* (1990)

Notes:
[a] AB[R], antibiotic resistance; Kan, kanamycin; Cat, chloramphenicol (acetyltransferase); Cm, chloramphenicol; Hyg, hygromycin.

illegitimate recombination has been reported. In *Mycobacterium tuberculosis* and BCG, electroporation of a pBR322-derived vector (containing the *aph* gene of Tn*5seq*1 as selectable marker) that could not replicate in these hosts still resulted in transformants (at a frequency 10^4 to 10^5 times lower than when autonomously replicating DNA was used), provided that DNA was linearized prior to electroporation. Independent transformants revealed different Southern hybridization patterns, indicating random integrating events (Kalpana, Bloom & Jacobs, 1991).

Similar observations were made in *R. fascians* upon introduction of replication-deficient pUC13- or pUC18-based vectors containing a selectable marker gene for this species. However, no linearization prior to electrotransformation was required to obtain frequencies of integration as high as those in *Mycobacterium*. The presence of intact ColE1 replicon and ampicillin resistance gene on the inserted sequences allowed straightforward cloning of the interrupted target DNA sequences. Furthermore, integration seemed to occur via specific positions on the plasmids, which were determined by sequence analysis to overlap the restriction enzyme *Nar*I site in pUC13 or pUC18. No homologies could be found among the sequences of different target sites nor between the recombining target and plasmid sequences,

indicating that recombination was illegitimate and essentially random along the genome. This latter property had already been suggested by the multitude of different phenotypes obtained (auxotrophies, pigmentation mutants, phytopathogenicity mutants). Although the mechanism by which integration of the plasmids occurs in *R. fascians* is unknown, available evidence seems to point in the direction of a linear recombination intermediate, generated from the incoming circular plasmid (Desomer, Crespi & Van Montagu, 1991). While transposition of movable genetic elements has been described in mycobacteria (Leskiw *et al.*, 1990; Martin *et al.*, 1990; McAdam *et al.*, 1990), none of these mutagenesis systems achieves the efficiencies reported above.

Examples of application

Pathobiology of *Rhodococcus fascians* infection in plants

Rhodococcus fascians is a nocardioform phytopathogen that infects a large range of plants (Tilford, 1936; Lacey, 1939) and causes the development of numerous adventitious shoots. Although in earlier publications (Murai *et al.*, 1980; Murai, 1981) phytopathogenic properties had been associated with the presence of large covalently closed circular cadmium resistance-encoding plasmids (a thesis contradicted by Lawson, Gantotti & Starr (1982)), recently, large linear extrachromosomal elements unique to virulent strains have been described (Crespi *et al.*, 1992). Transformation technology (Desomer *et al.*, 1990) has allowed insertion of a marker gene into the linear plasmid. By conjugation of these linear fasciation-inducing (Fi) plasmids from virulent to avirulent cured strains, it was demonstrated unequivocally that they carried essential fasciation-inducing genes (Crespi *et al.*, 1992). The random insertion of nonreplicating plasmids along the Fi plasmid has identified at least three loci, involved in fasciation induction, one encoding an isopentenyltransferase, the enzyme that catalyzes the first step in cytokinin biosynthesis.

The use of BCG for recombinant vaccines

The transformation procedures developed for mycobacteria have allowed the construction of recombinant 'bacille Calmette–Guérin' (BCG), expressing antigens of different pathogens (Stover *et al.*, 1991). BCG has a long history of use to immunize against tuberculosis and exhibits some unique features (oral administration, single inoculum requirement, innocuous persistence *in vivo*) that could allow development of a inexpensive multivaccine. Therefore, an integrative expression vector was developed based on *hsp*60 expression signals and the phage attachment site (*att*P) and integrase (*int*) gene of a temperate mycobacteriophage. Integrated vectors are stably maintained even without antibiotic selection (Lee *et al.*, 1991) and *hsp*60-driven expression of several antigens from a variety of pathogens was obtained (Stover *et al.*, 1991). In mice, inoculated by low titers of BCG, long-lasting humoral and cellular immune responses were elicited and persistent BCGs were recovered from host spleens. Heat-killed BCGs were not effective for immunization, supporting the view that the persisting immune responses with BCG are due to growth and persistence of the vaccine (Stover *et al.*, 1991).

Concluding remarks

Although significant progress in the development of molecular-genetic tools for coryneform and nocardioform bacteria has been achieved in the last decade, several improvements are still required for effective genetic engineering. A first drawback is the lack of incompatibility studies among the different replicons, making simultaneous introduction and stable inheritance of more than one plasmid (e.g. in genetic repressor/operator studies) unpredictable. Also, very little is known about stability of the used replicons without selective pressure. Available information is scant on the structure of expression signals in these bacteria, let alone regulation signals, and virtually nothing is known about secretion signals. Studies of this kind could increase the 'tool box' for successful manipulation of the specific properties of these bacterial genera.

References

Appel, M., Raabe, T. & Lingens, F. (1984). Degradation of *o*-toluidine by *Rhodococcus rhodochrous*. *FEMS Microbiology Letters*, **24**, 123–126.

Batt, C. A., Follettie, M. T., Shin, H. K., Yeh, P. & Sinskey, A. J. (1985). Genetic engineering of coryneform bacteria. *Trends in Biotechnology*, **3**, 305–310.

Best, G. R. & Britz, M. L. (1986). Facilitated protoplasting in certain auxotrophic mutants of *Corynebacterium glutamicum*. *Applied Microbiology and Technology*, **23**, 288–293.

Brownell, G. H., Saba, J. A., Denniston, K. & Enquist, L. W. (1982). The development of a *Rhodococcus*-actinophage gene cloning system. *Developments in Industrial Microbiology*, **23**, 287–298.

Cadenas, R. F., Martín, J. F. & Gil, J. A. (1991). Construction and characterization of promoter-probe vectors for Corynebacteria using the kanamycin-resistance reporter gene. *Gene*, **98**, 117–121.

Chang, A. C. & Cohen, S. (1979). High frequency transformation of *Bacillus subtilis* protoplasts by plasmid DNA. *Molecular and General Genetics*, **168**, 111–115.

Chassy, B. M., Mercenier, A. & Flickinger, J. (1988). Transformation of bacteria by electroporation. *Trends in Biotechnology*, **6**, 303–309.

Cook, A. M. & Hütter, R. (1986). Ring dechlorination of deethylsimazine by hydrolases from *Rhodococcus corallinus*. *FEMS Microbiology Letters*, **34**, 335–338.

Cooper, D. G., Akit, J. & Kosaric, N. (1982). Surface activity of the cells and extracellular lipids in *Corynebacterium fascians* CF15. *Journal of Fermentation Technology*, **60**, 19–24.

Crespi, M., Messens, E., Caplan, A. B., Van

Montagu, M. & Desomer, J. (1992). Fasciation induction by the phytopathogen *Rhodococcus fascians* depends upon a linear plasmid encoding a cytokinin synthase gene. *EMBO Journal*, 11, 795–804.

Dabbs, E. R. (1987). A generalised transducing bacteriophage for *Rhodococcus erythropolis*. *Molecular and General Genetics*, 206, 116–120.

Dabbs, E. R. & Sole, G. J. (1988). Plasmid-borne resistance to arsenate, arsenite, cadmium, and chloramphenicol in a *Rhodococcus* species. *Molecular and General Genetics*, 211, 148–154.

Dabbs, E. R., Gowan, B. & Andersen, S. J. (1990). Nocardioform arsenic resistance plasmids and construction of *Rhodococcus* cloning vectors. *Plasmid*, 23, 242–247.

Desomer, J., Crespi, M. & Van Montagu, M. (1991). Illegitimate integration of non-replicative vectors in the genome of *Rhodococcus fascians* upon electrotransformation as an insertional mutagenesis system. *Molecular Microbiology*, 5, 2115–2124.

Desomer, J., Dhaese, P. & Van Montagu, M. (1988). Conjugative transfer of cadmium resistance plasmids in *Rhodococcus fascians* strains. *Journal of Bacteriology*, 170, 2401–2405.

Desomer, J., Dhaese, P. & Van Montagu, M. (1990). Transformation of *Rhodococcus fascians* by high-voltage electroporation and development of *R. fascians* cloning vectors. *Applied and Environmental Microbiology*, 56, 2818–2825.

Desomer, J., Vereecke, D., Crespi, M. & Van Montagu, M. (1992). The plasmid-encoded chloramphenicol resistance protein of *Rhodococcus fascians* is homologous to the transmembrane tetracycline efflux proteins. *Molecular Microbiology*, 6, 2377–2385.

Dunican, L. K. & Shivnan, E. (1989). High frequency transformation of whole cells of amino acid producing coryneform bacteria using high voltage electroporation. *Bio/Technology*, 7, 1067–1070.

Ferreira, N. P., Robson, P. M., Bull, J. R. & van der Walt, W. H. (1984). The microbial production of 3α-H-4α-(3′-propionic acid)-5α-hydroxy-7αβ-methylhexahydro-indan-1-one-δ-lactone from cholesterol. *Biotechnology Letters*, 6, 517–522.

Furukawa, S., Azuma, T., Nakanishi, T. & Sugimoto, M. (1988). Breeding an L-isoleucine producer by protoplast fusion of *Corynebacterium glutamicum*. *Applied Microbiology and Biotechnology*, 29, 248–252.

Goodfellow, M. (1986). Genus *Rhodococcus* Zopf 1891, 28[al]. In *Bergey's Manual® of Systematic Bacteriology*, Vol. 2, ed. P. H. A. Sneath, N. S. Mair, M. E. Sharpe & J. G. Holt, pp. 1472–1481. Williams and Wilkins, Baltimore.

Goodfellow, M. & Minnikin, D. E. (1981). The genera *Nocardia* and *Rhodococcus*. In *The Prokaryotes*, Vol. II, ed. M. P. Starr, H. Stolp, H. G. Trüper,

A. Balows & H. G. Schlegel, pp. 2016–2027. Springer-Verlag, Berlin.

Gormley, E. P. & Davies, J. (1991). Transfer of plasmid RSF1010 by conjugation from *Escherichia coli* to *Streptomyces lividans* and *Mycobacterium smegmatis*. *Journal of Bacteriology*, 173, 6705–6708.

Gross, D. C., Vidaver, A. K. & Keralis, M. B. (1979). Indigenous plasmids from phytopathogenic *Corynebacterium* species. *Journal of General Microbiology*, 115, 479–489.

Hasegawa, S., Vandercook, C. E., Choi, G. Y., Herman, Z. & Ou, P. (1985). Limonoid debittering of citrus juice sera by immobilized cells of *Corynebacterium fascians*. *Journal of Food Science*, 50, 330–332.

Hashimoto, Y., Nishiyama, M., Yu, F., Watanabe, I., Horinouchi, S. & Beppu, T. (1992). Development of a host-vector system in a *Rhodococcus* strain and its use for expression of the cloned nitrile hydratase gene cluster. *Journal of General Microbiology*, 138, 1003–1010.

Haynes, J. A. & Britz, M. L. (1989). Electrotransformation of *Brevibacterium lactofermentatum* and *Corynebacterium glutamicum*: growth in Tween 80 increases transformation frequencies. *FEMS Microbiology Letters*, 61, 329–334.

Haynes, J. A. & Britz, M. L. (1990). The effect of growth conditions of *Corynebacterium glutamicum* on the transformation frequency obtained by electroporation. *Journal of General Microbiology*, 136, 255–263.

Hendrick, C. A., Haskins, W. P. & Vidaver, A. K. (1984). Conjugative plasmid in *Corynebacterium flaccumfaciens* subsp. *oortii* that confers resistance to arsenite, arsenate, and antimony(III). *Applied and Environmental Microbiology*, 48, 56–60.

Hermans, J., Boschloo, J. G. & de Bont, J. A. M. (1990). Transformation of *Mycobacterium aurum* by electroporation: the use of glycine, lysozyme and isonicotinic acid hydrazide in enhancing transformation efficiency. *FEMS Microbiology Letters*, 72, 221–224.

Hermans, J., Martin, C., Huijberts, G. N. M., Goosen, T. & de Bont, J. A. M. (1991). Transformation of *Mycobacterium aurum* and *Mycobacterium smegmatis* with the broad host-range Gram-negative cosmid vector pJRD215. *Molecular Microbiology*, 5, 1561–1566.

Hopwood, D. A. (1981). Genetic studies with bacterial protoplasts. *Annual Review of Microbiology*, 35, 237–272.

Husson, R. N., James, B. E. & Young, R. A. (1990). Gene replacement and expression of foreign DNA in mycobacteria. *Journal of Bacteriology*, 172, 519–524.

Jacobs, W. R. Jr, Tuckman, M. & Bloom, B. R. (1987). Introduction of foreign DNA into mycobacteria using a shuttle plasmid. *Nature (London)*, **327**, 532–535.

Kalkus, J., Reh, M. & Schlegel, H. G. (1990). Hydrogen autotrophy of *Nocardia opaca* strains is encoded by linear megaplasmids. *Journal of General Microbiology*, **136**, 1145–1151.

Kalpana, G. V., Bloom, B. R. & Jacobs, W. R. Jr (1991). Insertional mutagenesis and illegitimate recombination in mycobacteria. *Proceedings of the National Academy of Sciences, USA*, **88**, 5433–5437.

Katsumata, R., Ozaki, A., Oka, T. & Furuya, A. (1984). Protoplast transformation of glutamate-producing bacteria with plasmid DNA. *Journal of Bacteriology*, **159**, 306–311.

Keddie, R. M. & Jones, D. (1981). Saprophytic, aerobic coryneform bacteria. In *The Prokaryotes*, Vol. II, ed. M. P. Starr, H. Stolp, H. G. Trüper, A. Balows & H. G. Schlegel, pp. 1838–1878. Springer-Verlag, Berlin.

Kurane, R., Toeda, K., Takeda, K. & Suzuki, T. (1986). Culture conditions for production of microbial flocculant by *Rhodococcus erythropolis*. *Agricultural and Biological Chemistry*, **50**, 2309–2313.

Labidi, A., Dauget, C., Goh, K. S. & David, H. L. (1984). Plasmid profiles of *Mycobacterium fortuitum* complex isolates. *Current Microbiology*, **11**, 235–240.

Lacey, M. S. (1939). Studies on a bacterium associated with leafy galls, fasciations and 'cauliflower' disease of various plants. Part III. Further isolation, inoculation experiments and cultural studies. *Annals of Applied Biology*, **26**, 262–278.

Lawson, E. N., Gantotti, B. V. & Starr, M. P. (1982). A 78-megadalton plasmid occurs in avirulent strains as well as virulent strains of *Corynebacterium fascians*. *Current Microbiology*, **7**, 327–332.

Lazraq, R., Clavel-Sérès, S., David, H. L. & Roulland-Dussoix, D. (1990). Conjugative transfer of a shuttle plasmid from *Escherichia coli* to *Mycobacterium smegmatis*. *FEMS Microbiology Letters*, **69**, 135–138.

Lee, M. H., Pascopella, L., Jacobs, W. R. Jr & Hatfull, G. F. (1991). Site-specific integration of mycobacteriophage L5: integration-proficient vectors for *Mycobacterium smegmatis*, *Mycobacterium tuberculosis*, and bacille Calmette-Guérin. *Proceedings of the National Academy of Sciences, USA*, **88**, 3111–3115.

Leskiw, B. K., Mevarech, M., Barritt, L. S., Jensen, S. E., Henderson, D. J., Hopwood, D. A., Bruton, C. J. & Chater, K. F. (1990). Discovery of an insertion sequence, IS*116*, from *Streptomyces clavuligerus* and its relatedness to other transposable elements from actinomycetes. *Journal of General Microbiology*, **136**, 1251–1258.

Martin, C., Timm, J., Rauzier, J., Gomez-Lus, R., Davies, J. & Gicquel, B. (1990). Transposition of an antibiotic resistance element in mycobacteria. *Nature (London)*, **345**, 739–746.

Martín, J. F., Cadenas, R. F., Malumbres, M., Mateos, L. M., Guerrero, C. & Gil, J. A. (1990). Construction and utilization of promoter-probe and expression vectors in corynebacteria. Characterization of corynebacterial promoters. In *Proceedings of the 6th International Symposium on Genetics of Industrial Microorganisms*, ed. H. Heslot, J. Davis, J. Florent, L. Bobichon, G. Durand & L. Penasse, pp. 283–292. Société Française de Microbiologie, Paris.

Martín, J. F., Santamaría, R., Sandoval, H., del Real, G., Mateos, L. M., Gil, J. A. & Aguilar, A. (1987). Cloning systems in amino acid-producing corynebacteria. *Bio/Technology*, **5**, 137–146.

Mazodier, P. & Davies, J. (1991). Gene transfer between distantly related bacteria. *Annual Review of Genetics*, **25**, 147–171.

McAdam, R. A., Hermans, P. W. M., van Soolingen, D., Zainuddin, Z. F., Catty, D., van Embden, J. D. A. & Dale, J. W. (1990). Characterization of a *Mycobacterium tuberculosis* insertion sequence belonging to the IS*3* family. *Molecular Microbiology*, **4**, 1607–1613.

Meletzus, D. & Eichenlaub, R. (1991). Transformation of the phytopathogenic bacterium *Clavibacter michiganense* subsp. *michiganense* by electroporation and development of a cloning vector. *Journal of Bacteriology*, **173**, 184–190.

Miwa, K., Matsui, K., Terabe, M., Ito, K., Ishida, M., Takagi, H., Nakamori, S. & Sano, K. (1985). Construction of novel shuttle vectors and a cosmid vector for the glutamic acid-producing bacteria *Brevibacterium lactofermentum* and *Corynebacterium glutamicum*. *Gene*, **39**, 281–286.

Momose, H., Miyashiro, S. & Oba, M. (1976). On the transducing phages in glutamic acid-producing bacteria. *Journal of General Applied Microbiology*, **22**, 119–129.

Murai, N. (1981). Cytokinin biosynthesis and its relationship to the presence of plasmids in strains of *Corynebacterium fascians*. In *Metabolism and Molecular Activities of Cytokinins*, ed. J. Guern & C. Péaud-Lenoël, pp. 17–26. Springer-Verlag, Berlin.

Murai, N., Skoog, F., Doyle, M. E. & Hanson, R. S. (1980). Relationship between cytokinin production, presence of plasmids, and fasciation caused by strains of *Corynebacterium fascians*. *Proceedings of the National Academy of Sciences, USA*, **77**, 619–623.

Novick, R., Sanchez-Rivas, C., Gruss, A. & Edelman, I. (1980). Involvement of the cell envelope in plasmid maintenance: plasmid curing during the regeneration of protoplasts. *Plasmid*, **3**, 348–358.

Ozaki, A., Katsumata, R., Oka, T. & Furuya, A. (1984). Transfection of *Corynebacterium glutamicum* with temperate phage φCG1. *Agricultural and Biological Chemistry*, **48**, 2597–2601.

Ranes, M. G., Rauzier, J., Lagranderie, M.,

Gheorghiu, M. & Gicquel, B. (1990). Functional analysis of pAL5000, a plasmid from *Mycobacterium fortuitum*: construction of a 'mini' mycobacterium–*Escherichia coli* shuttle vector. *Journal of Bacteriology*, **172**, 2793–2797.

Rast, H. G., Engelhardt, G., Ziegler, W. & Wallnöfer, P. R. (1980). Bacterial degradation of model compounds for lignin and chlorophenol derived lignin bound residues. *FEMS Microbiology Letters*, **8**, 259–263.

Rauzier, J., Moniz-Pereira, J. & Gicquel-Sanzey, B. (1988). Complete nucleotide sequence of pAL5000, a plasmid from *Mycobacterium fortuitum*. *Gene*, **71**, 315–321.

Reyes, O., Guyonvarch, A., Bonamy, C., Salti, V., David, F. & Leblon, G. (1991). 'Integron'-bearing vectors: a method suitable for stable chromosomal integration in highly restrictive Corynebacteria. *Gene*, **107**, 61–68.

Sánchez, F., Peñalva, M. A., Patiño, C. & Rubio, V. (1986). An efficient method for the introduction of viral DNA into *Brevibacterium lactofermentum* protoplasts. *Journal of General Microbiology*, **132**, 1767–1770.

Sandoval, H., del Real, G., Mateos, L. M., Aguilar, A. & Martín, J. F. (1985). Screening of plasmids in non-pathogenic corynebacteria. *FEMS Microbiology Letters*, **27**, 93–98.

Santamaría, R., Gil, J. A., Mesas, J. M. & Martín, J. F. (1984). Characterization of an endogenous plasmid and development of cloning vectors and a transformation system in *Brevibacterium lactofermentum*. *Journal of General Microbiology*, **130**, 2237–2246.

Santamaría, R. I., Gil, J. A. & Martín, J. F. (1985). High-frequency transformation of *Brevibacterium lactofermentum* protoplasts by plasmid DNA. *Journal of Bacteriology*, **162**, 463–467.

Schäfer, A., Kalinowski, J., Simon, R., Seep-Feldhaus, A.-H. & Pühler, A. (1990). High-frequency conjugal plasmid transfer from Gram-negative *Escherichia coli* to various Gram-positive coryneform bacteria. *Journal of Bacteriology*, **172**, 1663–1666.

Schwarzer, A. & Pühler, A. (1991). Manipulation of *Corynebacterium glutamicum* by gene disruption and replacement. *Bio/Technology*, **9**, 84–87.

Sensfuss, C., Reh, M. & Schlegel, H. G. (1986). No correlation exists between the conjugative transfer of the autotrophic character and that of plasmids in *Nocardia opaca* strains. *Journal of General Microbiology*, **132**, 997–1007.

Serwold-Davis, T. M., Groman, N. & Rabin, M. (1987). Transformation of *Corynebacterium diphtheriae*, *Corynebacterium ulcerans*, *Corynebacterium glutamicum*, and *Escherichia coli* with the *C. diphtheriae* plasmid pNG2. *Proceedings of the National Academy of Sciences, USA*, **84**, 4964–4968.

Smith, M. D., Flickinger, J. L., Lineberger, D. W. & Schmidt, B. (1986). Protoplast transformation in coryneform bacteria and introduction of an α-amylase gene from *Bacillus amyloliquefaciens* into *Brevibacterium lactofermentum*. *Applied and Environmental Microbiology*, **51**, 634–639.

Snapper, S. B., Lugosi, L., Jekkel, A., Melton, R. E., Kieser, T., Bloom, B. R. & Jacobs, W. R. Jr (1988). Lysogeny and transformation in mycobacteria: stable expression of foreign genes. *Proceedings of the National Academy of Sciences, USA*, **85**, 6987–6991.

Snapper, S. B., Melton, R. E., Mustafa, S., Kieser, T. & Jacobs, W. R. Jr (1990). Isolation and characterization of efficient plasmid transformation mutants of *Mycobacterium smegmatis*. *Molecular Microbiology*, **4**, 1911–1919.

Stover, C. K., de la Cruz, V. F., Fuerst, T. R., Burlein, J. E., Benson, L. A., Bennett, L. T., Bansal, G. P., Young, J. F., Lee, M. H., Hatfull, G. F., Snapper, S. B., Barletta, R. G., Jacobs, W. R. Jr & Bloom, B. R. (1991). New use of BCG for recombinant vaccines. *Nature (London)*, **351**, 456–460.

Tilford, P. E. (1936). Fasciation of sweet peas caused by *Phytomonas fascians* n.sp. *Journal of Agricultural Research*, **53**, 383–394.

Tkachuk-Saad, O. & Prescott, J. (1991). *Rhodococcus equi* plasmids: isolation and partial characterization. *Journal of Clinical Microbiology*, **29**, 2696–2700.

Vogt Singer, M. E. & Finnerty, W. R. (1988). Construction of an *Escherichia coli* – *Rhodococcus* shuttle vector and plasmid transformation in *Rhodococcus* spp. *Journal of Bacteriology*, **170**, 638–645.

Yeh, P., Oreglia, J., Prévots, F. & Sicard, A. M. (1986). A shuttle vector system for *Brevibacterium lactofermentum*. *Gene*, **47**, 301–306.

Yeh, P., Oreglia, J. & Sicard, M. (1985). Transfection of *Corynebacterium lilium* protoplasts. *Journal of General Microbiology*, **131**, 3179–3183.

Yoshihama, M., Higashiro, K., Roa, E., Akedo, M., Shanabruch, W. G., Folletie, M., Walker, G. C. & Sinskey, A. J. (1985). Cloning vector system for *Corynebacterium glutamicum*. *Journal of Bacteriology*, **162**, 591–597.

3

Agrobacterium, Rhizobium, and Other Gram-negative Soil Bacteria

Alan G. Atherly

Introduction

The introduction of DNA into bacterial cells is essential for rapid genetic analysis, i.e., mapping, complementation analysis, and as a tool for introduction of plasmids that contain cloned fragments. Several different approaches are used with Gram-negative soil bacteria, including transformation (natural and electroporation assisted), transduction and conjugation. Protoplast fusion has not been successful in Gram-negative bacteria. Of all the procedures for DNA introduction, transformation is the most frequently used owing to the advent of gene cloning and the need to introduce plasmids (with cloned inserts) into bacteria. In this respect, introduction of plasmids is important for complementation analysis, generation of clone banks, integration of DNA sequences into the genome by recombination, and introduction of a desired gene into a cell on a stably maintained plasmid. Only a limited number of bacterial species permit natural transformation (e.g. *Haemophilus influenzae, Streptococcus pneumonia, Bacillus subtilis* and *Acinetobacter calcoaceticus*), but several bacteria have become amenable to transformation after treatment with special chemicals and/or heat cycles (*Escherichia coli, Rhizobium, Agrobacterium,* and *Bacillus cereus*). More recently, a wide variety of bacteria have become transformable by the application of high electric currents, which causes membrane permeability, allowing uptake of large DNA molecules. This process, termed electroporation, has had a dramatic impact on the ability to insert any desired DNA sequence into any bacterial species.

The genera *Rhizobium, Bradyrhizobium* and *Agrobacterium* are economically important soil bacteria; *Rhizobium* and *Bradyrhizobium* fix nitrogen in symbiosis with the family Leguminosae, and *Agrobacterium*, besides causing plant diseases, is used as a vector system for introduction of foreign DNA into dicotyledonous plants. *Rhizobium* and *Agrobacterium* are Gram-negative soil bacteria, possess considerable DNA sequence similarity (Gibbins & Gregory, 1972), and have similar growth characteristics and requirements. The slow-growing genus *Bradyrhizobium* is more distantly related (Elkan, 1981). Other important Gram-negative bacteria considered in this review are *Azospirillum, Erwinia* and *Xanthomonas*. Bacteria of the genus *Azospirillum* are diazotrophs associated with the roots of grasses and cause increased crop yields, and the genera *Erwinia* and *Xanthomonas* cause many different plant diseases.

Vectors and selectable markers

A considerable number of DNA cloning vectors have been developed over the years; however, only a limited number are useful in *Agrobacterium, Rhizobium* and other Gram-negative soil bacteria. Three groups of broad host range vectors have been developed and have been used in a large number of Gram-negative bacteria including *Klebsiella, Serratia, Pseudomonas, Acinetobacter, Xanthomonas, Erwinia, Azorhizobium,* as well as *Bradyrhizobium, Rhizobium* and *Agrobacterium* species. The first of these vectors (pRK290) was developed by Ditta *et al.* (1980) from RK2, a large (56 kb) P-1 incompatibility group plasmid. pRK290 is a relatively large cloning vector (20 kb) with a tetracycline resistance marker and lacks the *tra* genes for mobilization. pRK290 contains a single recognition site for restriction enzymes *Eco*RI and *Bgl*II that is suitable for cloning. This vector can be easily transformed into Gram-negative bacteria using electroporation or conjugated from *Escherichia coli* in conjunction with the helper plasmid pRK2013, which contains the RK2 transfer genes cloned onto a ColE1 replicon (Figurski & Helinski, 1979). A cosmid of pRK290 was created by inserting a *cos* sequence into the unique *Bgl*II site (pLAFR1, pVK100; Friedman *et al.*, 1982; Knauf & Nester, 1982). These plasmids/ cosmids have been widely used to support replication of recombinant molecules in *Agrobacterium* and *Rhizobium*. Other *Inc*P plasmids, R772 and pTJS75, have also been used but have technical limitations similar to those of pRK290, i.e. few unique restriction sites, and scarcity of antibiotic resistance markers. pTJS75 (Schmidhauser & Helinski, 1985) has a reduced size, but large inserts tend to be unstable and rearrange (R. K. Prakash & A. G. Atherly, unpublished data). Another large *Inc*P plasmid, pVS1, was isolated from *Pseudomonas aeruginosa* and is capable of replication in a wide variety of Gram-negative bacteria, but not in *E. coli* (Itoh *et al.*, 1984). However, when an 8 kb segment of pVS1 containing the replication, stability and mobilization regions was ligated into pBR325 the resulting plasmid, pGV910, was capable of replication in all tested Gram-negative bacteria including *E. coli* and simultaneously acquired compatability with other *Inc*P plasmids such as pRK290 (Van den Eede *et al.*, 1992). Thus, pGV910 can be used in combination with other P-type plasmids to form a novel binary vector system.

A second group of broad host range cloning vectors has been developed from the smaller 8.9 kb *Inc*Q/P4 plasmid RSF1010 (Guerry, van Embden & Falkow, 1974; Bagdasarian *et al.*, 1981). Chloramphenicol/streptomycin resistant derivatives of RSF1010 have been prepared (pJRD215, 10.2 kb; Davison *et al.*, 1987) and have considerable advantage over pRK290 as a wide range of unique cloning sites is available (*Eco*RI, *Sst*I, *Hind*III, *Xma*I, *Xho*I, *Sal*I, *Bam*HI, and *Cla*I) and the size of the vectors is greatly reduced, allowing larger inserts for transformation.

A third group of cloning vectors comes from the *Inc*W plasmid pSa, which contains a small (1.5 kb) well-characterized region of DNA that supports replication in a wide range of bacterial hosts. Tait *et al.* (1983), Leemans *et al.* (1982), Shaw *et al.* (1983) and Close, Zaitlin & Kado (1984) developed a set of relatively low copy number plasmids, ranging from 5.8 to 15 kb, which have wide application in Gram-negative soil bacteria. These pSa vectors have been modified by increasing the number of convenient cloning sites and phenotypic markers. In addition to the endogenous spectino-mycin and streptomycin resistance genes, newly constructed plasmids include kanamycin (pGV1106, pSa151 and pSa4), sulfonamide (pGV1113), tetracycline (pGV1122) and chloramphenicol (pGV1124, pSa4 and pSa152) resistance genes. pSa-derived vectors have been shown to replicate in *E. coli*, *Klebsiella*, *Serratia*, *Erwinia*, *Rhizobium*, *Agrobacterium*, *Pseudomonas* and *Alcaligenes* species (Tait *et al.*, 1983).

Other potential sources of cloning vectors for soil bacteria are the replication origins of stable endogenous plasmids, with the addition of selectable markers and cloning sites. Mozo, Cabrera & Ruiz-Argueso (1990) cloned the origin of DNA replication (about 5 kb) from a small cryptic plasmid from *Rhizobium* species *(Hedysarum)* and found it to be more stable than RK2 derivatives in a variety of *Rhizobium* and *Agrobacterium* strains, in the absence of selective pressure. In addition *Agrobacterium* plasmid replication regions have been cloned (Gallie *et al.*, 1984; Gallie, Hagiya & Kado, 1985a; Nishiguchi, Takanami & Oda, 1987; Tabata, Hooykaas & Oda, 1989) and some cloning vectors have been constructed using these replicator regions (Gallie, Novak & Kado, 1985b; Tabata *et al.*, 1989). Tabata *et al.* (1989) inserted the 6.8 kb replicator region of the Ti plasmid pTiB6S3 into a ColE1 replicon and found it to be stably maintained in *Agrobacterium*. Likewise, Gallie *et al.* (1985b) constructed a very stable low copy number plasmid/cosmid from the replicator region of pTAR of *A. tumefaciens* (pUCD1001 and pUCD2001). Some useful cloning vectors are summarized in Table 3.1.

Methods for introduction of DNA into bacteria

Transformation

Early reports on the transformation of *Agrobacterium* and *Rhizobium* occurred in the 1960s and have been extensively reviewed

Table 3.1. *Cloning vectors for genetic analysis of Gram-negative soil bacteria*

Plasmids	Size (kb)	Selectable markers and other characteristics	Reference
pRK290	20	Tc, RK2 *ori*, *Inc*P	Ditta *et al.* (1980)
pVK102	23	Tc, Km, *cos*, RK2 *ori*, *Inc*P	Knauf & Nester (1982)
pLAFR1	21.6	Tc, *cos*, RK2 *ori*, *Inc*P	Friedman *et al.* (1982)
pTJS75	7.5	Tc, RK2 *ori*, *Inc*P	Schmidhauser & Helinski (1985)
pJRD215	10.2	Km, Sp, *cos*, RFS1010 *ori*, *Inc*Q	Davison *et al.* (1987)
pUCD2	13	Km, Sp, Cb, Tc, ColE1 *ori*, pSa *ori*, *Inc*W	Close *et al.* (1984)
pUCD5	13	pUCD with *cos*	Close *et al.* (1984)
pUCS2001	10.4	Km, Tc, Ap, *cos*, ColE1 *ori*, pTAR *ori*	Gallie *et al.* (1985*b*)
pC22	17.5	Km, Cb, Sp, *cos*, pTAR *ori*, binary vector for *Agrobacterium*	Simoens *et al.* (1986)

Notes:
Tc, tetracycline; Sp, spectinomycin; Km, kanamycin; Cb, carbenicillin; Sm, streptomycin; *ori*, origin of replication; Ap, ampicillin; *cos*, phage λ cohesive ends.

(Schwinghamer, 1977; Kondorosi & Johnston, 1981). Many of the data from these early reports were not reproducible and gave widely varying results with respect to DNA concentration and frequency of transformants.

Gram-negative bacteria can take up and stably establish exogenous DNA. Uptake is dependent upon a transitory state of competence for transformation, which is generally related to both the conditions of growth and the circumstances under which the cells and DNA are combined. The presence of divalent cations often plays an essential role. *Escherichia coli* was the first nonnaturally competent Gram-negative bacteria where conditions were found that allowed uptake of DNA (Mandel & Higa, 1970) and Ca^{2+} is an absolute requirement. Subsequently, Holsters *et al.* (1978) developed a procedure for transformation of *Agrobacterium* that has been modified (Ebert, Ha & An, 1987) to improve the transformation frequency. This procedure has also been adapted for transformation of *Rhizobium meliloti* (Selvaraj & Iyer, 1981; Courtois, Courtois & Guillaume, 1988) and its close relatives (Bullerjahn & Benzinger, 1982), but not *Bradyrhizobium* species. Transformation frequencies for both *Rhizobium* and *Agrobacterium* are as high as 10^{-5} with large plasmids such as pRK290 and both require growth in special medium, a temperature shock at 0 °C and 37 °C, followed by a recovery period of 3–4 h at 28 °C, also in a special medium. Competent agrobacteria can be stored at −70 °C for use at a later time (Holgen & Willmitzer, 1988).

Transformation of *B. japonicum* poses a special problem, and high frequencies of transformation have been obtained only by electroporation (see Electroporation, below). Berry & Atherly (1981) reported the transformation of *B. japonicum* strains with RP1 after spheroplast formation and treatment with polyethylene glycol; however, the frequencies of transformation were very low (10^{-7}).

The soil bacterium *Acinetobacter calcoaceticus* (strain BD4), in contrast, has been shown to acquire high natural competence during normal growth (Juni & Janik, 1969) and is transformed by several plasmids (Singer, Van Tuijl & Finnerty, 1986). In fact, *A. calcoaceticus* is seemingly competent in natural environments as it is able to take up DNA in groundwater in the presence of divalent cations (Lorenz, Reipschläger & Wackernagel, 1992).

Electroporation

In 1973 it was discovered that high current pulses in a narrow range of intensity and duration result in a transient and reversible increase in plasma membrane permeability (Zimmerman, Schultz & Pilwat, 1973; Zimmerman, 1983). After a microsecond pulse of electric current the lifespan of the field-induced pores is strongly dependent on the temperature, very likely due to the fluidity of the proteins and lipids in the membrane. As a consequence, time, intensity of the current, and the temperature are very tightly regulated for each cell type. In practice, a highly concentrated suspension of cells in a nonconductive medium is exposed to a high voltage source for a few microseconds by the discharge of a capacitor through the system, the time of exposure to the field being determined by the time required for capacitor discharge. The cell membrane acts as an electrical insulator that is unable to pass current. The high voltage surge results in the formation of pores that are large enough to allow macromolecules such as double-stranded DNA molecules to pass into the cell. The closure of the pores is a natural decay process that is delayed by holding the cells at a low temperature, usually 0 °C.

Electroporation is simple and rapid and was first applied to many different eukaryotic cell types, some of which were refractory to transformation. These included many mammalian, fungal and plant cell types (Potter, Weir & Leder, 1984; Fromm, Taylor & Walbot, 1985; Shigekawa & Dower, 1988). Only recently has electroporation been applied to both Gram-positive and Gram-negative bacteria and has generally met with uniform success (Chassy & Flickinger, 1987; Dower, Miller & Ragsdale, 1988; Luchansky, Muriana & Klaenhammer, 1988). Correspondingly, the following Gram-negative soil bacteria are easily transformed using electroporation: *Rhizobium, Bradyrhizobium, Agrobacterium, Erwinia, Pseudomonas, Xanthomonas* and *Azospirillum* (Broek, van Gool & Vanderleyden, 1989; Jun & Forde, 1989; Mattanovich *et al.*, 1989; Wirth, Friesenegger & Fiedler, 1989; Guerinot, Morisseau & Kapatch, 1990; Hatterman & Stacey, 1990; Nagel *et al.*, 1990; White & Gonzolas, 1991). Electroporation-assisted transformation of DNA directly into soil bacteria has many advantages over conjugation from *E. coli*. For example, with *Agrobacterium* triparental matings from *E. coli* are routinely used to insert the binary Ti plasmids for eventual infection of plant cells, and this sometimes leads to plasmid rearrangements (An *et al.*, 1988) as well as necessitating that the recipient *Agrobacterium* strain possess an antibiotic resistance marker. Electroporation-assisted transformation also allows the introduction of DNA directly into *B. japonicum* strains, where plasmid introduction was previously restricted to conjugation (Guerinot *et al.*, 1990; Hatterman & Stacey, 1990). *Azospirillum* can now be transformed by means of electroporation, where previously no method was available for introducing DNA (White & Gonzolas, 1991).

The efficiency of electroporation-assisted transformation was initially low (Wirth *et al.*, 1989) but with growth media changes and optimization of electroporation conditions and equipment, the efficiency has increased to dramatic levels (10^6–10^8/µg DNA; Nagel *et al.*, 1990). Some factors that affect the efficiency of electroporation-assisted transformation include excretion of nucleases that degrade the added DNA, effective restriction endonuclease systems for digesting the unmodified DNA (Wirth *et al.*, 1989), and the presence of large quantities of extracellular polysaccharide (Regué *et al.*, 1992).

Electroporation can also assist in the transfer of plasmids directly between bacterial strains, without the necessity of purifying the DNA from the donor strain. This has been dubbed 'electroduction' (Pfau & Youderian, 1990). The strain possessing the donor plasmid is mixed with the recipient strain and the mixture is subjected to a typical electrical discharge for Gram-negative bacteria (15–20 kV/cm, 200 Ω, 25 µF in a 0.10 cm wide cuvette). The frequency of plasmid transfer is as high as 4×10^{-5} per recipient cell (Pfau & Youderian, 1990). This phenomenon is likely to be due to the diffusion of the plasmid out of the donor cells, via the generated pores, and subsequent uptake by the recipient cells, and not due to fusion of the two cell types.

Transformation frequencies dramatically decrease with increasing size of the double-stranded DNA molecule. In *E. coli* the transformation frequency is lower with plasmids larger than 50 kb. This restricts the preparation of genome libraries with large insert sizes. In contrast, the plasma membrane pores generated during electroporation of bacteria are seemingly quite big, as extremely large molecules of DNA can be transformed using electroporation. Mozo & Hooykaas (1991) were able to introduce DNA molecules as large as 250 kb (the pTiB6::R772 plasmid of *A. tumefaciens*) into *A. tumefaciens*, using electroporation, with low but reproducible efficiencies: 2.7×10^{-9} transformants/survivor versus 1.4×10^{-6} for a 20 kb plasmid with crude DNA preparations. High electroporation efficiencies were also obtained with *E. coli* cells using plasmids up to 136 kb in size (Leonardo & Sidivy, 1990).

Conjugation

Cloning vectors usually have *mob* or *tra* functions removed to prevent unwanted plasmid transfers into undesirable bacteria. For example, RK2 derivative plasmids lack the *tra* genes, thus conjugative transfer from a donor to a recipient strain is accomplished by supplying the *tra* genes *in trans* on a ColE1 plasmid (pRK600 or similar plasmids). The ColE1 plasmid replicates only in *E. coli*, not in *Agrobacterium* or *Rhizobium*, and the recipient strain is selected by growth on medium with an appropriate antibiotic.

Plasmids coding for symbiotic functions in some strains of *R. leguminosarum*, *R. trifolii* and *R. phaseoli* are self-transmissible (Johnson, Bibb & Beringer, 1978; Beynon, Beringer & Johnston, 1980; Hooykaas *et al.*, 1981; Rolfe *et al.*, 1981; Hooykaas, Snijdewint & Schilperoort, 1982*b*; Lamb, Hombrecher & Johnston, 1982; Scott & Ronson, 1982), but this is not the case in *R. meliloti*, *R. fredii* and *B. japonicum* (Kondorosi *et al.*, 1982; Appelbaum *et al.*, 1985; Atherly *et al.*, 1985). Several approaches have been devised to transfer entire plasmids from one strain to another. For example, Hooykaas, Den Dulk-Ras & Schilperoort (1982*a*) developed a helper plasmid (pRL180) that assists in the transfer of nontransmissible plasmids. Kondorosi *et al.* (1982) inserted the *mob* region of the P-1 type plasmid RP4 into the large plasmid of *R. meliloti* to obtain plasmid transfer. Mobilization of a nontransmissible plasmid of *R. trifolii* was achieved by cointegration with the transfer-proficient plasmid R68.45 (Scott & Ronson, 1982). A more generally used system utilizes a *mob*::Tn5 element present in a ColE1-based vector (Simon, 1984). Conjugation of this plasmid into Gram-negative soil bacteria, with selection for kanamycin, results in the random insertion of the *mob*::Tn5 element into the genome. This approach has been used for plasmid and chromosome transfer, utilizing *tra* genes *in trans*, with excellent success (Glazebrook & Walker, 1991), permitting high frequency of recombination (Hfr)-like mapping. The *mob*::Tn5 system has also been used to transfer plasmids from *Azospirillum lipoferum* into *A. tumefaciens* (Bally & Givaudan, 1988).

The Ti plasmids of *A. tumefaciens* are transmissible between strains (Kerr, Manigault & Tempé, 1977; Genetello *et al.*, 1977) as well as into other Gram-negative bacteria, such as *R. trifolii* (Hooykaas *et al.*, 1977). Transfer of plasmids between strains is seemingly dependent upon specific virulence (*vir*) gene functions as mutations in *vir*A, *vir*G, *vir*B, and *vir*E operons (Steck & Kado, 1990), and, in some cases, *vir*C (Gelvin & Habeck, 1990) greatly decreases plasmid transfers. The presence or absence of acetosyringone has no effect. In contrast, Beijersbergen *et al.* (1992) found that the nonconjugative plasmid, pKT230 (a derivative of *Inc*Q plasmid RSF1010), could be mobilized between *Agrobacterium* strains and mobilization was dependent upon the presence of acetosyringone, the *vir*A, *vir*G and *vir*B operons and *vir*D4. These data suggest that T-DNA (transferred-DNA) transfer from *Agrobacterium* to plants may occur in a manner similar to bacterial conjugation.

Transduction

Although little used, transduction is a valuable adjunct to plasmid cloning vectors and transformation in the genetic analysis of *Rhizobium*. Generalized transducing phages are available for *R. meliloti* (Casadesus & Olivaries, 1979; Sik, Hovath & Chatterjee, 1980; Finan *et al.*, 1984; Martin & Long, 1984; Williams, Klein & Signer, 1989; Novikova *et al.*, 1990; Glazebrook & Walker, 1991), *R. leguminosarum* and *R. trifolii* (Buchanan-Wollaston, 1979) and *B. japonicum* (Shah, Sousa & Modi, 1981). Phage can be used for fine-structure mapping, strain construction and for the transfer of large plasmids. The molecular size of most *Rhizobium* phage genomes is 150–200 kb, thus allowing transfer of encapsulated intact plasmids (Martin & Long, 1984; Glazebrook & Walker, 1991). Unfortunately, most *Rhizobium* phages are very strain specific.

Vector and DNA stability

Most cloning vectors now in use in Gram-negative soil bacteria are derived from broad host range replicons (Table 3.1) and many of these vectors are unstable during growth under nonselective conditions. For example, *Agrobacterium* and *Rhizobium* cells lose pRK290 and its derivatives at a rate ranging between 0.03% and 0.25%/generation (Ditta *et al.*, 1980; Close *et al.*, 1984). Plasmids derived from the *Inc*W origin from plasmid pSa are lost at an even more rapid rate, ranging between 2% and 13%/generation depending upon the construction (Close *et al.*, 1984). This rapid loss likely reflects the nontransmissibility of these plasmids as pSa, which is self-transmissible, shows no loss even after 40 generations (Close *et al.*, 1984).

A highly stable cloning vector is desirable for symbiotically expressed genes. In *Rhizobium* cells, RK2-derived vectors are relatively unstable during symbiosis with the host plant (Long, Buikema & Ausubel, 1982; Lambert *et al.*, 1987), although Weinstein, Roberts & Helinski (1992) stabilized RK2-derived plasmids by introducing a 3.2 kb DNA fragment (pTR102 and pMW708) or a 0.8 kb DNA fragment (pTR101 and pMW707) derived from the RK2 stabilization region. No plasmid loss was observed after 100 generations of nonselective growth, nor during nodule passage. Gallie *et al.* (1985*b*) constructed a series of cloning vectors using the origin of replication from the stable endogenous plasmid (pTAR) of *Agrobacterium* LBA4301. These vectors (pUCD2001 and pUCD1001) are extremely stable in *Agrobacterium* LBA4301 and *Rhizobium meliloti* RM102Z1 under nonselective conditions, showing no loss after 50 generations. The copy number was low, as expected from the presence of *par* gene functions, which are required for plasmid replication and partitioning. Similarly, Simoens *et al.* (1986) prepared a binary vector (pC22) utilizing the origin of replication from *A. rhizogenes* plasmid pRiHR1, which is very stable under nonselective conditions. Mozo *et al.* (1990) cloned the origin of DNA replication (about 5 kb) from a small cryptic plasmid from *Rhizobium* species *(Hedysarum)* and found it to be more stable than RK2 derivatives in a variety of *Rhizobium* and *Agrobacterium* strains in the absence of selective pressure. This plasmid sequence seemingly possesses *par* genes similar to *Agrobacterium* plasmids. The *Rhizobium* species *(Hedysarum)* origin of replication is more stable than RK2-derived plasmids in a variety of *Rhizobium* strains under symbiotic conditions, although foreign DNA inserts have not been tested in this plasmid or in pUCD2001 or pUCD1001.

Another approach for stably inserting a desired DNA sequence into *Rhizobium* is via homologous recombination, either into a stable endogenous plasmid or into the chromosome. Care must be taken that the insertion does not inactivate a gene with an unknown but desirable function, thus site-specific recombination is the safest approach. Legocki, Yun & Szalay (1984) inserted DNA sequences into *Rhizobium* species by homologous recombination, but relied upon randomly cloned

fragments to find a 'nonessential' region of the genome. In a related study (Yun, Noti & Szalay, 1986), *nif* genes were part of a genomic target, but this had the risk of creating undesired, symbiotically altered strains after integration. Acuña *et al.* (1987) constructed a vector (pRJ1035) that was shown to integrate unselected, cloned DNA into a specific and nonessential region of the *B. japonicum* genome. Unfortunately, this vector is restricted to use with *B. japonicum* strains, since recombination depends upon the presence of repeated sequences not found in other species.

Another important consideration is genomic rearrangements that result from insertion of a plasmid into a bacterial cell. Plasmid transfer into *Agrobacterium* and *Rhizobium* by conjugation frequently results in plasmid rearrangements (Berry & Atherly, 1984; An *et al.*, 1988; Shantharam, Engwall & Atherly, 1988). Although no conclusive studies have been done, introduction of DNA by transformation, either natural or electroporation assisted, seemingly produces fewer rearrangements of cloned fragments.

Applications

Some obvious uses of introduced DNA into Gram-negative soil bacterial cells include complementation analysis, introduction of a new, desirable gene or genes, and gene replacement by homologous recombination.

An example of complementation analysis is the experiments of Long *et al.* (1982), who were the first to clone a DNA fragment in *R. meliloti* that conferred nodulation ability. They mated a *R. meliloti* clone bank from *E. coli en masse* into a nod⁻ *R. meliloti* strain and selected for nod⁺ derivatives by direct selection on alfalfa plants (*Medicago sativa*). Similarly, Innes, Hirose & Kuempel (1988) constructed a clone from *R. trifolii* in a RK2-type plasmid and, after conjugation into a pSym plasmid-cured background, found a 32 kb clone that was able to confer nodulation and nitrogen fixation. Hahn & Hennecke (1988), using a deletion mutant of *B. japonicum* defective in nodulation, selected clones that complemented the inability to form nodules. Many examples exist, as complementation is a commonly used approach for cloning genes in mutant strains where the clone restores the original phenotype.

Introduction of new, desirable genes into

Rhizobium or other Gram-negative bacteria is less frequent, largely due to the lack of identified desirable genes. Lambert *et al.* (1987) introduced the gene for hydrogenase, originally cloned from *B. japonicum*, into *Hup⁻ B. japonicum, R. meliloti* and *R. leguminosarum* strains. It was believed that the added hydrogenase would increase the hydrogen recycling capability of these strains and improve the efficiency of nitrogen fixation; however, the effect is minimal. Even when and if desirable genes are found and stably introduced into *Rhizobium* strains, it is unlikely that the new strains could be utilized in a natural symbiosis in the soil, as endogenous strains out-compete introduced strains unless massive numbers of bacteria are added.

With the discovery of electroporation-assisted transformation, Gram-negative soil bacteria can be used for direct preparation of clone banks, bypassing the need to prepare the clone bank in *E. coli* and then conjugate the clones into the desired strain. Conjugation requires triparental mating, selectable markers on both the helper plasmid and the mated plasmid, and frequently causes rearrangements of inserted fragments (see above). Preparation of a clone bank directly in a desirable strain, or direct transformation for complementation analysis, bypasses these obstacles. In this respect, Simoens *et al.* (1986) utilized the stably replicating binary cloning vector pC22, which contains T-DNA border sequences, a kanamycin resistance gene and *Bam*HI and *Xba*I cloning sites between the border sequences, and a streptomycin resistance gene for plasmid selection to prepare a clone bank of *Arabidopsis thaliana* in *E. coli*. This clone bank was conjugated into an appropriate *Agrobacterium* strain that was used to introduce DNA fragments into plant cells. Unfortunately they found that a large percentage (60%) of the clones were unstable in the absence of selection, both in *E. coli* and in *Agrobacterium*. Rearrangements were likely due to the presence of repeated sequences, as Simoens *et al.* found a large number of rearrangements in clones containing ribosomal RNA sequences. In addition, they found that conjugation between *E. coli* and *Agrobacterium* preferentially selected against the more unstable DNA clones. Direct preparation of the library in the *Agrobacterium* host using electroporation-assisted transformation, however, would bypass this problem.

Another use of electroporation-assisted cloning into Gram-negative soil bacteria is the direct cloning of DNA replicator regions of Gram-negative replicons. Replicator regions of Gram-negative soil bacteria have made very stable cloning vectors and it is desirable to find others. A digest of total DNA from a desired strain would be cloned into a ColE1-based vector, which is incapable of replication in Gram-negative soil bacteria. The clone bank is transformed into a plasmidless strain, for example a cured *Agrobacterium* strain, where only clones possessing the replicator region replicate.

In summary, many stable cloning vectors are now available for a wide range of Gram-negative soil bacteria and with the advent of electroporation, it is relatively easy to introduce DNA sequences into any bacterial strain. It is very likely that we will see stable genetically engineered soil bacteria in general use in the USA in the near future, assuming that Federal regulatory agencies allow their introduction into the soil.

References

Acuña, G., Alverez-Morales, A., Hahn, M. & Hennecke, H. (1987). A vector for the site-directed, genomic integration of foreign DNA into soybean root-nodule bacteria. *Plant Molecular Biology*, **9**, 41–50.

An, G., Ebert, P. R. Mitra, A. & Ha, S. B. (1988). Binary vectors. In *Plant Molecular Biology Manual* A3 Kluwer Academic Publisher, Dordrecht, pp. 1–19.

Appelbaum, E. R., McLoughlin, T. J., O'Connell, M. & Chartrain, N. (1985). Expression of symbiotic genes of *Rhizobium japonicum* USDA191 in other rhizobia. *Journal of Bacteriology*, **163**, 385–388.

Atherly, A. G., Prakash, R. K., Masterson, R. V., Du Teau, N. B. & Engwall, K. S. (1985). The organization of genes involved in symbiotic nitrogen fixation on indigenous plasmids of *Rhizobium japonicum. Proceedings of the World Soybean Conference III*, ed. R. Shibles, pp. 229–300. Westview Press, Boulder, Co.

Bagdasarian, M., Lurz, R., Rüchert, B., Franklin, F. C. H., Bagdasarian, M. M., Frey, J. & Timmis K. N. (1981). Specific-purpose plasmid cloning vectors. II. Broad host range, high copy number, RSF1010-derived vectors, and a host-vector system for gene cloning in *Pseudomonas. Gene*, **16**, 237–247.

Bally, R. & Givaudan, A. (1988). Mobilization and transfer of *Azospirillum lipoferum* plasmid by the Tn5-*mob* transposon into a plasmid-free *Agrobacterium tumefaciens* strain. *Canadian Journal of Microbiology*, **34**, 1354–1357.

Beijersbergen, A., Dulk-Ras, A. D., Schilperoort, R. A. & Hooykaas, P. J. J. (1992). Conjugative transfer by the virulence system of *Agrobacterium tumefaciens*. *Science*, 256, 1324–1327.

Berry, J. O. & Atherly, A. G. (1981). Introduction of plasmids into *Rhizobium japonicum* by conjugation and spheroplast transformation. In *Proceedings of the 8th North American Rhizobium conference*, ed. R. W. Clark & J. H. W. Stephens, pp. 115–128. University of Manitoba Press, Winnipeg, Manitoba.

Berry, J. O. & Atherly, A. G. (1984). Induced plasmid genome rearrangements in *Rhizobium japonicum*. *Journal of Bacteriology*, 157, 218–224.

Beynon, J. L., Beringer, J. E. & Johnston, A. W. B. (1980). Plasmids and host range in *Rhizobium leguminosarum* and *Rhizobium phaseoli*. *Journal of Genetic Microbiology*, 120, 421–429

Broek, A. V., van Gool, A. & Vanderleyden, J. (1989). Electroporation of *Azospirillum brasilense* with plasmid DNA. *FEMS Microbiology Letters*, 61, 177–182.

Buchanan-Wollaston, V. (1979). Generalized transduction in *Rhizobium leguminosarum*. *Journal of Genetic Microbiology*, 112, 135–142.

Bullerjahn, G. S. & Benzinger, R. H. (1982). Genetic transformation of *Rhizobium leguminosarum* by plasmid DNA. *Journal of Bacteriology*, 150, 421–424.

Casadesus, J. & Olivaries, J. (1979). General transduction in *Rhizobium meliloti* by a thermosensitive mutant of bacteriophage DF2. *Journal of Bacteriology*, 139, 316–317.

Chassy, B. M. & Flickinger, J. L. (1987). Transformation of *Lactobacillus casei* by electroporation. *FEMS Microbiology Letters*, 44, 173–177.

Close, T. J., Zaitlin, D. & Kado, C. I. (1984). Design and development of amplifiable broad-host-range cloning vectors: analysis of the *vir* region of *Agrobacterium tumefaciens* plasmid pTiC58. *Plasmid*, 12, 111–118.

Courtois, J., Courtois, B. & Guillaume, J. (1988). High-frequency transformation of *Rhizobium meliloti*. *Journal of Bacteriology*, 170, 5925–5927.

Davison, J., Heusterspeute, M., Chevalier, N., Ha-Thi, V. & Brunel, F. (1987). Vectors with restriction site banks. v. pJRD215, a wide-host-range cosmid vector with multiple cloning sites. *Gene*, 51, 275–280.

Ditta, G., Stanfield, S., Corbin, D. & Helinski, D. R. (1980). Broad host range DNA cloning system for Gram-negative bacteria, construction of a gene bank of *Rhizobium meliloti*. *Proceedings of the National Academy of Sciences, USA*, 77, 7347–7351.

Dower, W. J., Miller, J. F. & Ragsdale, C. W. (1988). High efficiency transformation of *E. coli* by high voltage electroporation. *Nucleic Acid Research*, 16, 6127–6145.

Ebert, P. R., Ha, S. B. & An, G. (1987). Identification of an essential upstream element in the nopaline synthase promoter by stable and transient assays. *Proceedings of the National Academy of Sciences, USA*, 84, 5745–5749.

Elkan, G. H. (1981). The taxonomy of the Rhizobiaceae. In *The Biology of the Rhizobiaceae, International Review of Cytology* Suppl. 13, ed. K. L. Giles & A. G. Atherly, pp. 1–14. Academic Press, New York.

Figurski, D. & Helinski, D. R. (1979). Replication of an origin containing derivative of plasmid RK2 dependent on a plasmid function provided in trans. *Proceedings of the National Academy of Sciences, USA*, 76, 1648–1652.

Finan, T. M., Hartwieg, E., Lemieux, K., Bergman, K., Walker, G. C. & Signer, E. R. (1984). General transduction in *Rhizobium meliloti*. *Journal of Bacteriology*, 159, 120–124.

Friedman, A. M., Long, S. R., Brown, S. E., Buikema, W. J. & Ausubel, F. M. (1982). Construction of a broad host range cosmid cloning vector and its use in the genetic analysis of *Rhizobium* mutants. *Gene*, 18, 289–296.

Fromm, M., Taylor, L. P. & Walbot, V. (1985). Expression of genes transferred into monocot and dicot plant cells by electroporation. *Proceedings of the National Academy of Sciences, USA*, 82, 5824–5828.

Gallie, D. R., Hagiya, M. & Kado, C. I. (1985a). Analysis of *Agrobacterium tumefaciens* plasmid pTiC58 replication region with a novel high copy number derivative. *Journal of Bacteriology*, 161, 1034–1041.

Gallie, D. R, Novak, S. & Kado, C. I. (1985b). Novel high and low copy number stable cosmids for use in *Agrobacterium* and *Rhizobium*. *Plasmid*, 14, 171–175.

Gallie, D. R., Zaitlin, D., Perry, K. L. & Kado, C. I. (1984). Characterization of the replication and stability regions of *Agrobacterium tumefaciens* plasmid pTAR. *Journal of Bacteriology*, 157, 739–745.

Gelvin, S. B. & Habeck, L. L. (1990). *vir* genes influence conjugal transfer of the Ti plasmid of *Agrobacterium tumefaciens*. *Journal of Bacteriology*, 172, 1600–1608.

Genetello, C., Van Larebeke, N., Holsters, M., De Picker, A., Van Montagu, M. & Schell, J. (1977). Ti plasmids of *Agrobacterium* as conjugative plasmids. *Nature (London)*, 265, 561–563.

Gibbins, A. M. & Gregory, K. F. (1972). Relatedness among *Rhizobium* and *Agrobacterium* species determined by three methods of nucleic acid hybridization. *Journal of Bacteriology*, 111, 129–141.

Glazebrook, J. & Walker, G. C. (1991). Genetic techniques in *Rhizobium meliloti*. *Methods in Enzymology*, 204, 398–418.

Guerinot, M. L., Morisseau, B. A. & Kapatch, T. (1990). Electroporation of *Bradyrhizobium japonicum*.

Molecular and General Genetics, **221**, 287–290.

Guerry, P., Van Embden, J. & Falkow, S. (1974). Molecular nature of two non-conjugative plasmids carrying drug resistance genes. *Journal of Bacteriology*, **117**, 619–630.

Hahn, M. & Hennecke, H. (1988). Cloning and mapping of a novel nodulation region from *Bradyrhizobium japonicum* by genetic complementation of a deletion mutant. *Journal of Bacteriology*, **54**, 55–61.

Hatterman, D. R. & Stacey, G. (1990). Efficient transformation of *Bradyrhizobium japonicum* by electroporation. *Journal of Bacteriology and Applied Environmental Microbiology*, **56**, 833–836.

Holgen, R. & Willmitzer, L. (1988). Storage of competent cells for *Agrobacterium* transformation. *Nucleic Acids Research*, **16**, 9877.

Holsters, M., De Waele, D., Depicker, A., Messens, E., Van Montagu, M. & Schell, J. (1978). Transfection and transformation of *Agrobacterium tumefaciens*. *Molecular and General Genetics*, **163**, 181–187.

Hooykaas, P. J. J., Den Dulk-Ras, H. & Schilperoort, R. A. (1982*a*). Method for the transfer of large cryptic, non-self-transmissible plasmids: ex planta transfer of the virulence plasmid of *Agrobacterium rhizogenes*. *Plasmid*, **8**, 94–96.

Hooykaas, P. J. J., Klapwijk, P. M., Nuti, M. P., Schilperoort, R. A. & Rorsch, A. (1977). Transfer of the *Agrobacterium tumefaciens* Ti plasmid to avirulent agrobacteria and rhizobia ex planta. *Journal of General Microbiology*, **98**, 477–484.

Hooykaas, P. J. J., Snijdewint, F. G. M. & Schilperoort, R. A. (1982*b*). Identification of the Sym plasmid of *Rhizobium leguminosarum* strain 1001 and its transfer to and expression in other rhizobia and *Agrobacterium tumefaciens*. *Plasmid*, **8**, 73–82.

Hooykaas, P. J. J., Van Brussel, A. A. N., Den Dulk-Ras, H., Van Slogteren, G. M. S. & Schilperoort, R. A. (1981). Sym-plasmids of *Rhizobium trifolii* expressed in different rhizobial species and *Agrobacterium tumefaciens*. *Nature (London)*, **291**, 351–353.

Innes, R. W., Hirose, M. A. & Kuempel, P. L. (1988). Induction of nitrogen fixing nodules on clover requires only 32 kilobase pairs of DNA from the *Rhizobium trifolii* symbiosis plasmid. *Journal of Bacteriology*, **170**, 3793–3800.

Itoh, Y., Watson, J. M., Haas, D. & Leisinger, T. (1984). Genetic and molecular characterization of *Pseudomonas* plasmid pVS1. *Plasmid*, **11**, 206–220.

Johnson, A. W. B., Bibb, M. J. & Beringer, J. E. (1978). Tryptophan genes in *Rhizobium* – their organization and transfer to other bacterial genera. *Molecular and General Genetics*, **165**, 323–330.

Jun, S. W. & Forde, B. G. (1989). Efficient transformation of *Agrobacterium* spp. by high voltage electroporation. *Nucleic Acids Research*, **17**, 8385.

Juni, E. & Janik, A. (1969). Transformation of *Acinetobacter calcoaceticus*. *Journal of Bacteriology*, **98**, 281–288.

Kerr, A., Manigault, P. & Tempé, J. (1977). Transfer of virulence *in vivo* and *in vitro* in *Agrobacterium*. *Nature (London)*, **265**, 560–561.

Knauf, V. & Nester, E. W. (1982). Wide host range cloning vectors, a cosmid clone bank of an *Agrobacterium* Ti plasmid. *Plasmid*, **8**, 45–54.

Kondorosi, A. & Johnston, A. W. B. (1981). The genetics of *Rhizobium*. In *The Biology of the Rhizobiaceae, International Review of Cytology* Suppl. 13, ed. K. Giles & A. G. Atherly, pp. 191–224. Academic Press, New York.

Kondorosi, A., Kondorosi, E., Pankhurst, C. E., Broughton, W. J. & Banfalvi, Z. (1982). Mobilization of a *Rhizobium meliloti* megaplasmid carrying nodulation and nitrogen fixation genes into other rhizobia and *Agrobacterium*. *Molecular and General Genetics*, **188**, 433–440.

Lamb, J. W., Hombrecher, G. & Johnston, A. W. B. (1982). Plasmid-determined nodulation and nitrogen fixation abilities in *Rhizobium phaseoli*. *Molecular and General Genetics*, **186**, 449–452.

Lambert, G. R., Harker, A. R., Cantrell, M. A., Hanus, F. J., Russell, S. A., Haugland, R. A. & Evans H. J. (1987). Symbiotic expression of cosmid-borne *B. japonicum* hydrogenase genes. *Applied and Environmental Microbiology*, **53**, 422–428.

Leemans, J., Langenakens, J., De Greve, H., Deblaere, R., Van Montagu, M. & Schell, J. (1982). Broad-host-range cloning vectors derived from the W-plasmid Sa. *Gene*, **19**, 361–364.

Legocki, R. P., Yun, A. C. & Szalay, A. A. (1984). Expression of ß-galactosidase controlled by a nitrogenase promoter in stem nodules of *Aeschynomene scabra*. *Proceedings of the National Academy Sciences, USA*, **81**, 5806–5810.

Leonardo, E. D. & Sidivy, J. M. (1990). A new vector for cloning large eukaryotic DNA segments in *Escherichia coli*. *Bio/Technology*, **8**, 841–844.

Long, S. R., Buikema, W. & Ausubel, F. M. (1982). Cloning of *Rhizobium meliloti* nodulation genes by direct complementation of Nod⁻ mutants. *Nature (London)*, **198**, 495–498.

Lorenz, M. G., Reipschläger, K. & Wackernagel, W. (1992). Plasmid transformation of naturally competent *Acinetobacter calcoaceticus* in non-sterile soil extract and groundwater. *Archives of Microbiology*, **157**, 355–360.

Luchansky, J. B., Muriana, P. M. & Klaenhammer, T. R. (1988). Application of electroporation for transfer of plasmid DNA to *Lactobacillus, Lactococcus, Leuconostoc, Listeria, Pediococcus, Bacillus, Staphylococcus, Enterococcus, and Propionibacterium*. *Molecular Microbiology*, **2**, 637–646.

Mandel, M. & Higa, A. (1970). Calcium-dependent bacteriophage DNA infection. *Journal of Molecular Biology*, **53**, 159–162.

Martin, M. O. & Long, S. R. (1984). Generalized transduction in *Rhizobium meliloti*. *Journal of Bacteriology*, **159**, 125–129.

Mattanovich, D., Rucker, F., Machado, A. C., Lamer, M., Regner, F., Steinkelner, H., Himmler, G. & Katinger, H. (1989). Efficient transformation of *Agrobacterium* spp. by electroporation. *Nucleic Acids Research*, **17**, 6747.

Mozo, T., Cabrera, E. & Ruiz-Argueso, T. (1990). Isolation of the replication DNA region from a *Rhizobium* plasmid and examination of its potential as a replicon for Rhizobiaceae cloning vectors. *Plasmid*, **23**, 201–215.

Mozo, T. & Hooykaas, P. J. J. (1991) Electroporation of megaplasmids into *Agrobacterium*. *Plant Molecular Biology*, **16**, 917–918.

Nagel, R., Elliott, A., Mase., A., Birch, R. G. & Manners, J. M. (1990). Electroporation of binary Ti plasmid vector into *Agrobacterium tumefaciens* and *Agrobacterium rhizogenes*. *FEMS Microbiology Letters*, **67**, 325–328.

Nishiguchi, R., Takanami, M. & Oda, A. (1987). Characterization and sequence determination of the replicator region in the hairy-root inducing plasmid pRiA4b. *Molecular and General Genetics*, **206**, 1–8.

Novikova, N. I., Safronova, V. I., Lyudvikova, E. K. & Simarov, B. V. (1990). Production of recombinant transducing phage of *Rhizobium meliloti* with a broad spectrum of lytic activity. *Genetika*, **26**, 37–42.

Pfau, J. & Youderian, P. (1990). Transferring plasmid DNA between different bacterial species with electroporation. *Nucleic Acids Research*, **18**, 6165.

Potter, H., Weir, L. & Leder, P. (1984). Enhancer-dependent expression of human immunoglobin genes introduced into mouse pre-β-lymphocytes by electroporation. *Proceedings of the National Academy of Sciences, USA*, **81**, 7161–7165.

Regué, M., Enfedaque, J., Camprubi, S. & Tomás, J. M. (1992). The O-antigen lipopolysaccharide in the major barrier to plasmid DNA uptake by *Klebsiella pneumoniae* during transformation by electroporation and osmotic shock. *Journal of Microbiology Methods*, **15**, 129–134.

Rolfe, B. G., Djordjevic, M., Scott, K. F., Hughes, J. E., Badennoch-Jones, J., Gresshoff, P. M., Chen, Y., Dudman, W. G., Zurkowski, W. & Shine, J. (1981). Analysis of nodule formation in fast growing *Rhizobium* strains. In *Current Perspectives in Nitrogen Fixation*, ed. A. Gibson & A. Newton, pp. 142–145. Australian Academic Press, Canberra.

Schmidhauser, T. J. & Helinski, D. R. (1985). Regions of broad host range plasmid RK2 involved in replication and stable maintenance in nine species of Gram-negative bacteria. *Journal of Bacteriology*, **164**, 446–451.

Schwinghamer, E. A. (1977). Genetic aspects of nodulation and dinitrogen fixation by legumes, the microsymbiont. In *A Treatise on Dinitrogen Fixation*, ed. R. W. F. Hardy & W. S. Silver, Section III, pp. 577–622. Wiley (Interscience), New York.

Scott, D. B. & Ronson, C. W. (1982). Identification and mobilization by cointegrate formation of a nodulation plasmid in *Rhizobium trifolii*. *Journal of Bacteriology*, **151**, 36–43.

Selvaraj, G. & Iyer, I. V. (1981). Genetic transformation of *Rhizobium meliloti* by plasmid DNA. *Gene*, **15**, 279–283.

Shah, K., Sousa, S. & Modi, V. V. (1981). Studies on transducing phage M–1 for *Rhizobium japonicum*. *Archives of Microbiology*, **130**, 262–266.

Shantharam, S., Engwall, K. S. & Atherly, A. G. (1988). Symbiotic phenotypes of soybean root nodules associated with deletions and rearrangements in the symbiotic plasmid of *R. fredii* USDA 191. *Journal of Plant Physiology*, **132**, 431–438.

Shaw, C. H., Leemans, J., Shaw, C. H., Van Montagu, M. & Schell, J. (1983). A general method for the transfer of cloned genes to plant cells. *Gene*, **23**, 315–330.

Shigekawa, K. & Dower, W. J. (1988). Electroporation of eukaryotes and prokaryotes, a general approach to the introduction of macromolecules into cells. *BioTechniques*, **6**, 742–751.

Sik, T., Hovath, J. & Chatterjee, S. (1980). Generalized transduction in *Rhizobium meliloti*. *Molecular and General Genetics*, **178**, 511–516.

Simoens, C., Alliote, T., Mendel, R., Müller, A., Schiemann, J., Van Lijsebettens, M., Schell, J., Van Montagu, M. & Inzé, D. (1986). A binary vector for transferring genomic libraries to plants. *Nucleic Acids Research*, **14**, 8073–8090.

Simon, R. (1984). High frequency mobilization of Gram-negative bacterial replicons by the in vitro constructed Tn5-Mob transposon. *Molecular and General Genetics*, **196**, 413–420.

Singer, J. T., Van Tuijl, J. J. & Finnerty, W. (1986). Transformation and mobilization of cloning vectors in *Acinetobacter* spp. *Journal of Bacteriology*, **165**, 301–303.

Steck, T. R. & Kado, C. I. (1990). Virulence genes promote conjugative transfer of the Ti plasmid between *Agrobacterium* strains. *Journal of Bacteriology*, **172**, 2191–2193.

Tabata, S., Hooykaas, P. & Oda, A. (1989). Sequence determination and characterization of the replicator region in the tumor-inducing plasmid pTiB6S3. *Journal of Bacteriology*, **171**, 1665–1672.

Tait, R. C., Close, T. J., Lundquist, R. C., Haga, M., Rodriguez, R. L. & C. I. Kado. (1983).

Construction and characterization of a versatile broad host range DNA cloning system for Gram-negative bacteria. *Bio/Technology*, **1**, 269–275.

Van den Eede, G., Deblaere, R., Goethals, K., Van Montagu, M. & Holsters, M. (1992). Broad host range and promoter selection vectors for bacteria that interact with plants. *Molecular Plant–Microbe Interactions*, **5**, 228–234.

Weinstein, M., Roberts, R. C. & Helinski, D. R. (1992). A region of the broad-host-range plasmid RK2 causes stable in planta inheritance of plasmids in *Rhizobium meliloti* cells isolated from alfalfa root nodules. *Journal of Bacteriology*, **174**, 7486–7489.

White, T. J. & Gonzolas, C. F. (1991). Application of electroporation for efficient transformation of *Xanthomonas campestris* pv. *oryzae*. *Phytopathology*, **81**, 521–524.

Williams, M. V., Klein, S. & Signer, E. R. (1989).

Host restriction and transduction in *R. meliloti*. *Applied and Environmental Microbiology*, **55**, 3229–3230.

Wirth, R., Friesenegger, A. & Fiedler, S. (1989). Transformation of various species of Gram-negative bacteria belonging to 11 different genera by electroporation. *Molecular and General Genetics*, **216**, 175–177.

Yun, A. C., Noti, J. D. & Szalay, A. A. (1986). Nitrogenase promoter-*lacZ* fusion studies of essential nitrogen fixation genes in *Bradyrhizobium* I110. *Journal of Bacteriology*, **167**, 784–791.

Zimmermann, U. (1983). Electrofusion of cells, principles and industrial potential. *Trends in Biotechnology*, **1**, 149–155.

Zimmermann, U., Schultz, J. & Pilwat, G. (1973). Transcellular ion flow in *Escherichia coli* B and electrical sizing of bacteria. *Biophysics Journal*, **13**, 1005–1013.

4

Filamentous Fungi

Gustavo H. Goldman, Marc Van Montagu and Alfredo Herrera-Estrella

Introduction

Fungi are eukaryotic, heterotrophic organisms with an absorptive mode of nutrition. Most fungi are both unicellular and multinucleate, with rigid chitinous cell walls, and usually exhibit mycelial or yeast-like growth habit. Probably most of the biotechnologically important soil fungi have no stable recombination cycles in the laboratory. This creates problems for physiological and genetic studies in these species. Genetic manipulation using transformation and gene cloning provides the most logically directed approach to dissect and, eventually, alter the physiology of these filamentous fungi. Towards this end, there has been strong pressure to develop techniques of basic molecular biology suitable for these organisms. The most intensively studied fungi are the unicellular yeast *Saccharomyces cerevisiae* and the filamentous fungi *Neurospora crassa* and *Aspergillus nidulans*. The molecular genetic systems of these organisms have served as the basis for development of similar systems in less tractable but economically important fungal species (Timberlake & Marshall, 1989).

The first report of transformation of a fungal species mediated by DNA was published by Mishra & Tatum (1973). Growing cultures of an inositol-requiring mutant of *N. crassa* were transformed with total DNA of the wild-type together with calcium; from the conidia formed on such cultures it was possible to select prototrophic strains. This pioneering experiment was received with skepticism and only some years later, Hinnen, Hicks & Fink (1978) reported transformation of *S. cerevisiae* using protoplasts from a *leu*2 mutant by treatment with wild-type DNA in the presence of calcium chloride. The utilization of protoplasts was immediately applied to the filamentous fungi *N. crassa* and *A. nidulans* (Case *et al.*, 1979; Tilburn *et al.*, 1983). With the passing years, these techniques of transformation have been extended to other species of soil fungi (Table 4.1).

This chapter recapitulates the main technological developments, and the possibilities offered to biotechnology by the application of genetic transformation techniques to soil fungi. It is our aim to review the approaches being adopted and, when possible, to use examples with several soil fungi to illustrate these techniques. Transformation of soil yeasts will not be discussed in this chapter.

Procedures

Polyethylene glycol-mediated transformation

Until now, this technique has been the most exploited for fungal cells and is considered as the standard transformation system. However, its application involves delicate handling of cells because of the requirement for the use of protoplasts or osmotically sensitive cells (OSCs). As mentioned before, protoplast technology has been an important achievement for the elaboration of transformation protocols. Protoplasts or OSCs are produced by removing the cell wall of either germinating spores or hyphae with cell-wall-degrading enzymes. The most commonly used and successful product for this purpose is Novozym-234, a commercially available hydrolytic enzyme mixture secreted by the filamentous fungus *Trichoderma harzianum*, which is used alone or in

Table 4.1. *Examples of soil fungal species in which transformation has been achieved*

Species	Markers	Referene
A. NONPATHOGENIC		
Antibiotic producers		
Cephalosporium acremonium	*hph, niaD*	Skatrud *et al.* (1987); Whitehead *et al.* (1990)
Penicillium chrysogenum	*sul*I	Carramolino *et al.* (1989)
Enzyme producers		
Aspergillus oryzae	*pyr*G	Mattern *et al.* (1987)
Aspergillus niger	*arg*B	Buxton, Gwynne & Davis (1985)
Trichoderma reesei	*amd*S, *arg*B	Penttilä *et al.* (1987)
Trichoderma reesei	*ura*3, *ura*5	Bergès & Barreau (1991)
Trichoderma reesei	*pyr*4	Smith *et al.* (1991)
Trichoderma reesei	*hph*	A. Herrera-Estrella (unpublished data)
Trichoderma viride	*pyr*4	Cheng, Tsukagoshi & Udaka (1990)
Biocontrol agents		
Gliocladium roseum	*hph*	Thomas & Kenerly (1989)
Gliocladium viridens	*hph*	Thomas & Kenerly (1989)
Metarhizium anisopliae	*ben*A3	Goettel *et al.* (1990)
Trichoderma harzianum	*hph, bml*	Goldman *et al.* (1992); Herrera-Estrella *et al.* (1990); Uhloa, Vainstein & Pederby (1992)
Trichoderma viride	*hph, tub*2	Herrera-Estrella *et al.* (1990); Goldman *et al.* (1993)
Food processing (starter)		
Penicillium nalgiovense	*amd*S	Geisen & Leistner (1989)
Mycorrhiza		
Laccaria laccata	*hph*	Barrett, Dixon & Lemke (1990)
B. PHYTOPATHOGENIC		
Cochliobolus heterostrophus	*amd*S, *hph*	Turgeon, Garber & Yoder (1985, 1987)
Colletotrichum graminicola	*bml*R, *bml*R3	Panaccione, McKiernan & Hanau (1988)
Colletotrichum lindemuthianum	*hph, amd*S	Rodriguez & Yoder (1987)
Colletotrichum trifolii	*hph*, benomyl resistance	Dickman (1988)
Fusarium oxysporum	*hph*	Kistler & Benny (1988)
Fusarium sambucinum	*hph*	Salch & Beremand (1988)
Fusarium sporotrichiodes	*hph*	Turgeon *et al.* (1987)
Fulvia fulva	*hph*	Oliver *et al.* (1987)
Gaeumannomyces graminis	benomyl resistance	Henson, Blake & Pilgeraro (1988)
Leptosphaeria	*hph*	Farman & Oliver (1987); Turgeon *et al.* (1987)
Magnaporthe grisea (Pyricularia oryzae)	*arg*B	Parsons, Chumley & Valent (1987)
Nectria haematococca (Fusarium soloni f. sp. *pisi)*	*hph, arg*B	Dickman & Kolattukudy (1987); Turgeon *et al.* (1987); Rambosek & Leach (1987)
Phytophthora capsici	*hph*	Bailey, Mena & Herrera-Estrella (1991)
Phytophthora parasitica	*hph*	Bailey *et al.* (1991)
Septoria nodorum	*hph*	Cooley *et al.* (1988)
Ustilago hordei	*hph*	Holden, Wang & Leong (1988); Yoder & Turgeon (1985)
Ustilago maydis	*hph*	Wang *et al.* (1988); Yoder & Turgeon, (1985)
Ustilago nigra	*hph*	Holden *et al.* (1988)
Ustilago violacea	*hph*	Bej & Perlin (1988)

combination with other lytic enzymes such as β-glucuronidase and chitosanase. All protoplast preparations have to be protected by the presence of an osmotic stabilizer at all times. Although sorbitol, at concentrations between 0.8 and 1.2 M, has been commonly used and seems to be satisfactory for all fungi, mannitol, sodium chloride, and magnesium sulfate are used for some fungi (Fincham, 1989). Mycelia or germinative tubes are usually selected as the starting material for the production of protoplasts. When cell walls have been partially or totally removed, cells are treated with a mixture of calcium chloride, polyethylene glycol (PEG), and transforming DNA. The exogenous DNA molecules are apparently internalized while a PEG-induced protoplast fusion takes place (no transformation occurs when PEG is omitted; Timberlake & Marshall, 1989). After this treatment, the protoplasts are plated on appropriate regeneration medium that will allow selection for the expression of the phenotype encoded by the transforming DNA (see selection of transformants, below). While this phenomenon is always forced to occur by means of the combination of calcium chloride and PEG, the concentration of these chemicals, the type of PEG and the time during which the cells are exposed vary according to the cells to be transformed (Ballance, Buxton & Turner, 1983; Tilburn *et al.*, 1983).

The two major problems faced when one is trying to apply this technique are the efficiency with which protoplasts are formed and their regeneration, although special attention should also be paid to the time of recovery allowed to the cells before selection.

Alternative methods: lithium acetate, electroporation and particle bombardment

Because PEG-mediated transformation is time consuming and some fungi are not amenable to protoplast preparation and/or regeneration, alternative transformation methods have been developed. One of the alternative methods that avoids protoplast preparation is the use of high concentrations of lithium acetate in combination with PEG to induce permeability of intact cells to DNA. Initially developed for *S. cerevisiae* (Ito *et al.*, 1983), this method has been successfully used for *N. crassa* (Dhawale, Paietta & Marzluf, 1984) and *Coprinus lagopus* (Binninger *et al.*, 1987). In the filamentous fungi, germinating spores were exposed to the transforming DNA in the presence of 0.1 M lithium acetate. It is not known how lithium acetate assists the passage of DNA into intact cells, and this method is not frequently used for filamentous fungi.

Recently, electroporation has become a valuable technique for the introduction of nucleic acids into both eukaryotic and prokaryotic cells (Miller, Dower & Tompkins, 1988; Förster & Neumann, 1989). Intact cells, as well as cells treated with cell-wall-degrading enzymes, are amenable to electroporation (Fromm, Taylor & Walbot, 1985; Dower, Miller & Ragsdale, 1988; Shigekawa & Dower, 1988). When a cell is exposed to an electric field, the membrane components become polarized and a potential difference develops across it. If the voltage exceeds a threshold level, the membrane breaks down in localized areas and the cell becomes permeable to exogenous molecules (Shigekawa & Dower, 1988). The induced permeability is reversible provided the magnitude and/or duration of the electric field does not exceed a critical limit, otherwise the cell is damaged irreversibly (Shigekawa & Dower, 1988). There are also some reports about electroporation in filamentous fungi and the method is likely to become more popular, mainly due to its simplicity and reproducibility. Ward, Kodama & Wilson (1989) reported the transformation of protoplasts of *A. awamori* and *A. niger* by electroporation, obtaining transformation frequencies similar to those obtained with PEG. Richey *et al.* (1989) transformed protoplasts of *Fusarium solani* and *A. nidulans*. These authors reported transformation frequencies lower than those obtained by standard transformation methods. Goldman, Van Montagu & Herrera-Estrella (1990) reported electroporation of *T. harzianum* using a combination of OSCs and PEG. An important factor identified in this study was that the incubation time of the germinative tubes with the cell-wall-degrading enzymes greatly influenced the final outcome of the transformation. In all of these experiments, transformants could be obtained without the addition of PEG. However, Goldman *et al.* (1990) obtained a four-fold higher transformation frequency when 1% (w/v) PEG was present in the medium during the delivery of the electric shock. Intact cells have also been used for electroporation of filamentous fungi. Chakraborty, Patterson & Kapoor (1991) reported electroporation of germinative tubes of *N. crassa* and conidia of *Penicillium urticae* and Edman & Kwon-Chung (1990) have obtained similarly successful transformation of intact cells of *Cryptococcus neoformans*.

In most of these protocols, potentials between 2000 and 3000 V/cm with a time constant of 10 to 15 ms have been set up. In contrast with the calcium chloride–PEG-mediated transformation, there is no standard medium established for electroporation (Richey *et al.*, 1989; Ward *et al.*, 1989; Goldman *et al.*, 1990; Chakraborty *et al.*, 1991).

Thus, it is essential to take into account the following parameters when establishing the electroporation conditions for a new system: cell size, cell viability after shock delivery, and conductivity of the electroporation media.

Cells can also be transformed by literally shooting them with microprojectiles carrying nucleic acids, with the consequent expression of the introduced genes (Klein *et al.*, 1987; Sanford, Smith & Russell, 1993). The projectiles used can be metal particles made of colloidal gold or tungsten coated with nucleic acids, or other types of hardened particles such as bacteria and phages. This method is called particle bombardment or 'biolistic' (for biological ballistics) transformation, and has already been applied to plants (Klein *et al.*, 1988*a*,*b*,*c*; Wang, Holden & Leong, 1988), mammals (Williams *et al.*, 1991), bacteria (Shark *et al.*, 1991), and fungi (Armaleo *et al.*, 1990; A. Bailey, personal communication). According to Armaleo *et al.* (1990), this technique of delivering nucleic acids into cells is still in its infancy but has already demonstrated several advantages over preexisting transformation methods: (i) its simplicity of application allows the simultaneous targeting of hundreds of millions of cells; (ii) it is not limited to one particular species and has been employed successfully both in systems already transformable by other means as well as in less tractable systems; and (iii) it is the only technique so far described by which mitochondria and chloroplasts (Fox, Sanford & McMullin, 1988; Johnston *et al.*, 1988; Daniell *et al.*, 1990) can be transformed. In fungi, transformation by particle bombardment has been achieved for *S. cerevisiae, Schizosaccharomyces pombe, N. crassa* (Armaleo *et al.*, 1990), *Phytophthora capsici, P. citricola, P. cinnamomi, P. citrophthora* (A. Bailey, personal communication), and *Mucor circinelloides* (F. Gutierrez, personal communication). This technique has been more successful when self-replicating integrative vectors are used and is specially interesting when assaying for transient expression or one is titrating *trans*-acting factors. The future of this technique is promising in filamentous fungi, but, as in the case of electroporation, many biological and physical parameters underlying this process need to be better defined.

Selection of transformants

Marker genes

Selection of transformants from the background of nontransformed cells depends on the expression of genes conferring adequate dominant selectable phenotypes (Timberlake & Marshall, 1989). In the beginning, auxotrophic mutants were transformed to prototrophy; in these cases, the selection of the transformants was usually straightforward (Fincham, 1989). As mentioned above, the first report of transformation of a filamentous fungus was based on the complementation of an inositol-requiring *N. crassa* mutant strain with DNA from the inositol-independent parent (Mishra & Tatum, 1973). Later, Case *et al.* (1979) extended the types of selection for *N. crassa* by using the *qa-2* gene encoding the catabolic dehydroquinase from *N. crassa*. Subsequently, many different genes used as prototrophic markers in recipient auxotrophic mutants were used as selectable markers (for a review, see Fincham, 1989). An inconvenient consequence of relying on selection against auxotrophy in isolating transformants is the need for the appropriate auxotrophic mutation in the recipient strain. The isolation of certain auxotrophic mutants is sometimes facilitated by a positive selection procedure. For example, mutations resulting in the loss of orotidine 5'-monophosphate carboxylase (encoded by *ura*3 or *pyr*4), which is required for uridine biosynthesis, confer resistance to the normally inhibitory analog 5-fluoroorotic acid (Alani, Cao & Kleckner, 1987; Diez *et al.*, 1987; Smith, Bayliss & Ward, 1991).

A more versatile system is to use a dominant gene as the selectable marker. There are many genes already available for filamentous fungi from different origins, such as those conferring resistance to hygromycin B, G418, phleomycin, oligomycin, copper, the fungicide benomyl, and an *A. nidulans* gene, *amd*S, that enables the organism to grow on acetamide as sole carbon and nitrogen source (Austin, Hall & Tyler, 1990; Table 4.2). The existence of these markers means that most fungal species, even those not previously subjected to laboratory investigations, can now be genetically modified by transformation (Timberlake & Marshall, 1989). However, the problem still exists that many fungal species are naturally resistant to many of the antibiotics for which genes conferring resistance are available.

Table 4.2. *Some dominant selectable markers used for transformation in filamentous fungi*

Marker gene	Origin	Phenotype
hph	*E. coli*	Hygromycin B resistance
*npt*II	Bacterial Tn5	Kanamycin, G418 resistance
benA3, bml	*N. crassa, A. nidulans*	Benomyl resistance
*bml*R3	*C. graminicola*	Benomyl resistance
*oli*C^R	*A. niger*	Oligomycin resistance
*amd*S	*A. nidulans*	Acetamide utilization
ble	Bacterial Tn5	Bleomycin, phleomycin resistance
*sul*I	Enterobacteria	Sulfonamide resistance

Cotransformation

In cases where a transforming gene cannot directly be selected for, an option is to look for its assimilation along with a more readily selectable marker. Theoretically, there is a high probability that a cell that takes up one kind of DNA will also take up another, specially if the ratio of cotransforming DNA : transforming DNA is kept high. According to Fincham (1989), this phenomenon of cotransformation can be rationalized by supposing that not all protoplasts are equally prone to take up DNA and that those more competent to do so will tend to take up several molecules simultaneously. The frequencies of reported cotransformation are highly variable and probably dependent upon the organism and the transformation conditions.

Genetic purification of transformants

Generally, when a protoplast is transformed, the transformed nuclei will exist within a population of wild-type nuclei, since most of the protoplasts are multinucleate. Selection cannot be applied to individual nuclei because the fungal thallus is coencytial. The initially transformed colonies are likely to be heterokaryons, with some nuclei transformed and some not, or with a heterologous mixture of independently transformed nuclei. If a fungus forms uninucleate conidia, a heterokaryon can be resolved very simply into its components by plating conidia and by isolating single colonies. If the conidia are multinucleate, genetic purification can be achieved more laboriously by several successive rounds of re-isolation of the transformed phenotype from single conidial colonies. After three rounds the probability of stochastic loss of one or other nuclear component is high (Fincham, 1989).

Fate of transforming DNA

Autonomously replicating vectors

Although most of the plasmids used for transformation in filamentous fungi have no origin of replication, a few autonomously replicating vectors have been used. These vectors are derived from naturally occurring mitochondrial plasmids or vectors with 'autonomously replicating sequences' (ARS). Autonomously replicating vectors in *S. cerevisiae* have a higher frequency of transformation than do integrative vectors. This enhanced efficiency can be exploited to simplify the isolation and cloning of genes. The origin of replication from the *S. cerevisiae* nuclear plasmid, the 2 μ circle, is used extensively as a component in yeast episomal vectors (Futcher, 1988). Unfortunately, the 2 μ replicon and yeast chromosomal ARS do not promote autonomous replication of vectors in any filamentous fungus so far tested. Therefore, sequences from chromosomal DNA and from endogenous fungal plasmids have been screened in attempts to identify sequences that promote autonomous replication of vectors.

Most fungal plasmids are found within mitochondria, but, in some yeast species, plasmids are found within nuclei. Circular plasmids appear to be less common in fungi than linear ones and have been reported only in *N. crassa*, *Cochliobolus heterostrophus* and *Cephalosporum acremonium* (Taylor, Smolich & May, 1985; Samac & Leong, 1989). Plasmids with diverse characteristics have been described in a wide range of fungal species. Unfortunately, there are only few examples of partial success of incorporation of mitochondrial replicons into transformation vectors (Stahl *et al.*, 1982; Esser *et al.*, 1983; Stohl & Lambowitz, 1983; Kuiper & de Vries, 1985). The replicons of mitochondrial plasmids could be useful in constructing autonomously replicating vectors for filamentous fungi because mitochondrial plasmids are maintained at high copy number (Samac & Leong, 1989).

A more traditional approach to vector building has been the incorporation of ARS. These sequences allow autonomous replication of the hybrid plasmid in transformants and, thus, enhance the efficiency of transformation (Mishra, 1985). Integrated sequences are stable but replicating sequences can be lost because they lack the centromeres and telomere size needed to behave like true chromosomes. Buxton & Radford (1984) screened a library of 700 cloned *N. crassa* sequences in the size range 1 to 7 kb for the ability to improve the *N. crassa* transformation frequency when inserted into a plasmid that also included *pyr*4$^+$ or *qa*-2$^+$ as a selective marker. Only four were found to have significant effect. In an attempt to isolate an *Aspergillus* origin of replication, Ballance & Turner (1985) utilized the fact that the *N. crassa pyr*4 gene is weakly expressed in yeast. Fragments of *A. nidulans* genomic DNA were cloned into a plasmid containing the *N. crassa pyr*4 gene and this library was used to transform a yeast strain. Transformants that were mitotically unstable, i.e. lost the plasmid under nonselective conditions, were subsequently identified on the assumption that these contained autonomously replicating plasmids. These replicating plasmids, complete with the *A. nidulans* DNA fragment acting as ARS, were re-isolated from yeast and used to transform *A. nidulans*. One plasmid derived in this manner was capable of transforming *A. nidulans* at frequencies 50- to 100-fold higher than the original plasmid, yielding up to 5×10^3 transformants/μg DNA (Ward, 1991). The *A. nidulans* sequence that conferred this high frequency of transformation was termed ARS1. Further investigation of the function of ARS1 has led to the conclusion that it does not confer autonomous replication in *A. nidulans*, although the unproven possibility exists that autonomous replication occurs for a brief period immediately after transformation and is followed by stabilization due to integration. Mapping the chromosomal location of ARS1 has shown it to be tightly linked to the centromere of linkage group I (Cullen *et al.*, 1987). It has since been confirmed that many sequences from heterologous species that act as ARS in yeast are not necessarily ARS or origin of replication in their native host (Maundrell *et al.*, 1985).

Van Heeswijck (1986) and Roncero *et al.* (1989) presented evidence for the autonomous replication of recombinant plasmids in *Mucor circinelloides leu*$^+$ transformants. Plasmids consisting of a unique fragment of *Mucor* DNA inserted into YRp17 or pBR322 give a high frequency of transformation (up to 7800 *leu*$^+$ transformants/μg DNA), are mitotically unstable, can be re-isolated in an unmodified form from uncut transformant DNA, and are present as discrete extrachromosomal DNA molecules. Subcloning of the recombinant plasmids and analysis of subsequent *leu*$^+$ transformants show that autonomous replication is independent of the vector sequences and locates a *Mucor* ARS within a 4.4 kb *Pst*I fragment of the insert DNA. Recently, it has been shown that high frequency transformation and mitotically unstable transformants are obtained when *Ustilago maydis* is transformed with a vector incorporating a putative ARS from nuclear DNA of *U. maydis*. The vector is present at high copy number, approximately 25 molecules/cell, of vector DNA (Tsukuda *et al.*, 1988). The same vector has been successfully used for transformation of *Phytophthora parasitica* and *P. capsici* (M. Bailey, personal communication).

Integration of DNA into the chromosomes

Genetic modification of most filamentous fungi depends on the genomic incorporation of exogenously added DNA. Therefore, most of the plasmids used for transformation in filamentous fungi do not need a fungal origin of replication. Plasmids readily integrate into either homologous or heterologous sites. Genomic integration of circular plasmids occurs in several ways. According to the most commonly used classification, three types of integration event can be defined (Figure 4.1). Type I involves integration of the plasmid at a region of homology within the genome (Figure 4.1(*a*)) and it is usually called homologous recombination. In general, the relative frequency of homologous integration is highly variable from one organism to another: 100% in *S. cerevisiae* (Struhl, 1983), about 80% in *A. nidulans* (Yelton, Hamer & Timberlake, 1984) and only 1% to 5% in *N. crassa* (Case, 1986) or *Coprinus cinereus* (Binninger *et al.*, 1987). Type II transformants have the plasmid integrated into sites within the genome where no known homology exists (Figure 4.1(*b*)). The mechanism by which the DNA

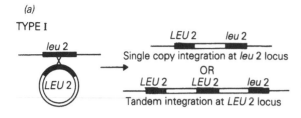

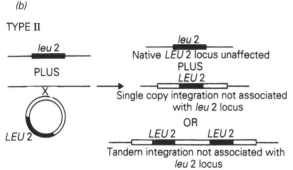

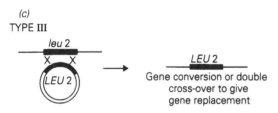

Figure 4.1. Patterns of plasmid integration in fungal transformants. (*a*) Type I involves integration of the plasmid at a region of homology within the genome and is usually called homologous recombination. (*b*) Type II transformants have the plasmid integrated into sites within the genome where no known homology exists (heterologous or ectopic recombination). (*c*) Type III, or a gene conversion event, no plasmid sequence would be detected within the genome but the gene (for example, *leu*) would apparently have been replaced by the introduced copy.

recombines in type II (heterologous recombination) has not been determined but there is presumably a requirement for short stretches of fortuitous sequence identity at the sites of apparently nonhomologous integration (Ward, 1991). In type III or a gene conversion event, no plasmid sequence would be detected within the genome, but the gene (*leu* in the example) would apparently have been replaced by the introduced copy (Figure 4.1(*c*)). Multiple plasmid copies integrated in tandem are a common feature of transformants. There are two obvious mechanisms through which this might come about: either

extrachromosomal plasmids first undergo homologous recombination with each other to form circular oligomers, which could then integrate by homology with the chromosomal, single copy; or a monomeric plasmid recombines with its chromosomal homolog and then tandem repeats arise through secondary integration (Fincham, 1989).

The use of linearized DNA has been shown to give higher efficiencies of integrative transformation in *S. cerevisiae* when compared to circular DNA (Suzuki *et al.*, 1983). In contrast, linearization of vector DNA does not seem to have any effect when transforming *N. crassa* (Huiet & Case, 1985) and it results in lower efficiencies when used for transformation in *Trichoderma* species (Herrera-Estrella, Goldman & Van Montagu, 1990). However, the use of linear molecules appears to increase significantly homologous recombination (Boylan *et al.*, 1987; Aramayo, Adams & Timberlake, 1989). Genomic integration of linear molecules often occurs by a process equivalent to a double cross-over event (Kinsey & Rambosek, 1984; Miller, Miller & Timberlake, 1985). However, circularization of linear molecules prior to integration can produce tandem duplications (Miller *et al.*, 1985).

Although integrated plasmids are mitotically stable, introduced DNA sequences are often meiotically unstable. In *A. nidulans*, tandemly repeated sequences are lost at variable, but readily detectable, frequencies after self-fertilization or out-crossing (Tilburn *et al.*, 1983). In *N. crassa*, duplicated sequences are eliminated at high frequency during the second phase by a process called 'repeat-induced point mutation' (RIP) (Selker *et al.*, 1987; Selker & Garret, 1988). Considering that most of the soil nonpathogenic fungi lack a sexual phase, the behavior of inserted DNA sequences in meiosis is irrelevant.

Use of transformation for analysis of gene function

Self-cloning in filamentous fungi

Although there are several reports on cloning of filamentous fungal genes in *Escherichia coli* and *S. cerevisiae* (Ballance, 1986; for a short review, see Schrank *et al.*, 1991; Goldman *et al.*, 1992), many fungal genes will not be expressed in these species and so cannot be isolated by interspecific complementation in these hosts. Additionally, any cloning method that involves isolation of the specific mRNA species depends on its abundance in the total mRNA population. This is not the case for many genes of interest, especially regulatory genes. Theoretically, the establishment of a transformation system could allow that fungal genes could be cloned by complementation of mutations, i.e. self-cloning. So, a genomic library of a particular fungal species would be constructed and used to transform the strain bearing a particular mutation. Transformants would be selected in which the mutation had been complemented and the plasmid bearing the sequence responsible for the complementation would be isolated again. According to Ward (1991), this type of self-cloning requires two features: (i) transformation frequencies must be high enough so that a gene bank can be screened using realistic quantities of DNA and protoplasts, and (ii) it must be possible to re-isolate or rescue any complementing plasmid. The lack of satisfactory shuttle vectors makes it somewhat more difficult to clone genes by complementation in filamentous fungi, but in some cases intact bacterial plasmids can be recovered from transformants. In cases in which there is no detectable free plasmid remaining, a transforming sequence can sometimes still be recovered by cleaving the transformant DNA with a restriction enzyme that cuts once, but no more than once, within the sequence duplicated through type I integration (see p. 39; Fincham, 1989). The fragments generated in this way are circularized with ligase and the reconstituted plasmid is selected by transformation of *E. coli* (Yelton *et al.*, 1984). Yelton, Timberlake & van den Hondel (1985) developed a more efficient way of cloning complementing *A. nidulans* genes by constructing a cosmid library – a plasmid with bacteriophage λ cos sequences – including *trpC*[+] for selection in *A. nidulans*, ampicillin and chloramphenicol resistance genes for selection in *E. coli*, and a *Bam*HI cloning site that would accept fragments cleaved with the 'tetra-cutting' endonuclease *Mbo*I or *Sau*3AI. A cosmid library carrying *A. nidulans* genomic fragments of 35 to 40 kb, was used to transform a *trpCyA* (yellow spored) double-mutant strain. DNA isolated from the *trp*+, green-spored selected transformants was packaged *in vitro* and the cosmids recovered by infection of a suitable *E. coli* strain, where they could be selected by ampicillin resistance. The identity of these clones was then confirmed by a new transformation experiment with the rescued *E. coli* plasmids.

Cosmid rescue has also been used to clone genes from one species by detecting their expression in a different species. For example, a disease-determinant gene, pisatin-demethylating ability (PDA) from the phytopathogen *Nectria haematococca* was cloned by detecting its expression in *A. nidulans* (Weltring *et al.*, 1988). It should be possible to use identical or related strategies to isolate genes of interest from any other fungal species. A recent example of this technology has been provided by the cellulolytic fungus *T. reesei*. Barreau *et al.* (1991) cloned an *A. niger* gene for invertase by transformation of the QM9A14 Ura5[-] strain of this species. First, the authors developed a very efficient transformation system for this species using pyrimidine auxotrophic *ura3* (*pyr*4) and *ura5* mutants (Bergès & Barreau, 1991). Then, a sibling (sib) selection (see below) procedure was carried out in order to clone the structural gene of an *A. niger* invertase by direct expression in *ura5*[+] transformants. Two cosmid clones were obtained and, using oligonucleotides derived from the N-terminus and internal tryptic peptides from the purified protein, one of the clones was identified as containing the gene for the secreted form of *A. niger* invertase.

A widely used method for gene identification in *N. crassa* is named sib selection (Akins & Lambowitz, 1985). Sib selection refers to the construction of an ordered gene bank in which *E. coli* clones, each containing a plasmid with a different *N. crassa* DNA fragment, are maintained separately. Transformation is originally by DNA extracted from pools of large numbers of these clones and subsequently by smaller and smaller subpools until the individual clone containing DNA capable of complementation of the *N. crassa* mutation of interest is identified.

An important point about cloning by complementation is the need to confirm that the cloned gene obtained is not really some other sequence that acts as a suppressor of the mutation being complemented. Initial tests to rule out this possibility are to show that the cloned gene will complement several different allelic mutations and, if available, deletions of that gene, or to show by several crossings that the cloned gene maps at the expected locus (provided the gene has been mapped in the first place) (Ward, 1991). In organisms that do not have a sexual cycle, these tests are not available and then, of course, confirmation of whether or not the cloned gene is a suppressor becomes more difficult.

Gene disruption and replacement

Homologous integration of transforming plasmids opens up the possibility of gene disruption and gene replacement techniques. It often happens that a cloned DNA sequence looks like a functional gene in that it is transcribed, contains an open reading frame, and perhaps has some interesting similarities to known genes in other organisms, but it cannot be assigned a function because no mutations have been identified in it. In such cases, the first step in understanding the gene is to use the clone to disrupt the equivalent sequence in the genome, thus creating a null mutant. In gene replacement, as opposed to gene disruption, the purpose is to retain gene activity but modify its product or its mode of transcription.

The most common methodology for gene disruption in filamentous fungi is the procedure called one-step disruption, originally described by Rothstein (1983) for *S. cerevisiae*. This procedure consists of inserting a copy of a selectable marker (e.g. *hph*, encoding resistance to the antibiotic hygromycin) into the cloned gene under investigation. This construction is then used to transform and convert a hygromycin-sensitive strain into a resistant one. There are many examples of successful gene disruption in *N. crassa* and *A. nidulans* (Miller *et al.*, 1985; Paietta & Marzluf, 1985). Hoskins *et al.* (1990) have carried out a gene disruption analysis in the antibiotic producer *C. acremonium*. In this case, the one-step disruption procedure using *hph* as a marker was inserted into the genomic region immediately upstream from *pcb*C (the gene coding for isopenicillin synthetase). Approximately 4% of the *C. acremonium* transformants obtained were unable to produce β-lactam antibiotics.

The first procedure for gene replacement in fungi was described for *S. cerevisiae* (Scherer & Davies, 1979). In this method, transformation was carried out with a plasmid containing both a modified form of the target gene and a separate selectable marker. Transformants with the plasmid integrated by homology into the target gene had tandemly arranged copies of both the target gene and the modified version that was to replace it, with the rest of the plasmid including the selec-

tive marker between them. After screening for subclones (after ten cycles of budding) that had lost the marker by crossing over between tandem gene copies, there was either restoration of the natural gene or its replacement, depending on whether the recombinant event occurred to the right or to the left of the sequence distinguishing the natural and modified gene copies (Scherer & Davies, 1979; Fincham, 1989). Miller *et al.* (1985), working with *A. nidulans*, replaced the *spo*C1C gene of the *spo*C1 (sporulation-specific) gene cluster with a partly deleted derivative following essentially the same method described above. A plasmid carrying the modified *spo*C1 sequence together with *trp*C$^+$ as a selectable marker was used to transform a *trp*C mutant strain. A transformant with *trp*C$^+$ integrated by crossing-over within *spo* tended to lose *trp*C$^+$ during vegetative growth by further crossing-over between the flanking *spo* sequences.

Titration of *trans*-acting gene products

The introduction by transformation of multiple copies of a *cis*-acting sequence that binds to the protein product of a *trans*-acting regulatory gene can give valuable information about the function of that gene (Fincham, 1989). Andrianopoulos & Hynes (1988) showed that multiple copies of some *cis*-acting regulatory sequences titrate away their corresponding *trans*-acting transcriptional regulators, leading to an inability to induce genes in the regulon. Titrations are not expected to be useful in those instances where there is a large excess of *trans*-acting factor or where the *trans*-active regulatory gene is autogenously controlled (Timberlake & Marshall, 1989).

Applications to biotechnology

One of the most obvious applications of transformation is the development of gene expression systems for filamentous fungi. These organisms are potentially attractive as host systems for heterologous gene expression because of their high secretory capacity. Many species of filamentous fungi, including *Aspergillus, Trichoderma, Achlya, Mucor, Penicillium* and *Cephalosporium* have been in commercial use for decades. Transformation is now being used to improve existing fungal strains by providing them with the genes to produce enzymes or antibodies. The introduction of 'new phenotypes' or overexpression of existing ones in a given species could theoretically improve the biological process carried out by it. Cross-species expression can also be exploited to study a number of biological and biochemical problems including structure/function relationships of a particular gene product in an isozyme-free background, evolutionary relationships between functionally equivalent genes, elements that control transcription, translation and post-translational modification of a gene product, and pathway engineering in fungi (Fowler & Berka, 1991).

Cross-species gene expression can be used to 'engineer' biosynthetic pathways by introducing cloned genes from one organism into another. For example, the successful cloning of a penicillin biosynthetic gene cluster from *Penicillium chrysogenum* has allowed the introduction of at least three linked genes that are necessary for penicillin biosynthesis into fungi that do not normally make penicillin (Smith *et al.*, 1990; Fowler & Berka, 1991). Increasing the copy number of a fungal gene through transformation usually leads to increased gene expression (Fowler & Berka, 1991). Recent examples of this concept include the expression of the genes for glucose oxidase (*gox*A) (Whittington *et al.*, 1990), glucoamylase (*gla*A) (Fowler, Berka & Ward, 1990), and prepro-polygalacturonidase II (*pga*II) (Bussink, Kester & Visser, 1990) in *A. niger* as well as β-glucosidase (*bgl*I) gene expression in *T. reesei* (Barnett, Berka & Fowler, 1991).

An intensive effort for increasing protein secretion is being carried out in the cellulolytic producer *Trichoderma reesei* (for a review, see Nevalainen *et al.*, 1991). This species is well known for its ability to produce large quantities of cellulase. Under the appropriate conditions, production strains secrete into the culture medium well over 50% of all the protein produced, which represents over 40 g/l (Knowles *et al.*, 1989). Another very attractive feature present in *T. reesei* is a protein glycosylation system that appears to modify extracellular proteins in a manner very similar to that found in mammalian cells (Salovouri *et al.*, 1987). Two interesting projects employing this species are increased cellulase production using genetic engineering and the use of strong promoters of the cellulase gene for heterologous gene expression. The former is being done by using cloned cellulase genes and strains with

completely different cellulase profiles (Nevalainen *et al.*, 1991). In this sense, transformation with individual cellulase genes would be useful and a simple tool to alter the quantitative pattern of cellulolytic enzymes produced by *T. reesei*. Kubicek-Pranz, Gruber & Kubicek (1991) have shown that, on increasing the copy number of one of the enzymes of the complex, cellobiohydrolase II, transformants exhibited an increased specific activity against crystalline cellulose *in vitro*.

To date, there is just one report of heterologous gene expression in *T. reesei*. Harkki *et al.* (1989) studied the ability of *T. reesei* to express and secrete bovine chymosin. They inserted chymosin complementary DNA (cDNA) sequences into expression units that included the promoter and terminator regions of the highly expressed cellobiohydrolase I (*cbh*1) gene. Several expression units were constructed that employed different configurations of *cbh*1 and chymosin signal peptides and they were introduced into *T. reesei*. In the resulting transformants, more than 90% of the chymosin that was synthesized was extracellular and yields of 40 mg/l were produced by some isolates. The chymosin produced by *T. reesei* was processed to an active form (Berka & Barnett, 1989).

Aspergillus niger and *A. awamori* also have the ability to secrete copious amounts of proteins in submerged culture. Since these strains are widely regarded as safe for the production of food-grade enzymes, harnessing even a portion of this capacity for the production of high value enzymes or pharmaceutical proteins could provide an economically significant advantage over more expensive approaches such as mammalian cell cultures. Extracellular yields of glucoamylase from *A. niger* could be improved by increasing the gene copy number (Berka & Barnett, 1989). Multiple copies of the *A. niger* glucoamylase (*gla*A) gene were introduced into both *A. niger* and *A. awamori*. Multiple, integrated *gla*A gene copies were found to be arranged in tandem repeats that were stable in the absence of selective pressure. Transformants that contained multiple copies of the *gla*A gene overproduced *gla*A-specific mRNA, resulting in increased enzyme synthesis. In a similar approach, *A. niger* was used for the production of hen egg-white lysozyme, which was correctly processed and folded, as shown by two-dimensional H-NMR. In this case, lysozyme was routinely produced at 10 mg/ml (Archer *et al.*, 1990). Also in

1990, Ward *et al.* selected a strain of *A. awamori* for the production of a glucoamylase–chymosin fusion protein. In their work, they reported the enzyme generated as being autocatalytically released from the fusion and active.

Fungi, however, are the most important class of plant pathogens and also affect plant productivity in positive ways. For example, symbiotic mycorrhizae increase the ability of plant roots to obtain limiting nutrients. Genetic engineering provides a new opportunity to study mechanisms controlling symbiosis and pathogenesis by allowing the identification of symbiosis- and pathogenicity-related genes.

Kolattukudy and colleagues provided an example of an enzyme as a pathogenicity determinant. In their work, they introduced the *Fusarium solani pisi* cutinase gene into *Mycosphaerella* species, a parasitic fungus that affects papaya fruits only if the fruit skin is mechanically breached before inoculation. The transformants obtained of the wound-requiring fungus were then capable of infecting intact papaya fruits (Dickman, Podilia & Kolattukudy, 1989).

Nectria haematococca, a fungal pathogen of pea, carries genes that encode pisatin demethylase (PDA), a cytochrome P-450 monooxygenase that detoxifies the phytoalexin pisatin. Because PDA is required by *N. haematococca* for pathogenicity on pea, pisatin helps to defend pea against *N. haematococca*. The possibility that pisatin is a general defense factor (i.e. PDA can confer pathogenicity to fungi not normally pathogenic on pea) was investigated by Schafer *et al.* (1989). Genes encoding PDA were transformed into and highly expressed in *Cochliobolus heterostrophus* (a fungal pathogen of maize but not of pea) and in *A. nidulans* (a saprophytic fungus), neither of which produces a significant amount of PDA. Recombinant *C. heterostrophus* was normally virulent on maize, but it also caused symptoms on pea, whereas recombinant *A. nidulans* did not affect pea.

In spite of all recent efforts to develop new techniques and approaches to study soil fungi, it is clear that, compared to the sophisticated molecular biological tools available for well-characterized organisms such as *E. coli* and *S. cerevisiae*, the development of gene expression systems for filamentous fungi is still at an early stage. Nevertheless, the potential advantages and commercial applications for expression of both homologous and heterologous genes continues to provide the impetus for intense research efforts.

Acknowledgements

The authors are grateful to Drs Allan Caplan, Dominique Van Der Straeten and June Simpson for critical reading of the manuscript, Martine De Cock for typing it, Karel Spruyt, Vera Vermaercke, and Stefaan Van Gijsegem for drawings and photographs.

References

Akins, R. A. & Lambowitz, A. M. (1985). General method for cloning *Neurospora crassa* nuclear genes by complementation of mutants. *Molecular and Cellular Biology*, 5, 2272–2278.

Alani, E., Cao, L. & Kleckner, N. (1987). A method for gene disruption that allows repeated use of *URA3* selection in the construction of multiply disrupted yeast strains. *Genetics*, **116**, 541–545.

Andrianopoulos, A. & Hynes, M. J. (1988). Cloning and analysis of the positively acting regulatory gene *amdR* from *Aspergillus nidulans*. *Molecular and Cellular Biology*, **8**, 3532–3541.

Aramayo, R., Adams, T. H. & Timberlake, W. E. (1989). A large cluster of highly expressed genes is dispensable for growth and development in *Aspergillus nidulans*. *Genetics*, **122**, 65–71.

Archer, D. B., Jeense, D. J., MacKenzie, D. A., Brightwell, G., Lambert, N., Lowe, G., Radford, S. E. & Dobson, C. M. (1990). Hen egg white lysozyme expressed in, and secreted from, *Aspergillus niger* is correctly processed and folded. *Bio/Technology*, **8**, 741–745.

Armaleo, D., Ye, G.-N., Klein, T. M., Shark, K. B., Sanford, J. C. & Johnston, S. A. (1990). Biolistic nuclear transformation of *Saccharomyces cerevisiae* and other fungi. *Current Genetics*, **17**, 97–103.

Austin, B., Hall, R. M. & Tyler, B. M. (1990). Optimized vectors and selection for transformation of *Neurospora crassa* and *Aspergillus nidulans* to bleomycin and phleomycin resistance. *Gene*, **93**, 157–162.

Bailey, A. M., Mena, G. L. & Herrera-Estrella, L. (1991). Genetic transformation of the plant pathogens *Phytophthora capsici* and *Phytophthora parasitica*. *Nucleic Acids Research*, **19**, 4273–4278.

Ballance, D. J. (1986). Sequences important for gene expression in filamentous fungi. *Yeast*, **2**, 229–236.

Ballance, J., Buxton, F. P. & Turner, G. (1983). Transformation of *Aspergillus nidulans* by the orotidine–5′-phosphate decarboxylase gene of

Neurospora crassa. *Biochemical and Biophysical Research Communications*, **112**, 284–289.

Ballance, D. J. & Turner, G. (1985). Development of a high-frequency transforming vector for *Aspergillus nidulans*. *Gene*, **36**, 321–331.

Barnett, C. C., Berka, R. M. & Fowler, T. (1991). Cloning and amplification of the gene encoding an extracellular β-glucosidase from *Trichoderma reesei*: evidence for improved rates of saccharinification of cellulosic substrates. *Bio/Technology*, **9**, 562–567.

Barreau, C., Boddy, L. M., Peberdy, J. F. & Berges, T. (1991). Direct cloning of an *Aspergillus niger* invertase gene by transformation of the QM9414 ura5⁻ strain of the cellulase producer *Trichoderma reesei*. EMBO Workshop on Molecular Biology of Filamentous Fungi, Berlin (FRG), 24–29 August, 1991, Abstract A3.

Barrett, V., Dixon, R. K. & Lemke, P. A. (1990). Genetic transformation of a mycorrhizal fungus. *Applied Microbiology and Biotechnology*, **33**, 313–316.

Bej, A. K. & Perlin, M. (1988). Apparent transformation and maintenance in basidiomycete mitochondria of a plasmid bearing the hygromycin B gene. *Genome*, **19** (Suppl. 1), 300.

Bergès, T. & Barreau, C. (1991). Isolation of uridine auxotrophs from *Trichoderma reesei* and efficient transformation with the cloned ura3 and ura5 genes. *Current Genetics*, **19**, 359–365.

Berka, R. M. & Barnett, C. C. (1989). The development of gene expression systems for filamentous fungi. *Biotechology Advances*, **7**, 127–154.

Binninger, D. M., Skrzynia, C., Pukkila, P. J. & Casselton, L. A. (1987). DNA-mediated transformation of the basidiomycete *Coprinus cinereus*. *EMBO Journal*, **6**, 835–840.

Boylan, M. T., Mirabito, P. M., Willett, C. E., Zimmerman, C. R. & Timberlake, W. E. (1987). Isolation and physical characterization of three essential conidiation genes from *Aspergillus nidulans*. *Molecular and Cellular Biology*, **7**, 3113–3118.

Bussink, H. J. D., Kester, H. C. M. & Visser, J. (1990). Molecular cloning, nucleotide sequence and expression of the gene encoding prepro-polygalacturonidase II of *Aspergillus niger*. *FEBS Letters*, **273**, 127–130.

Buxton, F. P., Gwynne, D. I. & Davies, R. W. (1985). Transformation of *Aspergillus niger* using the argB gene of *Aspergillus nidulans*. *Gene*, **37**, 207–214.

Buxton, F. P. & Radford, A. (1984). The transformation of mycelial spheroplasts of *Neurospora crassa* and the attempted isolation of an autonomous replicator. *Molecular and General Genetics*, **196**, 339–344.

Carramolino, L., Lozano, M., Pérez-Aranda, A., Rubio, V. & Sánchez, F. (1989). Transformation of *Penicillium chrysogenum* to sulfonamide resistance. *Gene*, **77**, 31–38.

Case, M. E. (1986). Genetical and molecular analyses of qa-2 transformants in *Neurospora crassa*. *Genetics*, **113**, 569–587.

Case, M. E., Schweizer, M., Kushner, S. R. & Giles, N. H. (1979). Efficient transformation of *Neurospora crassa* by utilizing hybrid plasmid DNA. *Proceedings of the National Academy of Sciences, USA*, **76**, 5259–5263.

Chakraborty, B. N., Patterson, N. A. & Kapoor, M. (1991). An electroporation-based system for high efficiency transformation of germinated conidia of filamentous fungi. *Canadian Journal of Microbiology*, **37**, 858–863.

Cheng, C., Tsukagoshi, N. & Udaka, S. (1990). Transformation of *Trichoderma viride* using the *Neurospora crassa* pyr4 gene and its use in the expression of a Taka-amylase A gene from *Aspergillus oryzae*. *Current Genetics*, **18**, 453–456.

Cooley, R. N., Shaw, R. K., Franklin, F. C. H. & Caten, C. E. (1988). Transformation of the phytopathogenic fungus *Septoria nodorum* to hygromycin B resistance. *Current Genetics*, **13**, 383–389.

Cullen, D., Wilson, L. J., Grey, G. L., Henner, D. J., Turner, G. & Ballance, D. J. (1987). Sequence and centromere proximal location of a transformation enhancing fragment ans1 from *Aspergillus nidulans*. *Nucleic Acids Research*, **15**, 9163–9175.

Daniell, H., Vivekananda, J., Nielsen, B. L., Ye, G. N. & Sanford, J. C. (1990). Transient foreign gene expression in chloroplasts of cultured tobacco cells after biolistic delivery of chloroplast vectors. *Proceedings of the National Academy of Sciences, USA*, **87**, 88–92.

Dhawale, S. S., Paietta, J. V. & Marzluf, G. A. (1984). A new, rapid and efficient transformation procedure for *Neurospora*. *Current Genetics*, **8**, 77–79.

Dickman, M. B. (1988). Whole cell transformation of the alfalfa fungal pathogen *Colletotrichum trifolii*. *Current Genetics*, **14**, 241–246.

Dickman, M. B. & Kolattukudy, P. E. (1987). Transformation of *Fusarium solani* f. sp. *pisi* using the cutinase promoter. *Phytopathology*, **77**, 1740.

Dickman, M. B., Podilia, G. K. & Kolattukudy, P. E. (1989). Insertion of cutinase gene into a wound pathogen enables it to infect intact host. *Nature (London)*, **342**, 446–448.

Diez, B., Alvarez, E., Cantoral, J. M., Barredo, J. L. & Martín, J. F. (1987). Selection and characterization of pyrG mutants of *Penicillium chrysogenum* lacking orotidine-5′-phosphate decarboxylase and complementation of the pyr4 gene of *Neurospora crassa*. *Current Genetics*, **12**, 277–282.

Dower, W. J., Miller, J. F. & Ragsdale, C. W. (1988). High efficiency transformation of *E. coli* by high voltage electroporation. *Nucleic Acids Research*, **16**, 6127–6145.

Edman, J. C. & Kwon-Chung, K. J. (1990). Isolation of the *URA5* gene from *Cryptococcus neoformans* var. *neoformans* and its use as a selective marker for transformation. *Molecular and Cellular Biology*, 10, 4538–4544.

Esser, K., Kuck, U., Stahl, U. & Tudzynski, P. (1983). Cloning vectors of mitochondrial origin for eukaryotes: a new concept in genetic engineering. *Current Genetics*, 13, 327–330.

Farman, M. L. & Oliver, R. P. (1988). The transformation of protoplasts of *Leptosphaeria maculans* to hygromycin B resistance. *Current Genetics*, 13, 327–330.

Fincham, J. R. S. (1989). Transformation in fungi. *Microbiology Reviews*, 53, 148–170.

Förster, W. & Neumann, E. (1989). Gene transfer by electroporation. A practical guide. In *Electroporation and Electrofusion in Cell Biology*, ed. E. Neumann, A. E. Sowers & C. A. Jordan, pp. 299–318. Plenum Press, New York.

Fowler, T. & Berka, R. M. (1991). Gene expression systems for filamentous fungi. *Current Opinion in Biotechnology*, 2, 691–697.

Fowler, T., Berka, R. M. & Ward, M. (1990). Regulation of the *glaA* gene of *Aspergillus niger*. *Current Genetics*, 18, 537–545.

Fox, T. D., Sanford, J. C. & McMullin, T. W. (1988). Plasmids can stably transform yeast mitochondria lacking endogenous mtDNA. *Proceedings of the National Academy of Sciences, USA*, 85, 7288–7292.

Fromm, M., Taylor, L. P. & Walbot, V. (1985). Expression of genes transferred into monocot and dicot plant cells by electroporation. *Proceedings of the National Academy of Sciences, USA*, 82, 5824–5828.

Futcher, A. B. (1988). The 2μ circle plasmid of *Saccharomyces cerevisiae*. *Yeast*, 4, 27–40.

Geisen, R. & Leistner, L. (1989). Transformation of *Penicillium nalgiovense* with the *amdS* gene of *Aspergillus nidulans*. *Current Genetics*, 15, 307–309.

Goettel, M. S., Leger, R. J. S., Bhairi, S., Jung, M. K., Oakley, B. R., Roberts, D. W. & Staples, R. C. (1990). Pathogenicity and growth of *Metarhizium anisopliae* stably transformed to benomyl resistance. *Current Genetics*, 17, 129–132.

Goldman, G. H., Demolder, J., Dewaele, S., Herrera-Estrella, A., Geremia, R. A., Van Montagu, M. & Contreras, R. (1992). Molecular cloning of the imidazole-glycerolphosphate dehydratase gene of *Trichoderma harzianum* by genetic complementation in *Saccharomyces cerevisiae* using a direct expression vector. *Molecular and General Genetics*, 234, 481–488.

Goldman, G. H., Temmerman, W., Jacobs, D., Contreras, R., Van Montagu, M. & Herrera-Estrella, A. (1993). A nucleotide substitution in one of the ß-tubulin genes of *Trichoderma viride* confers resistance to the antimitotic drug methyl benzimidazole–2-yl-carbamate. *Molecular and General Genetics*, 240, 73–80.

Goldman, G. H., Van Montagu, M. & Herrera-Estrella, A. (1990). Transformation of *Trichoderma harzianum* by high-voltage electric pulse. *Current Genetics*, 17, 169–174.

Harkki, A., Uusitalo, J., Bailey, M., Penttilä, M. & Knowles, J. K. C. (1989). A novel fungal expression system: secretion of active calf chymosin from the filamentous fungus *Trichoderma reesei*. *Bio/Technology*, 7, 596–603.

Henson, J. M., Blake, N. K. & Pilgeram, A. L. (1988). Transformation of *Gaeumannomyces graminis* to benomyl resistance. *Current Genetics*, 14, 113–117.

Herrera-Estrella, A., Goldman, G. H. & Van Montagu, M. (1990). High-efficiency transformation system for the biocontrol agents, *Trichoderma* spp. *Molecular Microbiology*, 4, 839–843.

Hinnen, A., Hicks, J. B. & Fink, G. R. (1978). Transformation of yeast. *Proceedings of the National Academy of Sciences, USA*, 75, 1929–1933.

Holden, D. W., Wang, J. & Leong, S. A. (1988). DNA-mediated transformation of *Ustilago hordei* and *Ustilago nigra*. *Physiological and Molecular Plant Pathology*, 33, 235–239.

Hoskins, J. A., O'Callaghan, N., Queener, S. W., Cantwell, C. A., Wood, J. S., Chen, C. J. & Skatrud, P. L. (1990). Gene disruption of the *pcb*AB gene encoding ACV synthetase in *Cephalosporium acremonium*. *Current Genetics*, 18, 523–530.

Huiet, L. & Case, M. (1985). Molecular biology of the *qa* gene cluster of *Neurospora*. In *Gene Manipulations in Fungi*, ed. J. W. Bennett & L. L. Lasure, pp. 229–244. Academic Press, Orlando, FL.

Ito, H., Fukuda, Y., Murata, K. & Kimura, A. (1983). Transformation of intact yeast cells treated with alkali cations. *Journal of Bacteriology*, 153, 163–168.

Johnston, S. A., Anziano, P. Q., Shark, K., Sanford, J. C. & Butow, R. A. (1988). Mitochondrial transformation in yeast by bombardment with microprojectiles. *Science*, 240, 1538–1541.

Kinsey, J. A. & Rambosek, J. A. (1984). Transformation of *Neurospora crassa* with the cloned *am* (glutamate dehydrogenase) gene. *Molecular and Cellular Biology*, 4, 117–122.

Kistler, H. C. & Benny, U. K. (1988). Genetic transformation of the fungal plant wilt pathogen, *Fusarium oxysporum*. *Current Genetics*, 13, 145–149.

Klein, T. M., Fromm, M., Weissinger, A., Tomes, D., Schaaf, S., Sletten, M. & Sanford, J. C. (1988*a*). Transfer of foreign genes into intact maize cells with high-velocity microprojectiles. *Proceedings of the National Academy of Sciences, USA*, 85, 4305–4309.

Klein, T. M., Gradziel, T., Fromm, M. E. & Sanford, J. C. (1988*b*). Factors influencing gene delivery into *Zea mays* cells by high-velocity microprojectiles. *Bio/Technology*, 6, 559–563.

Klein, T. M., Harper, E. C., Svab, Z., Sanford, J. C., Fromm, M. E. & Maliga, P. (1988c). Stable genetic transformation of intact *Nicotiana* cells by the particle bombardment process. *Proceedings of the National Academy of Sciences, USA*, **85**, 8502–8505.

Klein, T. M., Wolf, E. D., Wu, R. & Sanford, J. C. (1987). High-velocity microprojectiles for delivering nucleic acids into living cells. *Nature (London)*, **327**, 70–73.

Knowles, J., Penttilä, M., Harkki, A., Nevalainen, H., Teeri, T., Saloheimo, M. & Uusitalo, J. (1989). Applications of the molecular biology of *Trichoderma reesei*. In *Molecular Biology of Filamentous Fungi*, Foundation for Biotechnical and Industrial Fermentation Research, Vol. 6, ed. H. Nevalainen & M. Penttilä, pp. 113–118. Foundation for Biotechnical and Industrial Fermentation Research, Helsinki.

Kubicek-Pranz, E. M., Gruber, F. & Kubicek, C. P. (1991). Transformation of *Trichoderma reesei* with the cellobiohydrolase II gene as a means for obtaining strains with increased cellulase production and specific activity. *Journal of Biotechnology*, **20**, 83–94.

Kuiper, M. T. R. & de Vries, H. (1985). A recombinant plasmid carrying the mitochondrial plasmid sequence of *Neurospora intermedia* LaBelle yields new plasmid derivatives in *Neurospora crassa* transformants. *Current Genetics*, **9**, 471–477.

Mattern, I. E., Unkles, S., Kinghorn, J. R., Pouwels, P. H. & van den Hondel, C. A. M. J. J. (1987). Transformation of *Aspergillus oryzae* using the *A. niger pyrG* gene. *Molecular and General Genetics*, **210**, 460–461.

Maundrell, K., Wright, A. P. H., Piper, M. & Shall, S. (1985). Evaluation of heterologous ARS activity in *S. cerevisiae* using cloned DNA from *S. pombe*. *Nucleic Acids Research*, **13**, 3711–3722.

Miller, B. L., Miller, K. Y. & Timberlake, W. E. (1985). Direct and indirect gene replacements in *Aspergillus nidulans*. *Molecular and Cellular Biology*, **5**, 1714–1721.

Miller, J. F., Dower, W. J. & Tompkins, L. S. (1988). High-voltage electroporation of bacteria: genetic transformation of *Campylobacter jejuni* with plasmid DNA. *Proceedings of the National Academy of Sciences, USA*, **85**, 856–860.

Mishra, N. C. (1985). Gene transfer in fungi. *Advances in Genetics*, **23**, 73–178.

Mishra, N. C. & Tatum, E. L. (1973). Non-Mendelian inheritance of DNA-induced inositol independence in *Neurospora*. *Proceedings of the National Academy of Sciences, USA*, **70**, 3875–3879.

Nevalainen, H., Penttilä, M., Harkki, A., Teeri, T. & Knowles, J. (1991). The molecular biology of *Trichoderma* and its application to the expression of both homologous and heterologous genes. In

Molecular Industrial Mycology, ed. S. A. Leong & R. M. Berka, pp. 129–148. Marcel Dekker, New York.

Oliver, R. P., Roberts, I. N., Harling, R., Kenyon, L., Punt, P. J., Dingemanse, M. A. & van den Hondel, C. A. M. J. J. (1987). Transformation of *Fulvia fulva*, a fungal pathogen of tomato, to hygromycin B resistance. *Current Genetics*, **12**, 231–233.

Paietta, J. V. & Marzluf, G. A. (1985). Gene disruption by transformation of *Neurospora crassa*. *Molecular and Cellular Biology*, **5**, 1554–1559.

Panaccione, D. G., McKiernan, M. & Hanau, R. M. (1988). *Colletotrichum graminicola* transformed with homologous and heterologous benomyl-resistance genes retains expected pathogenicity to corn. *Molecular Plant–Microbe Interactions*, **1**, 113–120.

Parsons, K. A., Chumley, F. G. & Valent, B. (1987). Genetic transformation of the fungal pathogen responsible for rice blast disease. *Proceedings of the National Academy of Sciences, USA*, **84**, 4161–4165.

Penttilä, M., Nevalainen, H., Rättö, M., Salminen, E. & Knowles, J. (1987). A versatile transformation system for the cellulolytic filamentous fungus *Trichoderma reesei*. *Gene*, **61**, 155–164.

Rambosek, J. & Leach, J. (1987). Recombinant DNA in filamentous fungi: progress and prospects. *CRC Critical Reviews in Biotechnology*, **6**, 357–393.

Richey, M. G., Marek, E. T., Schardl, C. L. & Smith, D. A. (1989). Transformation of filamentous fungi with plasmid DNA by electroporation. *Phytopathology*, **79**, 844–847.

Rodriguez, R. J. & Yoder, O. C. (1987). Selectable genes for transformation of the fungal plant pathogen *Glomerella cingulata* f. sp. *phaseoli* (*Colletotrichum lindemuthianum*). *Gene*, **54**, 73–81.

Roncero, M. I. G., Jepsen, L. P., Strøman, P. & van Heeswijck, R. (1989). Characterization of a *leuA* gene and an *ARS* element from *Mucor circinelloides*. *Gene*, **84**, 335–343.

Rothstein, R. J. (1983). One-step gene disruption in yeast. In *Recombinant DNA*, part C, *Methods in Enzymology*, Vol. 101, ed. R. Wu, L. Grossman & K. Moldave, pp. 202–211. Academic Press, New York.

Salch, Y. P. & Beremand, M. N. (1988). Development of a transformation system for *Fusarium sambucinum*. *Journal of Cellular Biochemistry*, Suppl. 12C, 290 (No. Y 341).

Salovuori, I., Makarow, M., Rauvala, H., Knowles, J. & Kääriäinen, L. (1987). Low molecular weight high-mannose type glycans in a secreted protein of the filamentous fungus *Trichoderma reesei*. *Bio/Technology*, **5**, 152–156.

Samac, D. A. & Leong, S. A. (1989). Mitochondrial plasmids of filamentous fungi: characteristics and use in transformation vectors. *Molecular Plant–Microbe Interactions*, **2**, 155–159.

Sanford, J. C., Smith, F. D. & Russell, J. A. (1993).
Optimizing the biolistic process for different
biological applications. In *Recombinant DNA*, part
H, *Methods in Enzymology*, Vol. 217, ed. R. Wu,
pp 483–508. Academic Press, San Diego.

Schafer, W., Straney, D., Ciufetti, L., Van Etten,
H. D. & Yoder, O. C. (1989). One enzyme makes a
fungal pathogen, but not a saprophyte, virulent on a
new host plant. *Science*, **246**, 247–249.

Scherer, S. & Davis, R. W. (1979). Replacement of
chromosome segments with altered DNA sequences
constructed *in vitro*. *Proceedings of the National
Academy of Sciences, USA*, **76**, 4951–4955.

Schrank, A., Tempelaars, C., Sims, P. F. G., Oliver,
S. G. & Broda, P. (1991). The *trpC* gene of
Phanerochaete chrysosporium is unique in containing
an intron but nevertheless maintains the order of
functional domains seen in other fungi. *Molecular
Microbiology*, **5**, 467–476.

Selker, E. U., Cambareri, E. B., Jensen, B. C. &
Haack, K. R. (1987). Rearrangement of duplicated
DNA in specialized cells of Neurospora. *Cell*, **51**,
741–752.

Selker, E. U. & Garrett, P. W. (1988). DNA sequence
duplications trigger gene inactivation in *Neurospora
crassa*. *Proceedings of the National Academy of Sciences,
USA*, **85**, 6870–6874.

Shark, K. B., Smith, F. D., Harpending, P. R.,
Rasmussen, J. L. & Sanford, J. C. (1991). Biolistic
transformation of a procaryote, *Bacillus megaterium*.
Applied and Environmental Microbiology, **57**, 480–485.

Shigekawa, K. & Dower, W. J. (1988). Electroporation
of eukaryotes and prokaryotes: a general approach to
the introduction of macromolecules into cells.
BioTechniques, **6**, 742–751.

Skatrud, P. L., Queener, S. W., Carr, L. G. & Fisher,
D. L. (1987). Efficient integrative transformation of
Cephalosporium acremonium. *Current Genetics*, **12**,
337–348.

Smith, D. J., Burnham, M. K. R., Edwards, J., Earl,
A. J. & Turner, G. (1990). Cloning and
heterologous expression of the penicillin biosynthetic
gene cluster from *Penicillium chrysogenum*.
Bio/Technology, **8**, 39–41.

Smith, J. L., Bayliss, F. T. & Ward, M. (1991).
Sequence of the cloned *pyr4* gene of *Trichoderma
reesei* and its use as a homologous selectable marker
for transformation. *Current Genetics*, **19**, 27–33.

Stahl, U., Tudzynski, P., Kück, U. & Esser, K.
(1982). Replication and expression of a
bacterial–mitochondrial hybrid plasmid in the
fungus *Podospora anserina*. *Proceedings of the National
Academy of Sciences, USA*, **79**, 3641–3645.

Stohl, L. L. & Lambowitz, A. M. (1983).
Construction of a shuttle vector for the filamentous
fungus *Neurospora crassa*. *Proceedings of the National
Academy of Sciences, USA*, **80**, 1058–1062.

Struhl, K. (1983). The new yeast genetics. *Nature
(London)*, **305**, 391–397.

Suzuki, K., Imai, Y., Yamashita, I. & Fukui, S.
(1983). In vivo ligation of linear DNA molecules to
circular forms in the yeast *Saccharomyces cerevisiae*.
Journal of Bacteriology, **155**, 747–754.

Taylor, J. W., Smolich, B. D. & May, G. (1985). An
evolutionary comparison of homologous
mitochondrial plasmid DNAs from three *Neurospora*
species. *Molecular and General Genetics*, **201**,
161–167.

Thomas, M. D. & Kenerley, C. M. (1989).
Transformation of the mycoparasite *Gliocladium*.
Current Genetics, **15**, 415–420.

Tilburn, J., Scazzocchio, C., Taylor, G. G., Zabicky-
Zissman, J. H., Lockington, R. A. & Davies, R. W.
(1983). Transformation by integration in *Aspergillus
nidulans*. *Gene*, **26**, 205–221.

Timberlake, W. E. & Marshall, M. A. (1989). Genetic
engineering of filamentous fungi. *Science*, **244**,
1313–1317.

Tsukuda, T., Carleton, S., Fotheringham, S. &
Holloman, W. K. (1988). Isolation and
characterization of an autonomously replicating
sequence from *Ustilago maydis*. *Molecular and
Cellular Biology*, **8**, 3703–3709.

Turgeon, B. G., Garber, R. C. & Yoder, O. C. (1985).
Transformation of the fungal maize pathogen
Cochliobolus heterostrophus using the *Aspergillus
nidulans amdS* gene. *Molecular and General Genetics*,
201, 450–453.

Turgeon, B. G., Garber, R. C. & Yoder, O. C. (1987).
Development of a fungal transformation system
based on selection of sequences with promoter
activity. *Molecular and Cellular Biology*, **7**, 3297–3305.

Ulhoa, C. J., Vainstein, M. H. & Pederby, J. F.
(1992). Transformation of *Trichoderma* species with
dominant selectable markers. *Current Genetics*, **21**,
23–26.

Van Heeswijck, R. (1986). Autonomous replication of
plasmids in *Mucor* transformants. *Carlsberg Research
Communications*, **51**, 433–443.

Wang, J., Holden, D. W. & Leong, S. A. (1988). Gene
transfer system for the phytopathogenic fungus
Ustilago maydis. *Proceedings of the National Academy
of Sciences, USA*, **85**, 865–869.

Ward, M. (1991). *Aspergillus nidulans* and other
filamentous fungi as genetic systems. In *Modern
Microbial Genetics*, ed. U. N. Streips & R. E. Yasbin,
pp. 455–496. Wiley-Liss, New York.

Ward, M., Kodama, K. H. & Wilson, L. J. (1989).
Transformation of *Aspergillus awamori* and *A. niger*
by electroporation. *Experimental Mycology*, **13**,
289–293.

Ward, M., Wilson, L. J., Kodama, K. H., Rey, M. W.
& Berka, R. M. (1990). Improved production of
chymosin in *Aspergillus* by expression as a

glucoamylase–chymosin fusion. *Bio/Technology*, **8**, 435–440.

Weltring, K.-M., Turgeon, B. G., Yoder, O. C. & VanEtten, H. D. (1988). Isolation of a phytoalexin-detoxification gene from the plant pathogenic fungus *Nectria haematococca* by detecting its expression in *Aspergillus nidulans*. *Gene*, **68**, 335–344.

Whitehead, M. P., Gurr, S. J., Grieve, C., Unkles, S. E., Spence, D., Ramsden, M. & Kinghorn, J. R. (1990). Homologous transformation of *Cephalosporium acremonium* with the nitrate reductase-encoding gene (*niaD*). *Gene*, **90**, 193–198.

Whittington, H., Kerry-Williams, S., Bidgood, K., Dodsworth, N., Peberdy, J., Dobson, M., Hinchliffe, E. & Ballance, D. J. (1990). Expression of the *Aspergillus niger* glucose oxidase gene in *A. niger*, *A. nidulans* and *Saccharomyces cerevisiae*. *Current Genetics*, **18**, 531–536.

Williams, R. S., Johnston, S. A., Riedy, M., DeVit, M. J., McElligott, S. G. & Sanford, J. C. (1991). Introduction of foreign genes into tissues of living mice by DNA-coated microprojectiles. *Proceedings of the National Academy of Sciences, USA*, **88**, 2726–2730.

Yelton, M. M., Hamer, J. E. & Timberlake, W. E. (1984). Transformation of *Aspergillus nidulans* by using a *trpC* plasmid. *Proceedings of the National Academy of Sciences, USA*, **81**, 1470–1474.

Yelton, M. M., Timberlake, W. E. & van den Hondel, C. A. M. J. J. (1985). A cosmid for selecting genes by complementation in *Aspergillus nidulans*: selection of the developmentally regulated *yA* locus. *Proceedings of the National Academy of Sciences, USA*, **82**, 834–838.

Yoder, O. C. & Turgeon, B. G. (1985). Molecular bases of fungal pathogenicity to plants. In *Gene Manipulations in Fungi*, ed. J. W. Bennett & L. L. Lasure, pp. 417–448. Academic Press, Orlando, FL.

PART II TRANSFORMATION OF CEREAL CROPS

5

Rice Transformation: Methods and Applications

Junko Kyozuka and Ko Shimamoto

Introduction

Genetic transformation offers new approaches to many fundamental problems in plant biology. In particular, transgenic plants are essential tools in understanding *in vivo* functions of plant genes and molecular mechanisms of their regulation (Schell, 1987). In addition, genetic transformation provides novel approaches to crop improvement. Early examples are herbicide- (Comai *et al.*, 1985; De Block *et al.*, 1987), virus- (Powell-Abel *et al.*, 1986) and insect- (Vaeck *et al.*, 1987; Perlak *et al.*, 1990) resistant dicotyledonous crops such as tobacco and potato. Recently, to improve agronomically important traits, more sophisticated strategies have been designed and applied to a number of crop species (Fraley, 1992).

In contrast to dicotyledonous species, production of transgenic monocotyledonous (monocot) plants has been difficult despite extensive efforts over many years. Because of this, expression of genes derived from monocot species is often examined in transgenic dicot species (Lamppa, Nagy & Chua, 1985; Keith & Chua, 1986; Colot *et al.*, 1987; Ellis *et al.*, 1987). However, increasing evidence obtained from transgenic experiments suggests that there are differences in regulation of gene expression between monocot and dicot species. Monocot genes are not always expressed correctly or at all in transgenic dicot plants (Keith & Chua, 1986; Colot *et al.*, 1987; Ellis *et al.*, 1987). Moreover, anatomical differences between monocot and dicot species often make it difficult to evaluate accurately tissue/cell specific expression of monocot-derived genes in transgenic dicot plants (Schernthaner, Matzke &

Matzke, 1988; Lloyd *et al.*, 1991; Matsuoka & Sanada, 1991).

Transgenic monocot plants are therefore required for elucidation of monocot gene expression and of differences in the system of gene expression between monocot and dicot plants; a monocot species amenable to genetic engineering is needed as a model plant for studies in gene regulation of monocot genes. In addition, the development of techniques for routine generation of transgenic monocots is of critical importance in plant breeding, because they include major crop species such as rice, maize, wheat and barley.

Recently, routine generation of fertile transgenic rice plants (*Oryza sativa*) by direct DNA transfer to protoplasts has been achieved (Shimamoto *et al.*, 1989; Shimamoto, 1991). In addition to this, rice has a number of favorable features as a model monocot plant that are shared with *Arabidopsis thaliana*, a model dicot plant for plant molecular genetics (Meyerowitz, 1989). These features include well-developed linkage and restriction fragment length polymorphism (RFLP) maps (Kinoshita, 1984; McCouch *et al.*, 1988; Wang & Tanksley, 1989), and a small genome three to five times as big as that of *A. thaliana* that contains less repetitive DNA compared to other plant species (Zhao *et al.*, 1989; Nishibayashi, 1992)

In this chapter, we review the present status of rice transformation (Table 5.1), and discuss key techniques and unsolved problems in the methods for generation of transgenic rice plants. Furthermore, we describe results of recent research activities making use of transgenic rice and discuss the potential of rice as a model monocot species for

Table 5.1. *Transgenic rice plants*

Cultivar	Method	Selection	Introduced genes	Fertility	Reference
Protoplast transformation					
Yamahoushi	EP	G418	35S-*npt*II, 35S-*uid*A	NT	Toriyama *et al.* (1988)
Pi4, Taipei 309	PEG	—	maize *adh*1-*uid*A	NT	Zhang & Wu (1988)
Taipei 309	EP	Km^R	35S-*npt*II	NT	Zhang *et al.* (1988)
Yamahoushi	EP	Hm^R	35S-*hph*, *rol*C-*uid*A	NT	Matsuki *et al.* (1989)
Nipponbare	EP	Hm^R	35S-*hph*, 35S-*uid*A	+	Shimamoto *et al.* (1989)
Taipei 309	EP	Km^R, G418	35S-*npt*II, 35S-*uid*A	NT	Bauraw & Hall (1990)
Chinsurah Boro II	PEG	Hm^R	35S-*hph*	+	Datta *et al.* (1990)
Nipponbare, Taipei 309	PEG	Hm^R	35S-*hph*	NT	Hayashimoto & Murai (1990)
IR54	PEG	Km^R	35S-*npt*II, 35S-*uid*A	NT	Peng *et al.* (1990)
Norin 8, Sasanishiki	EP	Hm^R	35S-*hph*, 35S-*uid*A	+	Tada *et al.* (1990)
Nipponbare	EP	Hm^R	35S-*hph*, 35S-*uid*A	+	Terada & Shimamoto (1990)
Nipponbare	EP	Hm^R	35S-*hph*, maize *adh*1-*uid*A	NT	Kyozuka *et al.* (1990)
Taipei 309	EP	Km^R	35S-*npt*II	+	Davey *et al.* (1991)
Toridel	EP	Hm^R	35S-*hph*, maize	NT	Izawa *et al.* (1991)
Nipponbare	EP	Hm^R	35S-*hph*, maize *adh*1-*uid*A	+	Kyozuka *et al.* (1991)
Taipei 309	EP, PEG	Hm^R, Mtx^R	35S-*hph*, p1',2'-*uid*A, 35S-*dhfr*	NT	Meijer *et al.* (1991)
Nipponbare	PEG	Hm^R	35S-*hph*, maize *Ac*	NT	Murali *et al.* (1991)
Nipponbare	EP	Hm^R	35S-*hph*, rice *cab*-*uid*A	+	Tada *et al.* (1991)
Taipei 309	PEG	—	rice *actin*-*uid*A	+	Zhang *et al.* (1991)
Microprojectile bombardment					
Gulfmont		Hm, Bialaphos	35S-*hph*, 35S-*bar* 35S-*uid*A	+	Christou *et al.* (1991)
IR 54, IR 26 IR 36, IR 72					

Notes:
EP, electroporation; PEG, polyethylene glycol; G418, synthetic aminoglycoside antibiotic, geneticin; Km, kanamycin; Hm, hygromycin; Mtx, methotrexate; 35S, CaMV 35S RNA promoter; *adh*1, maize alcohol dehydrogenase gene 1 promoter; *rol*C, promoter from a pathogenesis-related gene of the TL-DNA of the *Agrobacterium rhizogenes* Ri plasmid; P1', 2', T-DNA 1' and 2' gene promoters from *Agrobacterium tumefaciens*; *npt*II, neomycin phosphotransferase II gene; *uid*A, β-glucuronidase gene; *hph*, hygromycin phosphotransferase gene; *dhfr*, dihydrofolate reductase gene; *bar*, phosphinothrin acetyltransferase gene; *Ac*, maize autonomous transposable element activator; NT, not tested.

studies in regulation of gene expression that is unique to monocot plants.

Methods for generation of transgenic rice plants

Protoplast transformation

Agrobacterium-mediated transformation is routinely used for production of transgenic dicot plants. However, monocot species, especially cereals, are at best weakly susceptible to *Agrobacterium tumefaciens*. Thus, many laboratories have sought alternative transformation methods for cereals. Among a number of methods studied, protoplast transformation and microprojectile bombardment have been successfully used to produce fertile transgenic plants in rice and maize (*Zea mays*) (Table 5.1).

In cereals, rice is exceptional in that plant regeneration from protoplasts has been well established (Abdullah, Cocking & Thompson, 1986; Toriyama, Hinata & Sasaki, 1986; Yamada, Yang & Tang, 1986; Kyozuka, Hayashi & Shimamoto, 1987). Therefore, rice transformation can be routinely performed with direct DNA transfer into protoplasts (Table 5.1) and subsequent plant regeneration. Production of transgenic maize (Rhodes *et al.*, 1988) and orchardgrass (*Dactylis glomerata*; Horn *et al.*, 1988) by protoplast transformation has also been achieved; however, their fertilities have yet to be demonstrated.

The 'quality' of embryogenic suspension cultures from which protoplasts are isolated, is one of the most critical factors for successful production of transgenic plants. Normally we use protoplasts isolated from embryogenic suspension cultures derived from calli originating from mature seeds. Such protoplasts show high plating efficiency (*ca* 10%). A detailed protocol for initiation and maintenance of embryogenic suspension cultures was described by Kyozuka, Shimamoto & Ogura (1989) and Kyozuka & Shimamoto (1991). Suspension cultures of most japonica cultivars produce protoplasts capable of high frequency colony formation 1–2 months after initiation of the suspension culture. Suspension cultures are renewed every 6 months to avoid loss of their morphogenic capacity during prolonged culture.

The efficiency of generating transgenic rice plants varies depending on cultivars, presumably due to differences in their adaptability to culture conditions *in vitro* and in their competence for accepting foreign DNA. Although transgenic plants have been obtained in some indica rice (Datta *et al.*, 1990), protoplast culture of indica varieties is generally more difficult than that of japonica varieties (Kyozuka, Otoo & Shimamoto, 1988). For the recalcitrant varieties, including some japonica varieties and most of indica varieties, further improvement of culture conditions is necessary to generate transgenic plants reproducibly.

Both electroporation and PEG treatment are used to introduce foreign DNA into rice protoplasts (Table 5.1). When conditions are optimized, the efficiency of DNA uptake by protoplasts does not seem to be much different between these two methods. When protoplast transformation was compared with the microprojectile bombardment, the former is more frequently used for production of transgenic calli and plants in a wide variety of cereal species, including maize (Rhodes *et al.*, 1988), barley (*Hordeum vulgare*; Lazzeri *et al.*, 1991) and orchardgrass (Horn *et al.*, 1988).

Recently, production of fertile transgenic indica and japonica rice plants by microprojectile bombardment has been reported; the scutellar tissue of immature embryo was bombarded and plants were regenerated from calli derived from the transformed tissue (Christou, Ford & Kofron, 1991). The main advantage of this technique is the minimum requirement of tissue culture, which makes the method relatively genotype independent. However, two steps need to be further examined: the frequency of the plant regeneration from transformed calli and the degree of chimera formation in primary transgenic plants. Further-

more, collection of a large number of immature embryos is laborious and requires a constant supply of mature plants. Regeneration of fertile transgenic maize plants using microprojectile bombardment has also been reported by two groups (Fromm *et al.*, 1990; Gordon-Kamm *et al.*, 1990). Both groups adopted embryogenic suspension cells as target materials. At present, the success is limited to a few genotypes and the frequency of transformation is relatively low. Therefore, the frequency of plant regeneration from transformed cells should be improved to establish the method for the routine production of transgenic maize plants.

Selectable markers

The effective selection of transformed cells depends greatly on selectable markers and the selection procedures used. Hygromycin (*hph*), kanamycin (*npt*II), and bialaphos (*bar*) resistance genes have been used as selectable marker genes for rice transformation (Table 5.1). However, effective selection of transformed cells is not always easy with the Km^R marker because rice cells generally have background resistance to kanamycin, and furthermore albino or sterile plants were often obtained from rice callus after selection with G418, another aminoglycoside inactivated by the NPT II enzyme (K. Shimamoto, unpublished data). Although Dekeyser *et al.* (1989) suggested that the *bar* gene encoding phosphinothricin acetyltransferase (PAT), which confers resistance to herbicide bialaphos, is an efficient selection marker gene in rice, its usefulness in rice transformation needs to be further examined.

In our procedure, selection with hygromycin B is started after 10 to 14 days from the initiation of protoplast culture and selection is repeated twice for 7–10 days each (Figure 5.1). Our preliminary experiments showed that transformation with linearized plasmid carrying the marker gene produced more Hm^R calli than that with circular plasmid. Hm^R calli thus selected are transferred to the regeneration medium and then shoots arise from these transformed calli within 2–6 weeks. According to our procedure, transformed rice plantlets can be obtained within 8–10 weeks after electroporation. The frequency of plant regeneration from Hm^R calli is approximately 60%–80% (for a detailed protocol, see Kyozuka &

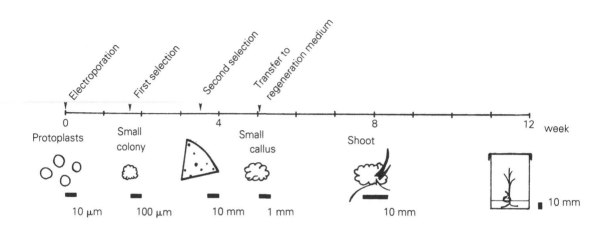

Figure 5.1. Steps in protoplast transformation in rice. Electroporated protoplasts are embedded in agarose blocks and cultured by mixed nurse culture method (Kyozuka *et al.*, 1987). After 10 days of the culture, the nurse cells are removed and agarose blocks are transferred to fresh medium containing 30 µg hygromycin B/ml (first selection). Hm-resistant colonies become visible (0.5–1.0 mm in diameter) at the end of the first selection; then agarose blocks are transferred on to solid medium with 30 µg hygromycin B/ml (second selection). At the end of the second selection (*ca* 5 weeks after electroporation), Hm-resistant calli become about 2 mm in diameter. Then the Hm-resistant (transformed) calli are transferred to regeneration medium without hygromycin. Shoots are obtained from 60%–80% of Hm-resistant calli within 1–2 months from the transfer to regeneration medium.

Shimamoto, 1991). Albino plants rarely appeared after selection with hygromycin, however, somaclonal variations seem to occur more often in transgenic plants than in plants regenerated from nontransformed protoplasts. Also, low fertility was observed in some transgenic plants probably due to somaclonal variations and ploidy changes.

Intron enhancement of gene expression

In cereal species, steady-state levels of mRNA and proteins are increased by the presence of introns in vector constructs. So far, the enhancing effect of introns has been found not only with introns from cereal species such as maize *adh*1 introns 1, 2, 3, 6, 8, and 9 (Kyozuka *et al.*, 1990; Callis, Fromm & Walbot, 1987; Oard, Paige & Dvorak, 1989; Mascarenhas *et al.*, 1990; Luehrsen & Walbot, 1991), maize *bz* intron 1 (Callis *et al.*, 1987), maize *sh*1 intron 1 (Vasil *et al.*, 1989), maize actin gene *(act)* intron 3 (Luehrsen & Walbot, 1991) and rice actin gene *(act)* intron 1 (McElroy *et al.*, 1990), but also with intron 1 of the catalase gene *(cat)* derived from the dicot species castor bean (Tanaka *et al.*, 1990). Stimulating effects of the first intron of maize *adh*1 gene (Kyozuka *et al.*, 1990), rice *act* gene intron 1 (McElroy *et al.*, 1990) and castor bean *cat*

intron 1 (Tanaka *et al.*, 1990) on the expression of the β-glucouronidase (*uid*A) reporter gene have been described in rice.

The degree of enhancement is dependent on the origin of the intron, the sequences flanking the intron, the reporter gene and the host cells used for experiments. The first intron of maize *adh*1 gene increased the β-glucuronidase (GUS) activity in rice protoplasts four- to six-fold when the 35S promoter of cauliflower mosaic virus (CaMV) was used, while 11- to 18-fold increases were observed with the maize *adh*1 promoter (Kyozuka *et al.*, 1990). These results are in agreement with the finding by Callis *et al.* (1987) that the first intron of maize *adh*1 gene enhanced chloramphenicol acetyltransferase (CAT) activity 16- to 112-fold in maize protoplasts when driven by the *adh*1 promoter, but 5- to 22-fold when the CaMV 35S promoter was used (Callis *et al.*, 1987). The strong enhancement of gene expression obtained by insertion of the first intron of castor bean *cat* gene in the N-terminal region of the coding sequence of the *uid*A gene driven by the 35S promoter was demonstrated in transiently, as well as stably, transformed rice cells (Tanaka *et al.*, 1990). Ten- to 90-fold increases in GUS activity were observed. Northern blot analysis showed that the increase in GUS activity was correlated

Table 5.2. *Frequency of cotransformation*

	No. of Hm^R calli	No. of Gus+ calli	%
Experiment 1	58	30	51.7
Experiment 2	38	15	39.5

Notes:
Protoplasts were cotransformed with 25 μg of plasmid DNA carrying 35S-*hph* gene and 25 μg of plasmid DNA carrying *uid*A gene.

with the increased level of mature mRNA and efficient splicing. When the same construct was introduced into tobacco cells, little increase of gene expression was observed (Ohta *et al.*, 1990) and the intron was not spliced efficiently (Tanaka *et al.*, 1990).

Cotransformation

As a means to introduce nonselectable genes into rice, cotransformation has been used successfully. In cotransformation, a plasmid carrying the nonselectable gene and another carrying the selectable marker gene are mixed and introduced into protoplasts. In our experiments, when a GUS plasmid is mixed with the Hm^R plasmid in the ratio of 1:1 (50 μg each/ml), the efficiency of cotransformation was 30%–50% at the level of expression (Table 5.2). The cotransformation frequency does not seem to vary, regardless of which nonselectable genes are used for each experiment. One clear advantage of this method is that construction of a composite plasmid carrying both the selectable marker gene and the nonselectable gene is not required. By using the cotransformation method, more than three different genes can be introduced into a protoplast by single treatment. Cotransformed genes can be segregated in the progeny when they are not linked, giving rise to transgenic plants carrying no selection marker gene.

Integration patterns of foreign DNA

Integration patterns of foreign DNA in the genome of transgenic rice were analyzed by Southern blot analysis using various fragments of Hm^R plasmids as hybridization probes (H. Fujimoto *et al.*, unpublished data). The results indicated that five out of six Hm^R transformants

contained one or two functional *hph* sequences and no other fragmented pieces of the *hph* sequence were detected. With respect to the copy number of the integrated *hph* gene, the results are generally in good accordance with those from transgenic cereals obtained by other groups, and the majority of the transgenic plants contained one or two copies of the transgene. Our study also revealed that there are three different patterns of integration of the Hm^R plasmid into rice genome. First, almost the entire unit of the plasmid DNA is integrated into a single site. Second, several fragmented pieces of the plasmid DNA are integrated into multiple sites of rice genome. Third, a tandem repeat of the entire plasmid is integrated (four- to five-copy) into one site. Further analysis of integrated foreign DNA in transgenic rice should be important for understanding the mechanism of DNA integration and for using transgenic plants effectively in the introduction of economically important genes.

Unsolved problems

Despite rapid progress in rice transformation based on protoplast culture, further improvements will be required to fully realize the potential of transgenic approaches in rice research. First, establishment of transformation protocols for some of the major indica varieties of rice are urgently required because this variety is grown in the majority of rice-growing countries in the world. Recent developments in indica rice transformation and plant regeneration from protoplasts indicates that careful identification of steps involved in generation of transgenic plants will eventually solve the problems and it will not be long before protoplast-mediated transformation becomes possible in most of the major indica varieties of rice. Second, controlled integration of foreign DNA will be desired to utilize transgenic rice plants for some research areas. Deleted or rearranged copies of the plasmid DNA are often detected in the chromosomes of transgenic rice plants and they may cause undesired mutations in addition to the somaclonal variations that occur inevitably through protoplast culture. At present, our understanding of factors influencing patterns and efficiency of integration of transgenes is limited. One possible approach is to establish a method for gene targeting using homologous recombination (Paszkowski *et al.*, 1988). The gene-targeting technique should allow the integration of foreign genes exclusively into specific sites on the chromosomes. Gene targeting may also be used to disrupt a gene of interest in order to reveal *in vivo* function where the actual role of the gene is unknown.

Regulated expression of genes revealed by *in vivo* GUS histochemical analysis

Analysis of regulation of gene expression in homologous transgenic system has many advantages because introduced genes exhibit their natural properties in the homologous host plants. However, for most cereals, such a system is not available, and none is as well developed as those for rice. Therefore, rice can be used as a host species to analyze *in vivo* functions not only of promoters of rice genes but also of promoters isolated from other cereal species.

The first promoter examined in transgenic rice plants is the 35S promoter of CaMV (Terada & Shimamoto, 1990) because it has been used extensively for expression of agronomically useful genes in dicot species. The quantitative and histochemical analysis of the 35S promoter expression in transgenic rice plants and their progeny demonstrated that it directs GUS expression in a number of tissues including root, leaf, flower and seeds, and that the level of GUS expression in leaf and root was comparable to that in transgenic tobacco plants. A general conclusion on the expression pattern of the 35S promoter is that it is a 'constitutive' promoter in rice and no major differences in its expression pattern between rice and tobacco are detected. Thus, the 35S promoter should be useful for introduction of agriculturally important genes such as coat protein genes of viruses (see p. 60).

Some studies demonstrated that the 35S promoter was less effective in cereal cells than in dicot cells (Hauptmann *et al.*, 1987). Non-optimal conditions of transformation and/or lack of competence of cells in DNA uptake used might have contributed to the low activity of the 35S promoter.

We examined the regulated expression of the *uid*A reporter gene fused with maize alcohol dehydrogenase 1 gene (*adh*1) promoter and a rice *rbc*S promoter (small subunits of Rubisco (ribulose-1,5-bisphosphate carboxylase oxygenase)) in transgenic rice plants (Plate 5.1). Maize *adh*1 is one of the best studied genes in higher plants. In maize plants, alcohol dehydrogenase (ADH) is present in pollen, embryo, endosperm and seedling roots. In transgenic rice plants, the maize *adh*1 promoter directs constitutive GUS expression in shoots, root caps of seedling roots and mature roots, anthers, anther filaments, pollen, scutellum and endosperm (Plate 5.1(*a*)–(*d*)). Although the histochemical GUS analysis with transgenic rice plants made more detailed observation of the expression possible, the overall spatial expression pattern of maize *adh*1 promoter-*uid*A gene in rice is similar to the distribution of ADH protein in maize plants except for the expression in shoots of transgenic rice. The most likely and the most attractive explanation for this difference in expression is that GUS expression conferred by the *adh*1 promoter in transgenic rice may be regulated by cellular factor(s) present or active only in rice shoots but not in maize shoots. This difference in the expression pattern between the two species may provide a suggestion to understanding the mechanism underlying species differentiation in expression of homologous genes.

Anaerobic induction of ADH proteins has been well documented in several plant species. Thus, anaerobic induction was carefully examined using 5- to 7-day-old seedlings derived from selfed progenies of primary transgenic plants. The expression was strongly (up to 81-fold) induced in roots of seedlings in response to anaerobic treatment for 24 h, concomitant with an increase in the level of *uid*A mRNA. Our results also indicate that induction in the expression by maize *adh*1 promoter takes place in specific regions of the root in transgenic rice plants and the spatial pattern of expression changes distinctly after induction.

These results indicate that the maize *adh*1-*uid*A fusion gene is expressed in a regulated manner that reflects the natural property of the promoter. As the maize *adh*1 promoter did not confer sufficient expression in transgenic tobacco plants, it is evident that transgenic rice plants are more appropriate hosts for studying the expression of monocot (cereal) genes.

Expression of genes encoding small subunits of Rubisco (*rbc*S) is leaf specific and regulated by light. To investigate the regulation of rice *rbc*S expression more precisely, the *rbc*S-*uid*A fusion gene was introduced into transgenic rice plants.

The histochemical study detected GUS activity in mesophyll cells of both leaf sheath and leaf blade but not in epidermis or vascular tissues (Plate 5.1(*e*)). No activity was detected in roots, flowers or seeds (Plate 5(*f*)–(*h*)). The levels of GUS expression examined in five independent transgenic plants ranged from 10 to 150 nmol 4-MU/min per mg protein, which is one order of magnitude higher than that obtained with the 35S promoter. Little difference in the GUS activity between leaf sheath and leaf blade was detected. The expression of the *rbc*S-*uid*A gene was induced by light in primary transformants as well as R1 progeny plants and the induction took place at the level of transcription.

Other examples of promoters whose functions have been studied in transgenic rice plants include the *rol*C of the Ri plasmid (Matsuki *et al.*, 1989), rice *cab* (Tada *et al.*, 1991) and rice actin gene (Zhang, McElroy & Wu, 1991). In leaves and roots of transgenic rice plants, the *rol*C promoter directed GUS activity only in vascular tissues. The activity of the rice *cab* gene promoter was detected in leaves, stems and floral organs but not in roots, and its expression was induced by light. The promoter of the rice actin gene is a useful constitutive promoter because it confers the high level of expression in all the tissues examined.

Examples examined so far of the regulated expression of promoters derived from rice as well as from other cereals clearly show that transgenic rice is a valuable host for studying the expression of monocot genes. Many interesting and important features of monocot gene expression will be revealed using transgenic rice plants in the near future. Among others, *cis*-elements responsible for tissue specific or developmentally regulated expression of monocot genes can be determined by using transgenic rice plants. Transgenic rice plants should become useful tools to elucidate *in vivo* functions of *trans*-acting factors that have been isolated from various plant species including monocots (Katagiri & Chua, 1992).

Introduction and expression of useful genes in transgenic rice plants

Genetic transformation provides a powerful tool for crop improvement. In some dicot species, agronomically useful traits such as resistance to herbicides (Comai *et al.*, 1985; De Block *et al.*, 1987), insects (Vaeck *et al.*, 1987; Perlak *et al.*, 1990) and viruses (Powell-Abel *et al.*, 1986) and also male sterility (Mariani *et al.*, 1990) have been conferred by genetic engineering. Novel approaches are being taken to alter other agronomically important traits of crops. However, this technique has not been applied to cereals, mainly due to the lack of reliable transformation strategies until recently. Establishment of an efficient transformation system in rice has made it possible to apply gene transfer methods to improve rice, one of the most important crops in the world.

Transposable elements

Transposon tagging has important uses in the isolation of new genes and a number of genes have been isolated by this method in maize and *Antirrhinum majus* (Wienand & Saedler, 1988; Carpenter & Coen, 1990). However, because active transposable elements have not been identified in the rice genome so far, the maize autonomous transposable element activator (*Ac*) and nonautonomous element dissociation (*Ds*) were introduced into rice to develop a transposon-tagging system.

In order to introduce *Ac*, a phenotypic assay for excision of the *Ac* element was employed (Izawa *et al.*, 1991; Mural, Kawagoe & Hayashimoto 1991). In this assay, excision of the *Ac* element is recognized by an Hm^R (hygromycin resistant) phenotype, since excision of the *Ac* element from the untranslated leader sequence of the *hph* gene reconstitutes a functional Hm^R gene. Excision and reintegration of the *Ac* element in the rice genome was examined by Southern blot analysis, demonstrating that the introduced *Ac* element was active in transformed calli and transgenic plants. Sequence analysis of excision sites indicated that the *Ac* element was excised in rice in a manner similar to that in maize.

As an alternative to the use of an autonomous *Ac* element, transposition of a nonautonomous *Ds* element was examined by cotransformation with the *Ac* transposase gene fused with the 35S promoter. This *Ac* transposase gene is not able to transpose because it lacks the ends of *Ac* necessary for transposition. The *Ds* element was inserted in the chimeric Hm^R gene in place of *Ac* and it was found that the *Ds* element can be excised from and reintegrated into the rice chromosome by the action of the transposase produced *in trans* by the

Ac transposase gene. Excision of the *Ds* element from the *hph* gene was monitored as the appearance of Hm^R cells. Transposition of the *Ds* element was examined by Southern blot analysis and by sequencing its excision sites and rice DNA flanking integrated *Ds* elements. The analysis indicated that the *Ds* element actively transposes in rice in the presence of the *Ac* transposase gene. Sequences of *Ds* excision sites are similar to those of *Ac* excision sites in transgenic rice and 8 bp duplication of target sequences observed in maize was also found in rice. These results suggested that a two-element (*Ac/Ds*) system can be also used for tagging genes in rice. Interestingly, most of the Hm^R calli did not contain integrated copies of the *Ac* transposase gene, suggesting that transiently expressed *Ac* transposase acted on the *Ds* element and caused its transposition into the rice genome. This gives us an interesting possibility of generating transgenic plants carrying only nonautonomous *Ds* elements scattered throughout the genome. Then, these *Ds* plants can be used for screening for possible mutations caused by *Ds* insertion in the next generation.

One characteristic of the *Ac* transposition is that *Ac* often transposes to sites linked to its original location. If it is also the case in rice, generation of transgenic rice plants carrying transposable elements on known chromosomes should greatly facilitate effective gene tagging by *Ds*. To map insertion sites of the elements on rice chromosomes, DNA sequences flanking integrated *Ds* elements can be isolated by the inverse polymerase chain reaction method (H. Hashimoto & K. Shimamoto, unpublished data).

waxy gene for controlling starch composition of the grain

Starch composition in endosperm is one of the economically important traits in rice. Starch of wild-type grain consists of 15%–30% amylose and 70%–85% amylopectin, whereas the starch in endosperm of the *waxy* mutant completely lacks amylose. Amylose content varies greatly among rice cultivars and affects grain qualities, grain tastes and cooking properties. Although several other factors contribute to the amylose contents of rice seeds, they are determined primarily by expression levels of the *waxy* gene.

A genomic region containing the rice *waxy* gene has been cloned and sequenced (Wang *et al.*,

1990). Therefore opportunities exist for manipulation of the amylose content in rice grains by genetic engineering techniques. In our laboratory, several modified *waxy* genes were introduced into rice. Expression of antisense *waxy* genes fused with two promoters that are known to express in developing endosperm indeed reduces amylose content of the seed. Also it was found that transgenic plants produced pollens that are not stained by iodine. Our preliminary analysis suggests that amylose content of the seed can be altered by introduction of modified *waxy* genes. Because other genes such as those for ADP-glucose pyrophosphorylase or a branching enzyme known to be involved in starch biosynthesis in cereal grains have been isolated, it should be feasible to alter a wide variety of starchs in rice in the future.

Coat protein gene of rice stripe virus

Coat protein (CP)-mediated protection against virus diseases has been applied with a number of dicot species since the first report showing that the CP of tobacco mosaic virus (TMV) expressed in transgenic plants conferred resistance to TMV (Powell-Abel *et al.*, 1986). Transgenic rice plants expressing the CP of rice stripe virus (RSV) was generated in our laboratory (Hayakawa *et al.*, 1992). RSV is a member of the Tehui virus group and is transmitted by small brown planthoppers. In Japan, Korea, China, Taiwan and the former USSR, RSV causes serious damage to rice. The CP expression vector used in the experiment consisted of the 35S promoter, the first intron of castor bean catalase gene, the coding sequence of the CP gene and polyadenylation site from the nopaline synthase gene. Western blot analysis using primary transgenic plants revealed that, out of 33 independent clones each of which gave rise to several plants, 19 expressed detectable levels of the CP. The amount of the CP produced in the rice leaves was estimated to be up to 0.5% of the total soluble protein. In the assay for viral resistance, transgenic plants expressing CP did not exhibit disease symptoms, whereas the nontransformed control plants as well as transformed plants not expressing the CP showed clear disease symptoms, indicating that the resistance to RSV depended on expression of the introduced CP gene. The CP gene was stably transmitted to the progeny of primary transgenic plants and CP expression and the viral resistance were observed in the progeny plants.

This study indicated that introduction of CP genes is a promising approach to introduce viral resistance in cereals. This strategy is applicable to other viruses such as Tungro virus, which is causing severe damage to indica rice in many Asian countries.

δ-Endotoxin gene of Bacillus thuringiensis

Genes encoding insect control proteins from *Bacillus thuringiensis* (*B.t.*) have been introduced into several crops for protection against insects. In rice, insect damage is one of the major agricultural problems in many Asian countries. Two lepidopterans, rice stem borer and rice leaf folder, are largely responsible for the insect damage to rice. Therefore, introduction of *B.t.* toxin genes should be a useful approach for protection against the pests.

Preliminary attempts to express a *B.t.* toxin gene in rice have been described (Yang *et al.*, 1989; Xie, Fan & Ni, 1990). In these studies the *B.t.* coding sequence was translationally fused with the *uid*A gene and introduced into rice by protoplast transformation. Southern blot analysis of resultant transgenic plants showed integration of the *B.t.-uid*A fusion gene in the rice genome. Furthermore, GUS expression has been detected in roots of the primary transgenic plants. Whether the transgenic plants exhibit resistance to any of rice pests has not been reported.

In order to use the *B.t.* gene effectively against major rice pests, a number of factors influencing protein expression needs to be considered. For instance, it has been known that the level of *B.t.* toxin genes in plants is very low and it is generally not sufficient to control insects in field conditions. One of the reasons for low level of expression is its codon usage which is substantially different from that of plant genes. Extensive modification of the sequence has dramatically improved expression level of the *B.t.* gene in plants (Perlak *et al.*, 1990). Thus, a similar modification may also be needed in order to apply the *B.t.* toxin for insect control of rice. Another consideration is the choice of promoters. Stemborers, for instance, enter the stems of plants and grow inside them. This should be taken into account when a *B.t.* expression vector is constructed. In the future, however, efficient expression of the *B.t.* gene in rice will be achieved and it should improve resistance of rice to insects.

Conclusions

Rice transformation is now well established and being used for genetic engineering and for studies in regulation of genes derived from monocotyledonous plants. Establishment of routine transformation in rice has depended on well-defined protoplast culture and subsequent plant regeneration. Nevertheless, further improvement is required in protoplast culture of major indica varieties and reduction of undesirable mutations generated during the process of production of transgenic plants. Understanding the mechanisms of nonhomologous recombination and competence of cells for DNA uptake should help to further improve techniques of rice transformation in the future.

Development of a nearly saturated RFLP map and use of transposable elements will provide valuable tools for identifying agriculturally useful genes of rice. These should expand our list of genes that can be introduced into rice after modification for better productivity.

In gene regulation, roles of specific sequences in promoter regions of various genes during development and differentiation will come from studies using transgenic plants. Furthermore, factors determining the spatial and temporal expression of genes should be better understood by making use of transgenic plants. As our understanding of these *cis* and *trans* factors involved in gene expression advances, chances of generating novel rice plants with various improved traits will considerably increase.

References

Abdullah, R., Cocking, E. C. & Thompson, J. A. (1986). Efficient plant regeneration from rice protoplasts through somatic embryogenesis. *Bio/Technology*, **4**, 1087–1090.

Battraw, M. J. & Hall, T. C. (1990). Histochemical analysis of CaMV 35S promoter-β-glucuronidase gene expression in transgenic rice plants. *Plant Molecular Biology*, **15**, 527–538.

Callis, J., Fromm, M. & Walbot, V. (1987). Introns increase gene expression in cultured maize cells. *Genes and Development*, **1**, 1183–1200.

Carpenter, R. & Coen, E. S. (1990). Floral homeotic mutations produced by transposon-mutagenesis in *Antirrhinum majus*. *Genes and Development*, **4**, 1483–1493.

Christou, P., Ford, T. L. & Kofron, M. (1991). Production of transgenic rice (*Oryza sativa* L.) plants from agronomically important indica and japonica varieties via electric discharge by particle acceleration of exogenous DNA into immature zygotic embryos. *Bio/Technology*, **9**, 957–962.

Colot, V., Robert, L. S., Kavanagh, T. A., Bevan, M. W. & Thompson, R. D. (1987). Localization of sequences in wheat endosperm protein genes which confer tissue-specific expression in tobacco. *EMBO Journal*, **6**, 3559–3564.

Comai, L., Gacciotti, D., Hiatt, W. R., Thompson, G., Rose, R. E. & Stalker, D. (1985). Expression in plants of a mutant *aroA* gene from *Salmonella typhimurium* confers tolerance to glyphosate. *Nature (London)*, **317**, 741–744.

Datta, S. K., Peterhans, A., Datta, K. & Potrykug, I. (1990). Genetically engineered fertile indica-rice recovered from protoplasts. *Bio/Technology*, **8**, 736–740.

Davey, M. R., Kothari, S. L., Zhang, H., Rech, E. L., Cocking, E. C. & Lynch, P. T. (1991). Transgenic rice: characterization of protoplast-derived plants and their seed progeny. *Journal of Experimental Botany*, **42**, 1159–1169.

De Block, M., Botterman, J., Vandewiele, M., Dockx, J., Thoen, C., Gossele, V., Movva, N. R., Thompson, C., Van Montagu, M. & Leemens, J. (1987). Engineering herbicide resistance in plants by expression of a detoxifying enzyme. *EMBO Journal*, **6**, 2513–2518.

Dekeyser, R., Claes, B., Marichal, M., Montague, M. C. & Caplan, A. (1989). Evaluation of selectable markers for rice transformation. *Plant Physiology*, **90**, 217–223.

Ellis, J. G., Llewellyn, D. J., Dennis, E. S. & Peacock, W. J. (1987). Maize *adh-1* promoter sequences control anaerobic regulation: addition of upstream promoter elements from constitutive genes is necessary for expression in tobacco. *EMBO Journal*, **6**, 11–16.

Fraley, R. (1992). Sustaining the food supply. *Bio/Technology*, **10**, 40–43.

Fromm, M., Morrish, F., Armstrong, C., Williams, R., Thomas, J. & Klein, T. M. (1990). Inheritance and expression of chimeric genes in the progeny of transgenic maize plants. *Bio/Technology*, **8**, 833–844.

Gordon-Kamm, W. J., Spencer, T. M., Mangano, M. L., Adams, T. R., Daines, R. J., Start, W. G., O'Brien, J. V., Chambers, S. A., Adams, W. R., Willetts, N. G., Rice, T. B., Mackey, C. J., Krueger, R. W., Kausch, A. P. & Lemaux, P. G. (1990). Transformation of maize cells and regeneration of fertile transgenic plants. *Plant Cell*, **2**, 603–618.

Hauptmann, R. M., Ozias-Akins, P., Vasil V., Tabaeizadeh, Z., Rogers, S. G., Horsch, R. B., Vasil, I. K. & Fraley, R. T. (1987). Transient expression of electroporated DNA in monocotyledonous and dicotyledonous species. *Plant Cell Reports*, **6**, 265–270.

Hayakawa, T., Zhu, Y., Itoh, X., Kimura, Y., Izawa, T., Shimamoto, K. & Toriyama, S. (1992). Genetically engineered rice resistant to rice stripe virus, an insect transmitted virus. *Proceedings of the National Academy of Science, USA*, **89**, 9865–9869.

Hayashimoto, A. & Murai, N. (1990). A polyethyleneglycol-mediated protoplast transformation system for production of fertile transgenic rice plants. *Plant Physiology*, **93**, 857–863.

Horn, M. E., Shillito, R. D., Conger, B. V. & Harms, C. T. (1988). Transgenic plants of orchardgrass (*Dactylis glomerata* L.) from protoplasts. *Plant Cell Reports*, **7**, 469–472.

Izawa, T., Miyazaki, C., Yamamoto, M., Terada, R., Iida, S. & Shimamoto, K. (1991). Introduction and transposition of the maize transposable element *Ac* in rice (*Oryza sativa* L.). *Molecular and General Genetics*, **227**, 391–396.

Katagiri, F. & Chua, N.-H. (1992). Plant transcription factors: present knowledge and future challenges. *Trends in Genetics*, **8**,. 22–27.

Keith, B. & Chua, N.-H. (1986). Monocot and dicot pre-mRNAs are processed with different efficiencies in transgenic tobacco. *EMBO Journal*, **5**, 2419–2425.

Kinoshita, T. (1984). Gene analysis and linkage map. In *Biology of Rice*, ed. S. Tsunoda & M. Takahashi, pp. 187–274. Japan Scientific Society Press, Tokyo.

Kyozuka, J., Fujimoto, H., Izawa, T. & Shimamoto, K. (1991). Anaerobic induction and tissue-specific expression of maize *Adh1* promoter in transgenic rice plants and their progeny. *Molecular and General Genetics*, **228**, 40–48.

Kyozuka, J., Hayashi, Y. & Shimamoto, K. (1987). High frequency plant regeneration from rice protoplasts by novel nurse culture methods. *Molecular and General Genetics*, **206**, 408–413.

Kyozuka, J., Izawa, T., Nakajima, M. & Shimamoto, K. (1990). Effect of the promoter and the first intron of maize Adh1 on foreign gene expression in rice. *Maydica*, **35**, 353–357.

Kyozuka, J., Otoo, E. & Shimamoto, K. (1988). Plant regeneration from protoplasts of indica rice: genotypic differences in culture response. *Theoretical and Applied Genetics*, **76**, 887–890.

Kyozuka, J. & Shimamoto, K. (1991). Transformation and regeneration of rice protoplasts. *Plant Tissue Culture Manual*, ed K. Lindsey, pp. 1–17. Kluwer Academic Publishers, Amsterdam.

Kyozuka, J., Shimamoto, K. & Ogura, H. (1989). Regeneration of plants from rice protoplasts. *Biotechnology in Agriculture and Forestry*, Vol. 8, *Plant Protoplasts and Genetic Engineering* I, ed. Y. P. S. Bajaj, pp. 109–203. Springer-Verlag, Berlin Heidelberg, New York.

Lamppa, G., Nagy, F. & Chua, N.-H. (1985). Light-regulated and organ-specific expression of wheat *Cab* gene in transgenic tobacco. *Nature (London)*, **316**, 750–752.

Lazzeri, P. A., Brettschneider, R., Luhrs, R. & Lörz, H. (1991). Stable transformation of barley via PEG-induced direct DNA uptake into protoplasts. *Theoretical and Applied Genetics*, **81**, 437–444.

Lloyd, J. C., Raines, C. A., John, U. P. & Dyer, T. A. (1991). The chloroplast FBPase gene of wheat: structure and expression of the promoter in photosynthetic and meristematic cells of transgenic tobacco plants. *Molecular and General Genetics*, **225**, 209–216.

Luehrsen, K. R. & Walbot, V. (1991). Intron enhancement of gene expression and the splicing efficiency of introns in maize cells. *Molecular and General Genetics*, **225**, 81–93.

Mariani, C., De Beuckeleer, M., Truettner, J., Leemans, J. & Goldberg, R. B. (1990). Induction of male sterility in plants by a chimaeric ribonuclease gene. *Nature (London)*, **347**, 737–741.

Mascarenhas, D., Mettler, I. J., Pierce, D. A. & Lowe, H. W. (1990). Intron-mediated enhancement of heterologous gene expression in maize. *Plant Molecular Biology*, **15**, 913–920.

Matsuki, R., Onodera, H., Yamauchi, T. & Uchimlya, H. (1989). Tissue-specific expression of the *rolC* promoter of the Ri plasmid in transgenic rice plants, *Molecular and General Genetics*, **220**, 12–16.

Matsuoka, M. & Sanada, Y. (1991). Expression of photosynthetic genes from the C4 plant, maize, in tobacco. *Molecular and General Genetics*, **225**, 411–419.

McCouch, R., Kochert, G., Yu, Z. H., Wang, Z. Y., Khush, G. S., Coffman, W. R. & Tanksley, S. D. (1988). Molecular mapping of rice chromosomes. *Theoretical and Applied Genetics*, **76**, 815–829.

McElroy, D., Zhang, W., Cao, J. & Wu, R. (1990). Isolation of an efficient actin promoter for use in rice transformation. *Plant Cell*, **2**. 163–171.

Meijer, E. G. M., Schilperoort, R. A., Rueb, S., Os-Ruygrok, P. E. & Hensgens, L. A. M. (1991). Transgenic rice cell lines and plants: expression of transferred chimeric genes. *Plant Molecular Biology*, **16**, 807–820.

Meyerowitz, E. M. (1989). *Arabidopsis*, a useful weed. *Cell*, **56**, 263–269.

Murai, N., Li, Z., Kawagoe, Y. & Hayashimoto, A. (1991). Transposition of the maize *Activator* element in transgenic rice plants. *Nucleic Acids Research*, **19**, 617–622.

Nishibayashi, S. (1992). Is genome of rice small? *Rice Genetics Newsletters*, **8**, 152–154.

Oard, J. H., Paige, D. & Dvorak, J. (1989). Chimeric gene expression using maize intron in cultured cells of breadwheat. *Plant Cell Reports*, **8**, 156–160.

Ohta, S., Mita, S., Hattori, T. & Nakamura, K. (1990). Construction and expression in tobacco of β-glucuronidase (GUS) reporter gene containing the intron within the coding sequence. *Plant Cell Physiology*, **31**, 805–815.

Paszkowski, J., Baur, M., Bogucki, A. & Potrykus, I. (1988). Gene targeting in plants. *EMBO Journal*, 7, 4021–4026.

Peng, J., Lyznik, L. A., Lee, L. & Hodges, T. (1990). Transformation of indica rice protoplasts with *gus*A and neo genes. *Plant Cell Reports*, **9**, 168–172.

Perlak, F. J., Deaton, R. W., Armstrong, T. A., Fuchs, R. L., Sims, S., Greenplate, J. T. & Fischhoff, D. A. (1990). Insect resistant cotton plants. *Bio/Technology*, **8**, 939–943.

Powell-Abel, P., Nelson, R. S., De, B., Hoffmann, N., Rogers, S. G., Fraley, R. T. & Beachy, R. N. (1986). Delay of disease development in transgenic plants that express the tobacco mosaic virus coat protein gene. *Science*, **232**, 738–743.

Rhodes, C. A., Pierce, D. A., Mettler, I. J., Mascarenhas, D. & Detmer, J. J. (1988). Genetically transformed maize plants from protoplasts. *Science*, **240**, 204–207.

Schell, J. (1987). Transgenic plants as tools to study the molecular organization of plant genes. *Science*, **237**, 1176–1183.

Schernthaner, J. P., Matzke, M. A. & Matzke, A. J. M. (1988). Endosperm-specific activity of a zein gene promoter in transgenic tobacco plants. *EMBO Journal*, 7, 1249–1255.

Shimamoto, K. (1991). Transgenic rice plants. In *Molecular Approaches to Crop Improvement*, ed. E. S. Dennis & D. J. Llewellyn, pp. 1–15. Springer, Wien.

Shimamoto, K., Terada, R., Izawa, T. & Fujimoto, H. (1989). Fertile transgenic rice plants regenerated from transformed protoplasts. *Nature (London)*, **338**, 274–276.

Tada, Y., Sakamoto, M. & Fujimura, T. (1990). Efficient gene introduction into rice by electroporation and analysis of transgenic plants: use of electroporation buffer lacking chloride ions. *Theoretical and Applied Genetics*, **80**, 475–480.

Tada, Y., Sakamoto, M., Matsuoka, M. & Fujimura, T. (1991). Expression of a monocot LHCP promoter in transgenic rice. *EMBO Journal*, **10**, 1803–1808.

Tanaka, A., Mita, S., Ohta, S., Kyozuka, J., Shimamoto, K. & Nakamura, K. (1990). Enhancement of foreign gene expression by a dicot intron in rice but not in tobacco is correlated with an increased level of mRNA and an efficient splicing of the intron. *Nucleic Acids Research*, **18**, 6767–6770.

Terada, R. & Shimamoto, K. (1990). Expression of CaMV 35S–GUS gene in transgenic rice plants. *Molecular and General Genetics*, **220**, 389–392.

Toriyama, K., Arimoto, Y., Uchimiya, H. & Hinata,

K. (1988). Transgenic rice plants after direct gene transfer into protoplasts. *Bio/Technology*, **6**, 1072–1074.

Toriyama, K., Hinata, K. & Sasaki, T. (1986). Haploid and diploid plant regeneration from protoplasts of anther callus in rice. *Theoretical and Applied Genetics*, **73**, 6–19.

Vaeck, M., Reynaerts, A., Hofte, H., Jensens, S., De Beuckeleer, M., Dean, C., Zabeau, M., Van Montagu, M. & Leemans, J. (1987). Transgenic plants protected from insect attack. *Nature (London)*, **328**, 33–37.

Vasil, V., Clancy, M., Ferl, R. J., Vasil, I. K. & Hannah, L. C. (1989). Increased gene expression by the first intron of maize *Shrunken-1* locus in grass species. *Plant Physiology*, **91**, 1575–1579.

Wang, Z. Y. & Tanksley, S. D. (1989). Restriction fragment length polymorphism in *Oryza sativa* L. *Genome*, **32**, 1113–1118.

Wang, Z. Y., Wu, Z. L., Xing, Y. Y., Zheng, F. G., Guo, X. I., Zhang, W. G. & Hong, M. M. (1990). Nucleotide sequence of rice *waxy* gene. *Nucleic Acids Research*, **18**, 5898–5899.

Wienand, U. & Saedler, H. (1988). Plant transposable elements: unique structures for gene tagging and gene cloning. In *Plant DNA Infectious Agents*, ed. T. Hohn & J. Schell, pp. 205–228. Springer, Wien.

Xie, D. X., Fan, Y. L. & Ni, P. C. (1990). Transgenic rice plant obtained by transferring the *Bacillus thuringiensis* toxin gene into a Chinese rice cultivar Zhonghua 11. *Rice Genetics Newsletters*, 7, 147–148.

Yamada, Y., Yang, Z. Q. & Tang, D. T. (1986). Plant regeneration from protoplast-derived callus of rice (*Oryza sativa* L.). *Plant Cell Reports*, **5**, 85–88.

Yang, H., Guo, S. D., Li, J. X., Chen, X. J. & Fan, Y. L. (1989). Transgenic rice plants produced from protoplasts following direct uptake of *Bacillus thuringiensis* δ-endotoxin protein gene. *Rice Genetics Newsletters*, **6**, 159–160.

Zhang, W., McElroy, D. & Wu, R. (1991). Analysis of rice *Act1* 5′ region activity in transgenic rice plants. *Plant Cell*, **3**, 1155–1165.

Zhang, W. & Wu, R. (1988). Efficient regeneration of transgenic rice plants from protoplasts and correctly regulated expression of foreign genes in plants. *Theoretical and Applied Genetics*, **76**, 835–840.

Zhang, H. M., Yang, H., Reach, E. L., Golds, T. J., Davis, A. S., Mulligan, B. J., Cocking, E. C. & Davey, M. R. (1988). Transgenic rice plants produced by electroporation-mediated plasmid uptake into protoplasts. *Plant Cell Reports*, 7, 379–383.

Zhao, X., Wu, T., Xie, Y. & Wu, R. (1989). Genome-specific repetitive sequences in the genus *Oryza*. *Theoretical and Applied Genetics*, **78**, 201–209.

6

Maize

H. Martin Wilson, W. Paul Bullock, Jim M. Dunwell, J. Ray Ellis*,
Bronwyn Frame, James Register III and John A. Thompson

Introduction

Genes advantageous for pathogen or insect resistance, or for enhancing yield, are continuously being sought by plant breeders. Many of these genes exist outside the species targeted for improvement. This fact has led researchers to contemplate ways of overcoming the biological barriers to gene transfer. More than a quarter of a century has passed since the first attempt was made to introduce DNA into maize (*Zea mays*) through physical intervention (Coe & Sarkar, 1966). Although this attempt, which involved direct injection of DNA into apical meristems of seeds, was not successful, it did serve to identify a principal problem – namely that 'The cell wall, a massive barrier to large structures, may have to be disrupted mechanically or chemically, or otherwise circumvented'. Another observation made in 1966 by Coe & Sarkar was that 'the number of meristem cells actually penetrated and observed may be too small to have included a fortuitous, observable transformation'. Over two decades later the efficient combination of DNA delivery and transformed cell selection has proved to be the key to production of transgenic plants of all species and particularly of maize, which as a cereal of major commercial importance has attracted substantial attention in recent years.

Maize transformation, defined as the production of transgenic plants that are fertile and produce transgenic seed, was achieved for the first time in 1990 (Fromm *et al.*, 1990; Gordon-Kamm *et al.*, 1990). Since then a number of commercial research groups have conducted field trials of transgenic maize, demonstrating the attainment of more widespread success.

In this review we briefly consider some of the approaches that have produced interesting preliminary results – *Agrobacterium*-mediated DNA delivery, whisker-mediated DNA delivery, germ-line transformation targets (pollen, ovules and embryos), and microinjection – and then concentrate on the three currently most successful approaches, namely direct gene transfer to protoplasts, particle bombardment of regenerable cell cultures, and, most recently, electroporation of immature embryos.

Interesting preliminary results

Agrobacterium-mediated delivery

Agrobacterium tumefaciens is the most widely used vector for transformation of dicotyledonous (dicot) species but its utility in maize transformation remains to be demonstrated. Monocotyledons (monocots) are rarely natural hosts for *Agrobacterium* (De Cleene & De Ley, 1976; De Cleene, 1985) and therefore are not expected to be susceptible to gene transfer mediated by *Agrobacterium*. This expectation has been challenged, however, by the demonstration that *Asparagus officinalis* (Liliaceae) could be transformed with DNA delivered by *Agrobacterium* to yield fertile transgenic plants (Hooykaas-Van Slogteren & Hooykaas, 1984; Bytebier *et al.*, 1987).

---

* We dedicate this review to the memory of our colleague Ray Ellis, who died in April, 1993.

Experimental inoculation of maize seedlings with wild-type *Agrobacterium* strains led to claims that nopaline or octopine synthase activities, encoded by transferred DNA (T-DNA) from the Ti plasmid, could be detected in seedling tissues and in progeny obtained from the inoculated plants (Graves & Goldman, 1986). However, the synthesis of opines cannot be unequivocally ascribed to T-DNA transfer, and these results on maize remain to be substantiated. More recently, a potent inhibitor of *Agrobacterium* virulence induction has been found in maize seedlings (Sahi, Chilton & Chilton, 1990).

A breakthrough in demonstrating the potential of *Agrobacterium* as a gene vector for maize was the unambiguous transfer of maize streak virus DNA from *Agrobacterium* to inoculated plants by a procedure commonly termed 'agroinfection' (Grimsley *et al.*, 1987). To effect the transfer, tandemly repeated or partially repeated copies of the viral genome were inserted between the T-DNA borders of the vector plasmid and the shoot meristem of seedlings inoculated with the engineered *Agrobacterium* strain (Grimsley *et al.*, 1988). Transfer of the T-DNA was detected by the presence of replicative forms of the viral genome and the development of viral disease symptoms in the host plant. The success of agroinfection in demonstrating T-DNA transfer to maize results from the sensitive detection system, since the marker DNA becomes highly amplified by viral replication and systemic spread.

The agroinfection technique has not produced stably transformed maize plants because the replicating viral genome is not transmitted to germline cells and does not become integrated into the nuclear genome. Attempts to detect T-DNA transfer using sensitive but nonreplicating marker genes, e.g. *uid*A, have not been successful (N. Grimsley, personal communication), indicating that *Agrobacterium* delivers T-DNA to very few initial target cells. Despite these limitations, the agroinfection procedure has permitted analysis of the physical and biological factors necessary for *Agrobacterium*-mediated gene transfer to maize and other cereals. For example, the nonrequirement of virulence inducers, the utility of binary vectors, the relative efficiency of C58 strains and the developmental competence of shoot meristems have been demonstrated (Grimsley *et al.*, 1987; Boulton *et al.*, 1989; Grimsley, 1990; Schlappi & Hohn, 1992). The inefficacy of octopine strains to infect maize has been ascribed to the inhibitory action of the *vir*F gene on the Ti plasmid (Jarchow, Grimsley & Hohn, 1991).

Integration of T-DNA into the genome of rice has been reported (Ranieri *et al.*, 1990). In addition, the production of transgenic maize plants from cultured shoot apices inoculated with *Agrobacterium* was reported by Gould *et al.* in 1991. However, molecular and genetic evidence for integration of T-DNA into the progeny remains inconclusive.

Agrobacterium clearly has potential as a gene vector for maize, but the efficiency of the bacterium–plant interaction must be improved for this approach to compete with direct gene transfer.

Germline transformation targets

Introduction of cloned genes directly into the germline is an attractive goal for maize transformation technology, since this would obviate the need to regenerate from totipotent cells via tissue culture. However, germline transformation has not been convincingly demonstrated in maize, nor in any other plant species.

From a practical point of view, pollen is the most amenable germline cell because of its abundance and accessibility. In the mature maize pollen grain (the male gametophyte) the germline is represented by the two sperm cells that lie within the vegetative cell and are surrounded by the complex pollen wall. Although numerous attempts to transform maize pollen have been published, and success has been claimed, in no case is convincing molecular and genetic evidence available. Most of these studies relied on poorly defined morphological traits, encoded by genomic DNA, as genetic markers. However, use of well-defined molecular markers cloned on plasmids has failed to produce transformants (Roeckel *et al.*, 1988; Booy, Krens & Huizing, 1989) or failed to provide the crucial molecular and genetic evidence (De Wet *et al.*, 1988).

Experimental procedures have generally been designed to permit DNA uptake into hydrating pollen grains or into the pollen tubes of germinating grains by imbibition (De Wet *et al.*, 1985; Sanford, Skubik & Reisch, 1985; De Wet *et al.*, 1986; Ohta, 1986; Waldron, 1987). Proven obstacles to DNA uptake include potent nuclease activity (Roeckel *et al.*, 1988; Booy *et al.*, 1989) and pollen wall impermeability (Kranz & Lörz, 1990). As an alternative, attempts have been made to introduce DNA into immature spikelets of maize tassels in the hope of inducing gene uptake by developing microspores (Bennetzen *et al.*, 1988). Efforts have also been made to target the growing pollen tube as a means of carrying marker DNA along the silks to the embryo sac (G. Feix, personal communication; Langridge *et al.*, 1992). The demonstration of endophytic bacteria in maize plants has emphasized how readily arte-factual transformation results can be obtained when nonaxenic target materials are used (Konstantinov, Mladenovic & Denic, 1991).

Delivery of marker genes directly to the egg cell has rarely been attempted because of its inaccessibility within the ovary. Two approaches under investigation are injection of DNA into the predetermined position of the egg cell in the ovule (Wagner, Dumas & Mogensen, 1990) or the enzymatic isolation of the embryo sac and egg cell followed by gene delivery and fertilization *in vitro* (Kranz, Bautor & Lörz, 1991).

Embryos also have been investigated as recipients for gene transfer by imbibition. Here, the target cells are those in the shoot meristem that will ultimately give rise to gametocytes. Both zygotic and somatic embryos have been used as DNA recipients, but no transgenic plants have been produced (Töpfer *et al.*, 1989; Lupotto & Lusardi, 1990). Nevertheless, transient expression of marker genes taken up by this approach has been obtained using embryos isolated from a range of cereal species (Töpfer *et al.*, 1989, 1990).

Despite the lack of success in targeting germline cells for transformation to date, these cells are likely to remain an option for experimentation because of their inherent potential benefits. The availability of marker genes that have been proven to function in transgenic maize should now allow earlier claims of germline transformation to be rigorously evaluated.

Microinjection

As an approach to maize transformation, micro-injection has received limited attention compared to other delivery systems. The considerable technical skill and expensive, sophisticated equipment required have contributed to the slow progress in experimentation. The main advantage of micro-injection over other delivery methods is the preci-sion with which DNA can be delivered into defined target cells (Neuhaus & Spangenberg, 1990). While microinjection leads to high transformation frequencies in suitable single cell systems (up to 20% in tobacco protoplasts; Schnorf *et al.*, 1991), the rate of injection is slow, and there are currently no single cell targets in maize that can be both efficiently injected and cultured at low density.

Attention has therefore turned with the cereals to multicellular systems as potential DNA recipients via microinjection (Potrykus, 1990). Success in the production of transgenic *Brassica napus* plants following microinjection of microspore-derived embryos (Neuhaus *et al.*, 1987) prompted the use of zygotic cereal embryos in transformation studies. Zygotic embryos can be considered an attractive target system for transformation, given the reduced effect of genotype, somaclonal variation and culture-related infertility compared to long-term cell cultures. However, it is clear that an inevitable consequence of transforming structures with many cells, such as zygotic embryos, is the production of chimeras. Since the fate of individual cells in an embryo cannot currently be described, it is necessary to analyze sexual offspring for the presence of introduced genes.

In collaboration with Gunter Neuhaus at ETH, Zurich, we have performed an extensive evaluation of the microinjection of zygotic proembryos as a route to maize transformation. Culture conditions were established for proembryos of an elite, inbred maize genotype. Microinjection was performed into the meristematic region of the embryos using a DNA construct carrying the *npt*II gene. The embryos were then grown in a micro-culture system and the resulting plants transferred to soil. Analysis of these plants was restricted to polymerase chain reaction (PCR) analysis of tassel tissue. Three individual tassels out of over 100 analyzed were identified and independently verified as being PCR positive for the presence of the *npt*II gene. The progeny of a further 340 plants derived from microinjected embryos were then screened for the presence and expression of the *npt*II marker gene, using PCR analysis and kanamycin selection. Despite the large number of plants analyzed (> 40 000), no evidence was found for transgene transmission to the progeny of microinjected individuals.

In light of these data and the lack of any other contradictory data in plants it has become clear that the possibility exists that cells within zygotic embryos lack the competence for transformation, as discussed by Potrykus (1990). The difficulty of targeting injections to germline progenitor cells in maize embryos large enough to be cultured may also dramatically reduce the efficiency of transformation. However, in view of the extensive screening of injected material described above, it seems likely there is a fundamental problem in effecting transformation of meristematic cells. This issue requires more basic research if further progress is to be made.

In vitro fertilized gametes offer an additional opportunity for microinjection (Kranz *et al.*, 1991). However, the development of a transformation system based on these target cells will require considerable further effort, and the degree of technical sophistication needed will prevent widespread experimentation.

Whisker-mediated transformation

A simple, rapid and inexpensive means of delivering naked DNA to nonregenerable maize suspension culture cells (cv. Black Mexican Sweet – BMS) without the need for sophisticated equipment has been demonstrated by Kaeppler *et al.* (1990). The method involves vortex treatment of suspension cells with silicon carbide whiskers and plasmid DNA. The rapid mixing resulting from this treatment appears to lead to whisker penetration of cells and DNA delivery. BMS cells were found to produce transient β-glucuronidase (GUS) activity at a frequency of 140 blue color-forming units per 250 μl (packed cell volume) of cells.

More recently, stable transformation of BMS using this approach has been reported by the same group (Kaeppler *et al.*, 1992). Cells were treated with a plasmid carrying the *bar* and *uid*A genes (pBARGUS) and selected on an herbicide (active ingredient phosphinothricin). An average of 40 GUS-expressing units and 3.4 stably transformed callus lines were recovered from a 300 μl packed cell volume of BMS cells.

Approaches proven to generate transformed maize plants

Three approaches have proved successful to-date and will be considered in detail below. These are (i) direct gene transfer to protoplasts, (ii) particle bombardment and (iii) electroporation of immature zygotic embryos and type 1 callus. In common to all three are the use of target cells from which whole plants can be regenerated and the direct delivery of DNA to these cells. To offer utility in a transformation system, transformed cultured cells must be capable of periods of sustained growth and division (for selection), and full expression of totipotency (for regeneration of fertile plants). This has proved an elusive formula in maize (and other cereals) and has delayed success for several years.

Direct gene transfer to protoplasts

Isolated protoplasts offer advantages over other recipient cell systems for DNA delivery since, using protoplasts, it is possible to define precise conditions for efficient DNA uptake into a large population of cells (Negrutiu *et al.*, 1987). Procedures for direct gene transfer involving polyethylene glycol (PEG) treatment or electroporation are well established for many species and have been used with a range of cereals, including maize. In addition, there is speculation that the process of enzymatic protoplast isolation might constitute the trigger in cereals for something akin to a 'wound response', thus increasing cell competence for transformation and regeneration (Potrykus, 1990). Multicellular structures resulting from protoplast culture are generally of single-cell origin (Thompson, Abdullah & Cocking, 1986), so the generation of chimeras is avoided.

Although protoplasts are attractive as a transformation target, maize protoplast culture remains technically difficult. As a result, slow progress has been made in the production of transgenic maize plants from protoplasts.

Protoplasts isolated directly from maize plant tissue are, as with other cereal species, extremely recalcitrant in culture. Numerous attempts to induce sustained cell division in such protoplasts have been made without success (Potrykus, Harms & Lörz, 1976). At best, cereal leaf protoplasts can be induced to undergo only a few cell divsions (Hahne, Lörz & Hahne, 1990). However, protoplasts of maize leaf or root origin are of value in transient expression experiments where information on regulation of promoter function may be obtained (Junker *et al.*, 1987).

Protoplasts from nonregenerable BMS suspension cultures can be induced to undergo sustained cell division (Chourey & Zurawski, 1981). However, the plating efficiency of such cultures remains low (at best 10%) compared to dicot species, despite improvements in division frequency obtained through, for example, the use of nurse cultures (Ludwig *et al.*, 1985) or conditioned medium (Somers *et al.*, 1987).

Stable transformation of BMS protoplasts has been achieved following either electroporation (Fromm, Taylor & Walbot, 1986) or PEG-mediated DNA uptake (Armstrong *et al.*, 1990). Information on the cotransformation of unlinked genes has been obtained in BMS using *npt*II/*uid*A (Lyznik *et al.*, 1989), *npt*II/*aph*IV (Armstrong *et al.*, 1990) and *npt*II/*cat* (ICI Seeds; J. A. Thompson *et al.*, unpublished data). Cotransformation occurs at frequencies of 40% – 80%, as described for other species (Saul & Potrykus, 1990). Cationic methods have also been used to transform maize protoplasts. Kanamycin-resistant BMS transformants were recovered after protoplast transfection with genomic DNA from a line previously transformed with the *npt*II gene (Antonelli & Stadler, 1990).

Protoplasts isolated from cell lines such as BMS, and from those of endosperm origin, have also been used widely in studies of transient gene expression and provide useful information on construct function, for example intron and enhancer effects (Callis, Fromm & Walbot, 1987; Last *et al.*, 1991; Quayle & Feix, 1992).

Embryogenic maize cultures offer the best possibility for the recovery of plants from protoplasts. The development of embryogenic suspension cultures capable of yielding division-competent cells is, however, a time-consuming and poorly understood process. Rhodes, Lowe & Ruby (1988*a*) succeeded in regenerating plants of A188 and B73 from embryogenic suspension-derived protoplasts of immature embryo origin, using feeder layers to stimulate protoplast division. Unfortunately all the regenerated plants were sterile, probably as a consequence of the age of the cultures used for protoplast isolation (from 18 months to 4 years). Transgenic plants of A188 were produced by the same group (Rhodes *et al.*, 1988*b*) following electroporation and kanamycin selection. Thirty-eight plants were regenerated from ten different cell lines. However, these plants were also sterile.

Several attempts have been made to develop an efficient system for the production of fertile plants from immature embryo-derived embryogenic maize suspension protoplasts that would be useful for transformation studies. Shillito *et al.* (1989) succeeded in regenerating fertile plants from embryogenic suspension protoplasts of a B73-derived inbred line despite the occurrence of many morphological and reproductive abnormalities. Again, feeder layers of nurse suspension cells were used to enhance protoplast division frequency in this work and in that of Prioli & Sondahl (1989), where fertile plants were regenerated from protoplasts of a Cateto inbred line adapted to tropical conditions. A wide range of morphological abnormalities were also observed in the regenerants described in this report.

Complex interactions between protoplast and feeder cell lines were noted by Petersen, Sulc & Armstrong (1992), where plants with only tassels, or plants with only ears, were regenerated from protoplasts of crosses involving A188, B73 and BMS. Protoplast-derived regenerants were used as embryo donors to develop lines with improved protoplast response.

Microspore-derived cultures have also been used to develop embryogenic suspension cultures from which regenerable protoplasts were isolated (Sun, Prioli & Sondahl, 1989; Mitchell & Petolino, 1991). No fertile dihaploid plants were described in these reports and again a range of phenotypic variants were noted in the regenerated plants.

In contrast, Morocz *et al.* (1990) found relatively low frequencies of developmental abnormalities in the plants regenerated from embryogenic suspension protoplasts of the complex synthetic maize genotype He/89. An auxin-autotrophic culture (patent applied for, EP application 0465875A1) derived from this genotype was shown to yield division-competent protoplasts as early as 10 weeks after culture initiation. More than 500 plants were obtained from the resulting callus and of these 60%–70% set seed when selfed or sibbed. The use of young cultures from a tissue culture-adapted genotype undoubtedly contributes to the high plant regeneration frequency and fertility seen in this work.

This efficient regeneration system was successfully used to achieve stable transformation following PEG-mediated DNA uptake and subsequent selection on phosphinothricin (Donn, Nilges & Morocz, 1990). Plants were regenerated from the resulting transformed callus, with the majority being fertile. This therefore represents the first report of fertile transgenic plant production from maize protoplasts.

It remains to be seen whether elements of these recent reports (Donn *et al.*, 1990; Morocz *et al.*, 1990) can be used to bring more general success in the production of transgenic maize plants from protoplasts. At present, genotype constraints and the reduced vigor and fertility of plants regenerated from protoplasts probably outweigh the benefits of protoplasts as recipients for the integration of foreign DNA. Transient expression in protoplasts will continue to provide valuable information on the functional analysis of genes.

Particle bombardment

The use of high velocity microprojectiles to carry DNA into living cells is now routine in many laboratories following the pioneering work of Sanford, Wolf, Klein and Allen at Cornell University (Klein *et al.*, 1987; Sanford *et al.*, 1987). Transient expression of an introduced gene was first reported in maize in 1988 by Klein *et al.* Stable transformation of cultured maize cells (Klein *et al.*, 1989) and regeneration of fertile transgenic maize plants from transformed cultured cells (Fromm *et al.*, 1990; Gordon-Kamm *et al.*, 1990) were subsequently reported. Particle bombardment is currently used in a number of commercial laboratories, including our own, to generate transformed maize plants. Plates 6.1–6.9 illustrate various stages in particle bombardment-mediated maize transformation.

Early successes with maize (e.g. see Fromm *et al.*, 1990; Gordon-Kamm *et al.*, 1990) were achieved using DNA coated on to tungsten particles that were accelerated by means of gunpowder-driven macroprojectiles or macrocarriers. More recent particle bombardment work (Russell, Roy & Sanford, 1992a,b; Sanford, Smith & Russell, 1993) has shown that acceleration of DNA-coated gold particles driven by helium gas provides a more efficient means for generation of transient and stable transformants in a number of biological systems. The use of gold particles and a helium-driven device (the Dupont PDS-1000) is now routine in many laboratories, including most of those working in the area of maize transformation.

A number of different protocols detailing methods of introducing genes into regenerable corn cells through particle bombardment have been described (Fromm *et al.*, 1990; Gordon-Kamm *et al.*, 1990). The protocol used at ICI Seeds is as follows. Gold particles (3 mg of 1 μm-sized particles) are washed and coated with plasmid DNA (5 μg) as described in the Dupont PDS-1000 manual. Ten microliters of gold particle/DNA suspension are then loaded on to each macrocarrier and thoroughly dried in a desiccator prior to bombardment. A fine wire mesh (150 μm) screen (after Gordon-Kamm *et al.* (1990), who used a 100 μm screen), placed between the macrocarrier holder and the target tissue, has been found to increase gold particle dispersion and reduce tissue death. Suspension cells (2 days post-subculture), or unorganized callus tissue, are vacuum filtered on to filter paper (Whatman no. 4) discs (0.25 ml PCV/filter), cultured overnight on the surface of N6 medium (suspension cells) or MS medium (callus), and then bombarded (900 p.s.i., 1/16 gap distance) the next day. Transformed clone selection is accomplished by including 1 mg bialaphos/l in the culture medium. An osmotic pretreatment of unorganized embryogenic callus (culture on 0.5 M osmoticum for 30 min) just prior to bombardment can enhance both transient expression and stably transformed clone recovery (Russell *et al.*, 1992b).

DNA-coated particles can be introduced into virtually any plant cell accessible to bombardment. This fact has led to consideration of what constitutes a suitable target cell for the generation of fertile transformed plants. To date, only cells capable of giving rise to cultures from which plants can be regenerated have proved useful for this purpose. Claims purporting to represent stable transformation of differentiated cells with subsequently normal development have not been substantiated (in any plant species). Target cells most widely used in corn are embryogenic cell suspension cultures (e.g. see Gordon-Kamm *et al.*, 1990) and callus (e.g. see Fromm *et al.*, 1990).

The great majority of cells in target tissue do not receive DNA during bombardment. Sanford (1990) has estimated a rate of one transient transformant per 1000 to 10 000 bombarded target cells, and rates of stable transformation at 2%–5% of the rate of transient transformation. These estimates suggest that stable transformation, on average, occurs at a rate of one event per hundreds of thousands, if not millions, of bombarded cells. The need is therefore plain for a means to select out the transformed cells.

Kanamycin selection has been used to select out

maize transformants following particle bombardment with the *npt*II gene (J. Register, unpublished data). There does, however, appear to be endogenous resistance to the antibiotic, and growth of nontransformed tissues ('escapes') is common. Similarly, attempts to use the selective agent Chlorsulfuron, and a mutant form of the *als* gene have been successful (Fromm *et al.*, 1990) but again escapes frequently occur (ICI Seeds, unpublished data). Hygromycin has also been used as a selective agent following bombardment with the gene for hygromycin phosphotransferase (Walters *et al.*, 1992). Here, growth on media containing hygromycin first at 15 mg/l and then at 60 mg/l was required for more than 9 weeks before sectors unequivocally resistant to the aminoglycoside could be identified. Recently, glyphosate has been used successfully as a selective agent in conjunction with a mutant enolpyruvyl shikimate-3-phosphate (EPSP) synthase gene (report from Monsanto, 1992 World Congress on Cell and Tissue culture, Washington, DC). To date, however, the selection system most widely used following particle bombardment-mediated transformation is expression of the *bar* gene (isolated from *Streptomyces hygroscopicus;* Thompson *et al.*, 1987) conferring resistance to phosphinothricin (PPT). PPT is an analog of glutamine that inhibits the amino acid biosynthetic enzyme glutamine synthase of bacteria and plants. Bialaphos is a tripeptide precursor of PPT produced by some strains of *Streptomyces,* in which two alanine residues are linked to the PPT moiety; the active PPT moiety is released intracellularly by peptidase activity (see Mazur & Falco, 1989). Although PPT has been used successfully as the selective agent in corn protoplast systems (Donn *et al.*, 1990; and see above), attempts to select on PPT following particle bombardment-mediated transformation with the *bar* gene have generally failed (ICI Seeds, unpublished data). However, for reasons that are not understood, if bialaphos, in the range 1–5 mg/l, is used instead of PPT, very effective selection for embryogenic suspension cells transformed with the *bar* gene can be achieved (Gordon-Kamm *et al.*, 1990; Spencer *et al.*, 1990). Escapes with the bialaphos/*bar* system following particle bombardment of embryogenic suspension cells are rare.

There are two areas where limitations in efficiency are encountered with bombardment of embryogenic cell cultures. First, the number of independent transgenic events from which plants can be regenerated is relatively low, and, second, transgenic regenerants often show poor fertility.

It is possible to produce large numbers (600 to 800) of stably transformed cell lines from single bombardments of certain plant cell cultures (e.g. NT1 of tobacco; Russell *et al.*, 1992*b*). However, with embryogenic corn cultures the average frequency of stable transgenic clone generation is closer to one or two per bombardment in most of the successful experiments where numbers have been reported (e.g. Gordon-Kamm *et al.*, 1991). This difference may relate to the nature of growth in embryogenic cell cultures compared to that in the populations of relatively homogeneous cells found in friable tobacco cultures such as NT1. Friable embryogenic maize cultures can be characterized as aggregates of cells organized into proembryonic structures at various stages of development. Division to generate new structures occurs from a subset of these cells; other cells within structures divide in the course of organized growth. The net result is that the number of cells from which stably transformed, selectable colonies can arise post-bombardment is relatively small. In the case of a culture such as NT1 tobacco, nearly all cells are capable of giving rise to stably transformed colonies.

Regeneration is not possible from all stably transformed clones and when plants can be produced they are not always fertile. Where numbers are given, it is clear that seed production on transformed regenerants is very low, with Gordon-Kamm *et al.* (1991) reporting a total of 176 seeds from 76 plants of one cell line and a total of 268 seeds from 219 plants of another cell line. Corn plants grown from seed in the greenhouse typically set between 300 and 400 seeds per plant (depending, of course, on genotype). The cause of low fertility in transgenic regenerants does not appear to be related in any general way to the presence of transgenes because in subsequent generations transgenic plants usually show normal fertility (Gordon-Kamm *et al.*, 1991; ICI Seeds, unpublished data). In addition, transgenic regenerants often appear 'stressed', i.e. they display premature senescence, poor tassel and ear development and stunted growth, whereas progeny of these plants display none of these symptoms and appear normal in all respects.

The causes of poor fertility and phenotypical abnormality in transgenic regenerants are unknown. These problems are encountered to varying degrees with regenerated corn plants in general and are thought to be related to the culture process. The phenomenon appears more marked in plants recovered from selection *in vitro*. Keeping the periods spent in culture and under selection to a minimum may be the most practical short-term solution to overcoming these problems.

We have shown (ICI Seeds, unpublished data) that improvements in the efficiency with which transgenic clones can be produced are often accompanied by a loss of regenerability, or by a loss in fertility of the plants that are produced. Deterioration in culture productivity in terms of fertile transgenic plant generation has also been suggested to occur as a consequence of culture age (Fromm *et al.*, 1990). One practical approach to overcome this problem is to maintain a bank of cryopreserved, productive cultures from which target tissue can be withdrawn when required.

Substantial progress will be required in understanding the control and dynamics of embryogenic culture development and plant regeneration before these culture systems can be fully exploited in the generation of fertile transgenic plants through particle bombardment. There has, unfortunately, been little fundamental work aimed at addressing the inefficiencies of culture initiation and fertile plant regeneration in cereal species to date.

Wounding and electroporation of immature zygotic embryos and type 1 callus

Recently, D'Halluin *et al.* (1992) described an electroporation method for the production of transgenic maize plants. This method utilizes as target tissue immature zygotic embryos of a size from which callus can be readily initiated (1.0–1.5 mm, typically about 10–14 days' post-pollination) or finely chopped type 1 callus. The freshly isolated embryos are treated for 1–2 min with an enzyme solution (0.3% (w/v) macerozyme, salts, buffer and 10% (w/v) mannitol) and then electroporated (after Dekeyser *et al.*, 1990) in the presence of linearized DNA. Following selection on kanamycin-containing medium, plants are regenerated from actively growing embryogenic tissue.

The authors of this method claim that wounding and/or degrading of intact tissue is an important prerequisite for success with DNA uptake.

This wounding may be achieved by physical means, e.g. cutting, or by chemical means such as an enzyme treatment for 1–2 min. Earlier work showed that electroporation could be used to introduce DNA into cells of intact rice leaf bases (Dekeyser *et al.*, 1990). Successful uptake was confirmed by transient gene expression assays. In this latter study, wounding – other than the dissecting out of the leaf bases – was not implicated.

In one example of the method, corn of the genotype H99 was transformed with a plasmid containing an *npt*II cassette and the barnase gene (Hartley, 1988) under the control of the tapetal-specific promoter TA29 from tobacco. It has been shown that expression of the barnase gene in tapetal tissue of tobacco (*Nicotiana tabacum*) and oilseed rape (*Brassica napus*) results in failure of normal anther development and, as a consequence, male sterility. The transgenic corn plants produced by the method of D'Halluin *et al.* (1992) were male sterile, but otherwise phenotypically normal. Progeny from these plants showed a 1:1 segregation for the *npt*II gene. The *npt*II negative progeny were male fertile, while the *npt*II positive progeny were male sterile. The barnase gene was detected by Southern analysis in all of the male sterile plants.

Further work to understand more fully the mechanism for DNA uptake into maize cells mediated by electroporation is clearly required. Nevertheless, the approach has yielded fertile, transgenic maize plants and has the promise of providing a means for testing transgenes with only the briefest exposure to growth *in vitro*.

Applicability of successful approaches to commercial germplasm

Reliance on a culture phase limits application of the above approaches to those genotypes from which the requisite response can be obtained. In the case of PEG-mediated DNA uptake into protoplasts, success has been restricted to one complex tissue culture-adapted genotype. Bombardment efforts to date have focused on the use of friable, embryogenic target tissue developed, for the most part, from immature F1 hybrid embryos of the genotype A188 × B73 (immature embryos dissected out of A188 plants pollinated with B73 pollen). Claims of success with elite maize inbreds have been made (stiff stalks of the

B73 type) but until very recently there had been no published reports. Koziel *et al.* (1993), however, have reported the introduction of a synthetic gene coding for an insecticidal protein into an elite inbred of Lancaster parentage. This was achieved by bombardment of immature zygotic embryos followed by culture initiation, selection of embryogenic callus and plant regeneration. Success following electroporation of immature zygotic embryos or type 1 callus has been achieved only with H99 and Pa91, two maize inbred lines known to be highly responsive in culture initiation experiments.

Transgenes can be introduced into any maize line from transgenic culture-responsive lines by backcrossing but clearly, from a commercial point of view, direct transformation of elite inbreds has major advantages. Transgene expression and agronomic performance can be assessed immediately, backcrossing is not required, and product development time is reduced. Several approaches are being employed in attempts to overcome genotype restrictions. Media improvements (Duncan *et al.*, 1985), and improved culture systems (selection of a certain cell type, culture maintenance protocols; Shillito *et al.*, 1989), have been tested with some success. An advance more readily reproducible and effective appears to be that of Vain, Flament & Soudain (1989*a*) and Vain, Yean & Flament (1989*b*), who showed that silver nitrate and/or aminoethoxyvinyl-glycine (AVG) could enhance friable embryogenic response from cultured immature embryos of A188. Songstad, Armstrong & Petersen (1991) subsequently showed that enhanced rates of response could be obtained with B73 and derivatives. This result has also been observed at ICI Seeds (W. P. Bullock *et al.*, unpublished data). Further exploitation of the encouraging lead identified by Vain and colleagues may result in success with other elite maize genotypes.

A different approach has been taken by Armstrong, Romero-Sevenson & Hodges (1992), who have attempted to identify the genetic components in A188 for culture responsiveness and introgress it into elite germplasm. However, from a practical point of view it would appear to be quicker and easier to introgress a transgene from A188 × B73 germplasm into elite lines than to introgress a complex trait such as culture responsiveness into elite lines as they are continually being developed by breeders.

Molecular characterization of transgenic maize

Transgene integration

The methods that have been used successfully to transform maize all employ direct DNA transfer. In general, transgene integration in maize (Spencer *et al.*, 1992) follows the trends observed with direct DNA delivery in other plant species (Riggs & Bates, 1986; Schocher *et al.*, 1986; Klein *et al.*, 1988; Bellini *et al.*, 1989; Tomes *et al.*, 1990). These trends are illustrated in Table 6.1, which shows a summary of results obtained in our laboratory from Southern analysis of over 100 independently transformed callus lines produced by particle bombardment of embryogenic A188 × B73 suspension cells (J. Register, unpublished data). Here, the selectable and nonselectable genes were on the same introduced plasmid.

The data in Table 6.1 illustrate several points. First, the absence of selection for gene expression correlates with a greater incidence of rearrangements (indicated by the presence of fragments of unexpected length) and deletions (nonselectable gene versus selectable gene). Second, integration of multiple gene copies (more than five) occurs with a frequency approaching 40%, although in around 90% of these cases the transgenes appear to be integrated at a single site (data not shown). This is a frequently observed phenomenon with introduction of transgenes through direct DNA delivery. Third, similar copy numbers of intact selectable and nonselectable genes were present in individual transformants.

At the whole plant level, transgene transmission has been studied through meiosis (Fromm *et al.*, 1990; Gordon-Kamm *et al.*, 1990, 1991; Spencer *et al.*, 1992; Walters *et al.*, 1992). Generally, transgenes are transmitted with fidelity from the primary transformed cell through to progeny of transgenic plants. There are, however, exceptions. Some transgene instability through meiosis was reported by Potrykus *et al.* (1985), following direct DNA delivery in tobacco. Similarly, in corn, transformants have been noted where transgene transmission has apparently occurred only through female gametes (Register *et al.*, 1992; Spencer *et al.*, 1992; Walters *et al.*, 1992).

As shown in Table 6.1, direct gene transfer tends to introduce multiple gene copies into cells and these gene copies often display complex inte-

Table 6.1. *Characterization of DNA integration after particle bombardment-mediated direct DNA delivery to embryogenic maize cells*

	Frequency of integration (%)	
	Selectable gene (*npt*II)	Nonselectable gene (*uid*A)
DNA rearrangements		
Unrearranged gene copies	54	27
Rearranged gene copies	7	19
Unrearranged and rearranged gene copies	39	42
Deleted genes	—	12
Copy number estimates		
1–4	59	62
5–10	22	22
>10	19	16

Notes:
Copy numbers were estimated from reconstructions and are expressed per haploid genome.

gration patterns (e.g. see Bellini *et al.*, 1989; Tomes *et al.*, 1990; Spencer *et al.*, 1992). Although direct gene transfer and *Agrobacterium*-mediated transformation have been compared head-to-head only once (Czernilofsky *et al.*, 1986*a,b*), it is clear that complex integration patterns occur more frequently following direct gene transfer. It remains to be shown definitively whether these integration characteristics have any functional significance but indirect evidence indicates that transgene instability may also be greater following direct gene transfer (for example, compare Chyi *et al.*, 1986; Muller *et al.*, 1987; and Heberle-Bors *et al.*, 1988; with Bellini *et al.*, 1989; Tomes *et al.*, 1990; and Spencer *et al.*, 1992).

Transgene expression

Study of transgene expression in primary maize transformants indicates that transgenes (driven by the 35S promoter) that confer resistance to a selective agent at the callus stage usually express at the whole plant level (e.g. see Gordon-Kamm *et al.*, 1991). Expression of nonselectable transgenes (e.g. *uid*A) introduced either on the same plasmid as the selectable marker gene (Fromm *et al.*, 1990; Register *et al.*, 1992; ICI Seeds, unpublished data) or on a separate plasmid (Gordon-Kamm *et al.*, 1990) is typically more variable. We have observed many cases in which expression of an apparently intact nonselectable gene occurs in fewer plants (regenerated from a single callus) than the selectable gene (Register *et al.*, 1992; ICI Seeds, unpublished data).

Inheritance of transgene expression has been studied for the most part in transgenic individuals produced by *Agrobacterium*-mediated transformation. The first study of transgene expression in plants transformed via direct DNA delivery was reported in 1985 by Potrykus *et al.* These authors noted that kanamycin-resistant tobacco transformants gave progeny that showed – with few exceptions – the expected segregation ratios for a single dominant trait. Later studies confirmed this result and also documented occasional ratios consistent with segregation of multiple loci and loss of gene expression (Morota & Uchimiya, 1988; Bellini *et al.*, 1989; Tomes *et al.*, 1990; Scheid, Paskowski & Potrykus, 1991). In transgenic maize, bialaphos or PPT resistance conferred by the *bar* gene has behaved as a single dominant trait in most of the progenies studied (Gordon-Kamm *et al.*, 1991; Register *et al.* 1992; Spencer *et al.*, 1992; Walters *et al.*, 1992).

There are few studies of the inheritance of expression of nonselectable genes in plants following direct DNA delivery. Results to date suggest that both Mendelian and non-Mendelian segregation can occur (Morota & Uchimiya, 1988; Tomes *et al.*, 1990). There are only two such reports in maize (Fromm *et al.*, 1990; Koziel *et al.*,

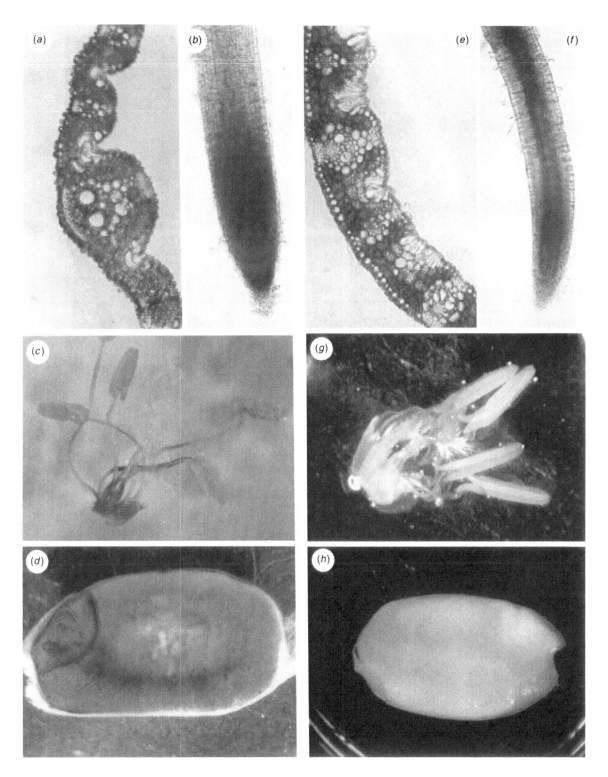

Plate 5.1. *In vivo* histochemical GUS analysis by using transgenic rice plants. (*a*)–(*d*) Expression of maize *adh*1-*uid*A gene. (*e*)–(*h*) Expression of rice *rbc*S-*uid*A gene. (*a*),(*e*) Expression in a leaf. (*b*),(*f*) Expression in a root. (*c*),(*g*) Expression in a flower. (*d*),(*h*) Expression in a seed.

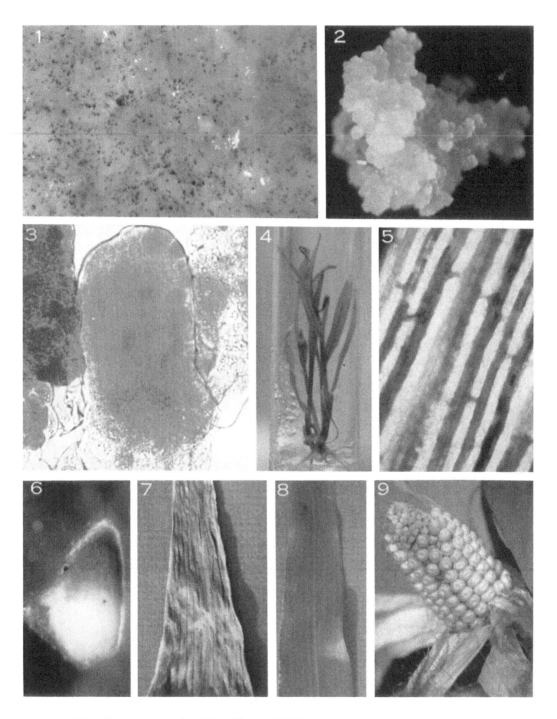

Plates 6.1–6.9. 1: Transient expression of the *uid*A gene in friable embryogenic callus tissue. 2: Transformed embryogenic callus clone selected on 1 mg bialaphos/l. 3: Transformed somatic embryo expressing ß-glucuronidase (GUS) activity. 4: Transformed plantlet during regeneration phase. 5: GUS activity in a leaf of a transgenic plant. Note vascular-specific expression. 6: GUS activity in a root tip of a transgenic plant. 7: Control plant reaction 7 days after leaf painting with Ignite (herbicide containing phosphinothricin as active ingredient). 8: Leaf from a transgenic plant carrying the *bar* gene and expressing resistance to Ignite. 9: Seed production on a regenerated transgenic plant.

1993), and in these instances expression of the transgenes segregated as single dominant traits.

A number of reports on plant transformation have shown stable inheritance of transgenes, with loss of expression. This has been observed following introduction of a native gene (e.g. see Napoli, Lemieux & Jorgenson, 1990; Van der Krol *et al.*, 1990) or multiple transgenes (e.g. Matzke *et al.*, 1989; Goring, Thomson & Rothstein, 1990), or following self-pollination of transformants (Scheid *et al.*, 1991; de Carvalho *et al.*, 1992). We have made similar observations in progeny of transgenic corn following self-pollinations (Register *et al.*, 1992). A number of mechanisms for transgene silencing (also referred to as cosuppression or *trans* inactivation) have been proposed (Matzke *et al.*, 1989; Napoli *et al.*, 1990; Grierson *et al.*, 1991; Mol, VanBlokland & Kooter, 1991; Scheid *et al.*, 1991). The only factor linking reports of gene silencing in transgenic plants is the presence of multiple copies of the silenced gene (or parts of the gene). It will be interesting to see if transgenic corn produced via direct DNA delivery (with its attendant increase in copy number and frequency of rearrangement) is particularly susceptible to this kind of epigenetic modification.

Although an in-depth consideration of factors affecting transgene expression is outside the scope of this review, it is worth briefly considering two of the most important. These are: regulatory sequences of the introduced gene, and transgene insertion site. Evaluation of regulatory sequences can be made in transient expression systems using particle bombardment or protoplasts (e.g. see Bodeau & Walbot, 1992). Presently, the only practical approach to addressing the insertion site is to produce a number of independent transformants and to select those in which the desired type and level of transgene expression is achieved. Site-specific recombination or creation of chromosome 'domains' (e.g. Hall *et al.*, 1991; Breyne *et al.*, 1992) to achieve expression independent of the transgene insertion site are approaches that could allow construct testing with the minimum number of transgenic individuals.

Concluding remarks

Progress in maize transformation is being made at what seems to be an ever-increasing pace. An efficiency sufficient for selection and recovery of transformed plants can be achieved by three approaches: (i) direct gene transfer to protoplasts, (ii) particle bombardment, and (iii) electroporation of zygotic embryos. There is little doubt that over the coming years these transformation techniques will become increasingly simple, more efficient and applicable to a wider range of genotypes.

The commercial goal with maize transformation is, in the first instance, to introduce performance-enhancing transgenes without disrupting the highly selected elite germplasm used for hybrid production. Small-scale field trials of transgenic maize plants carrying various single gene traits (e.g. for insect, virus and herbicide resistance) have already been conducted in the USA and Europe. In the longer term, transformation may become a tool that will allow the directed manipulation and exchange of genes between elite lines, as well as the introduction of complex multigene traits such as those involved in yield or those controlling recombination.

Recent developments

Recently, the use of silicon carbide whiskers (see whisker-mediated transformation, above) together with embryogenic maize suspension cultures resulted in the production of fertile, transgenic maize plants (Frame *et al.*, 1994). Transformation efficiency was estimated as one fifth to one tenth that achieved with microprojectile bombardment. In addition, stably transformed cell lines were obtained from whisker treatment of embryogenic callus derived from an elite stiff stalk maize inbred variety, indicating that the method can be extended to target tissues other than suspension cells.

References

Antonelli, N. M. & Stadler, J. (1990). Genomic DNA can be used with cationic methods for highly efficient transformation of maize protoplasts. *Theoretical and Applied Genetics*, **80**, 395–401.

Armstrong, C. L., Peterson, W. L., Buchholz, W. G., Bowen, B. A. & Sulc, S. L. (1990). Factors affecting PEG-mediated stable transformation of maize protoplasts. *Plant Cell Reports*, **9**, 335–339.

Armstrong, C. L., Romero-Sevenson, J. & Hodges, T. K. (1992). Improved tissue culture response of an elite maize inbred through backcross breeding and identification of chromosomal regions important for regeneration by RFLP analysis. *Theoretical and Applied Genetics*, **84**, 755–762.

Bellini, C., Guerche, P., Spielmann, A., Goujard, J., Lesaint, C. & Caboche, M. (1989). Genetic analysis of transgenic tobacco plants obtained by liposome-mediated transformation: absence of evidence for the mutagenic effect of inserted sequences in sixty characterised transformants. *Journal of Heredity*, **80**, 361–367.

Bennetzen, J. L., Lin, C., McCormick, S. & Staskawicz, B. J. (1988). Transformation of *Adh* null pollen to *Adh+* by 'macroinjection'. *Maize Genetics Cooperative Newsletter*, **62**, 113–114.

Bodeau, J. P. & Walbot, V. (1992). Regulated transcription of the maize *Bronze-2* promoter in electroporated protoplasts requires the C1 and R gene products. *Molecular and General Genetics*, **233**, 379–387.

Booy, G., Krens, F. A. & Huizing, H. J. (1989). Attempted pollen-mediated transformation of maize. *Plant Physiology*, **135**, 319–324.

Boulton, M. I., Buchholz, W. G., Marks, M. S., Markham, P. G. & Davies, J. W. (1989). Specificity of *Agrobacterium*-mediated delivery of maize streak virus DNA to members of the Gramineae. *Plant Molecular Biology*, **12**, 31–40.

Breyne, P., Van Montagu, M., Depicker, A. & Gheysen, G. (1992). Characterisation of a plant scaffold attachment region in a DNA fragment that normalises transgene expression in tobacco. *Plant Cell*, **4**, 463–471.

Bytebier, B., Deboeck, F., Greve, H., Van Montagu, M. & Van Hernalsteens, J. P. (1987). T-DNA organisation in tumor cultures and transgenic plants of the monocotyledon *Asparagus officinalis*. *Proceedings of the National Academy of Sciences, USA*, **84**, 5345–5349.

Callis, J., Fromm, M. & Walbot, V. (1987). Introns increase gene expression in cultured maize cells. *Genes and Development*, **1**, 1183–1200.

Chourey, P. S. & Zurawski, D. B. (1981). Callus formation from protoplasts of a maize cell culture. *Theoretical and Applied Genetics*, **59**, 341–344.

Chyi, Y.-S., Jorgenson, R. A., Goldstein, D., Tanksley, S. D. & Figueroa, F. (1986). Locations and stability of *Agrobacterium*-mediated T-DNA insertions in the *Lycopersicon* genome. *Molecular and General Genetics*, **204**, 64–69.

Coe, E. H. & Sarkar, K. R. (1966). Preparation of nucleic acids and a genetic transformation attempt in maize. *Crop Science*, **6**, 432–435.

Czernilofsky, A. P., Hain, R., Herrera-Estrella, L., Lörz, H., Goyvaerts, E., Baker, B. & Schell, J. (1986*a*). Fate of selectable marker DNA integrated into the genome of *Nicotiana tabacum*. *DNA*, **5**, 101–113.

Czernilofsky, A. P., Hain, R., Baker, B. & Wirtz, U. (1986*b*). Studies of the structure and functional organisation of foreign DNA integrated into the genome of *Nicotiana tabacum*. *DNA*, **5**, 473–482.

D'Halluin, K., Bonne, E., Bossut, M., De Beuckeleer, M. & Leemans, J. (1992). Transgenic maize plants by tissue electroporation. *Plant Cell*, **4**, 1495–1505.

de Carvalho, F., Gheysen, G., Kushnir, S., Van Montagu, M. C., Inzé, D. & Castreana, C. (1992). Suppression of β–1,3-glucanase transgene expression in homozygous plants. *EMBO Journal*, **11**, 2595–2602.

De Cleene, M. (1985). The susceptibility of monocotyledons to *Agrobacterium tumefaciens*. *Phytopathologie Zeitschrift*, **113**, 81–89.

De Cleene, M. & De Ley, J. (1976). The host range of crown gall. *Botanical Reviews*, **42**, 389–466.

De Wet, J. M. J., Bergquist, R. R., Harlan, J. R., Brink, D. E., Cohen, C. E., Newell, C. A. & De Wet, A. E. (1985). Exogenous gene transfer in maize (*Zea mays*) using DNA-treated pollen. In *The Experimental Manipulation of Ovule Tissues*, ed. G. P. Chapman, S. H. Mantell & R. W. Daniels, pp. 197–209. Longman, New York.

De Wet, J. M. J., Berthaud, J., Cubero, J. I. & Hepburn, A. G. (1988). Genetic transformation of cereals. In *Biotechnology in Tropical Crop Improvement: Proceedings of the International Biotechnology Workshop 12–15 January 1987*, pp. 27–32. ICRISAT Center, Patancheru, A. P. 502 324, India.

De Wet, J. M. J., De Wet, A. E., Brink, D. E., Hepburn, A. G. & Woods, J. A. (1986). Gametophyte transformation in maize (*Zea mays*, Gramineae). In *Biotechnology and Ecology of Pollen*, International Conference on Biotechnology and Ecology of Pollen, ed. D. L. Mulcahy, G. B. Mulcahy & E. Ottaviano, pp. 59–64. Springer-Verlag, New York.

Dekeyser, R. A., Claes, B., De Ryke, R. M. U., Habets, M. E., Van Montagu, M. C. & Caplan, A. B. (1990). Transient gene expression in intact and organised rice tissues. *Plant Cell*, **2**, 591–602.

Donn, G., Nilges, M. & Morocz, S. (1990). Stable transformation of maize with a chimeric, modified phosphinothricin acetyl transferase gene from *Streptomyces viridochromogenes*. In *Abstracts of VIIth International Congress on Plant Cell and Tissue Culture*, IAPTC, A2–38, p. 53.

Duncan, D. R., Williams, M. E., Zehr, B. E. & Widholm, J. M. (1985). The production of callus capable of plant regeneration from immature embryos of numerous *Zea mays* genotypes. *Planta*, **165**, 322–332.

Frame, B., Drayton, P., Bagnall, S., Lewnau, C., Bullock, P., Wilson, M., Dunwell, J., Thompson, J. & Wang, K. (1994). Production of fertile transgenic maize plants by silicon carbide whisker-mediated transformation. *In Vitro Cellular and Developmental Biology*, **30A**, 35.

Fromm, M. E., Taylor, L. P. & Walbot, V. (1986). Stable transformation of maize after gene transfer by electroporation. *Nature (London)*, **319**, 791–793.

Fromm, M. E., Morrish, F., Armstrong, C. L., Williams, R., Thomas, J. & Klein, T. M. (1990). Inheritance and expression of chimeric genes in the progeny of transgenic maize plants. *Bio/Technology*, **8**, 833–839.

Gordon-Kamm, W. J., Spencer, T. M., Mangano, M. L., Adams, T. R., Daines, R. J., Start, W. G., O'Brien, J. V., Chambers, S. A., Adams, W. R., Willetts, N. G., Rice, T. B., Mackey, C. J., Krueger, R. W., Kausch, A. P. & Lemaux, P. G. (1990). Transformation of maize cells and regeneration of fertile transgenic plants. *Plant Cell*, **2**, 603–618.

Gordon-Kamm, W. J., Spencer, T. M., O'Brien, J. V., Start, W. G., Daines, R. J., Adams, T. R., Mangano, M. L., Chambers, S. A., Zachweija, S. J., Willetts, N. G., Adams, W. R., Mackey, C. J., Krueger, R. W., Kausch, A. P. & Lemaux, P. G. (1991). Transformation of maize using particle bombardment: an update and perspective. *In Vitro Cellular and Developmental Biology*, **27P**, pp. 21–27, Tissue Culture Association.

Goring, D. R., Thomson, L. & Rothstein, S. J. (1990). Transformation of a partial nopaline synthase gene into tobacco suppresses the expression of a resident wild-type gene. *Proceedings of the National Academy of Sciences, USA*, **88**, 1770–1774.

Gould, J., Devey, M., Hasegawa, O., Ulian, E. C., Peterson, G. & Smith R. H. (1991). Transformation of *Zea mays* L. using *Agrobacterium tumefaciens* and the shoot apex. *Plant Physiology*, **95**, 426–434.

Graves, A. C. F. & Goldman, S. L. (1986). The transformation of *Zea mays* seedlings with *Agrobacterium tumefaciens*. *Plant Molecular Biology*, **7**, 43–50.

Grierson, D., Fray, R. G., Hamilton, A. J., Smith, C. J. S. & Watson, C. F. (1991). Does co-suppression of sense genes in transgenic plants involve antisense RNA? *Trends in Biotechnology*, **9**, 122–123.

Grimsley, N. (1990). Agroinfection. *Physiologia Plantarum*, **79**, 147–153.

Grimsley, N., Hohn, T., Davis, J. W. & Hohn, B. (1987). *Agrobacterium*-mediated delivery of infectious maize streak virus into maize plants. *Nature (London)*, **325**, 177–179.

Grimsley, N., Ramos, C., Hein, T. & Hohn, B. (1988). Meristematic tissues of maize plants are most susceptible to agroinfection with maize streak virus. *Bio/Technology*, **6**, 185–189.

Hahne, B., Lörz, H. & Hahne, G. (1990). Oat mesophyll protoplasts – their response to various feeder cultures. *Plant Cell Reports*, **8**, 590–593.

Hall, G., Allen, G. C., Loer, D. S., Thompson, W. F.

& Spiker, S. (1991). Nuclear scaffolds and scaffold-attachment regions in higher plants. *Proceedings of the National Academy of Sciences, USA*, **88**, 9320–9324.

Hartley, R. W. (1988). Barnase and Barstar expression of its cloned inhibitor permits expression of a cloned ribonuclease. *Journal of Molecular Biology*, **202**, 913–915.

Heberle-Bors, E., Charvat, B., Thompson, D., Schernthaner, J. P., Barta, A., Matzke, A. J. M. & Matzke, M. A. (1988). Genetic analysis of T-DNA insertions into the tobacco genome. *Plant Cell Reports*, **7**, 571–574.

Hooykaas-Van Slogteren, G. M. S. & Hooykaas, P. J. J. (1984). Expression of Ti plasmid genes in monocotyledonous plants infected with *Agrobacterium tumefaciens*. *Nature (London)*, **311**, 763–764.

Jarchow, E., Grimsley, N. H. & Hohn, B. (1991). *virF*, the host-range-determining virulence gene of *Agrobacterium tumefaciens*, affects T-DNA transfer to *Zea mays*. *Proceedings of the National Academy of Sciences, USA*, **88**, 10426–10430.

Junker, B., Zimny, J., Luehrs, R. & Lörz, H. (1987). Transient expression of chimeric genes in dividing and non-dividing cereal protoplasts after PEG-induced DNA uptake. *Plant Cell Reports*, **6**, 329–332.

Kaeppler, H. F., Gu, W., Somers, D. A., Rines, H. W. & Cockburn, A. F. (1990). Silicon carbide fiber-mediated DNA delivery into plant cells. *Plant Cell Reports*, **9**, 415–418.

Kaeppler, H. F., Somers, D. A., Rines, H. W. & Cockburn, A. F. (1992). Silicon carbide fiber-mediated stable transformation of plant cells. *Theoretical and Applied Genetics*, **84**, 560–566.

Klein, T. M., Fromm, M., Weissinger, A., Tomes, D., Schaaf, S., Sletten, M. & Sanford, J. C. (1988). Transfer of foreign genes into intact maize cells using high velocity microprojectiles. *Proceedings of the National Academy of Sciences, USA*, **85**, 4305–4309.

Klein, T. M., Kornstein, L., Fromm, M. E. & Sanford, J. C. (1989). Genetic transformation of maize cells by particle bombardment. *Plant Physiology*, **91**, 440–444.

Klein, T. M., Wolf, E. D., Wu, R. & Sanford, J. C. (1987). High velocity microprojectiles for delivery of nucleic acids into living cells. *Nature (London)*, **327**, 70–73.

Konstantinov, K., Mladenovic, S. & Denic, M. (1991). Recombinant DNA technology in maize breeding. v. Plant and indigenous bacterial strains. Transformation by the gene controlling kanamycin resistance. *Genetika (Beograd)*, **23**, 121–135.

Koziel, M. G., Beland, G. L., Bowman, C., Carozzi, N. B., Crenshaw, R., Crossland, L., Dawson, J., Desai, N., Hill, M., Kadwell, S., Launis, K., Lewis,

K., Maddox, D., McPherson, K., Meghji, M. R., Merlin, E., Rhodes, R., Warren, G. W., Wright, M. & Evola, S. V. (1993). Field performance of elite transgenic maize plants expressing an insecticidal protein derived from *Bacillus thuringiensis*. *Bio/Technology*, **11**, 194–200.

Kranz, E., Bautor, J. & Lörz, H. (1991). In vitro fertilisation of single, isolated gametes of maize mediated by electrofusion. *Sexual Plant Reproduction*, **4**, 12–16.

Kranz, E. & Lörz, H. (1990). Micromanipulation and in vitro fertilization with single pollen grains of maize. *Sexual Plant Reproduction*, **3**, 160–169.

Langridge, P., Brettschneider, R., Lazzeri, P. & Lörz, H. (1992). Transformation of cereals via *Agrobacterium* and the pollen pathway: a critical assessment. *Plant Journal*, **2**, 631–638.

Last, D. I., Brettell, R. I. S., Chamberlain, D. A., Chaudhury, A. M., Larkin, P. J., Marsh, E. L., Peacock, W. J. & Dennis, E. S. (1991). pEMU: an improved promoter for gene expression in cereal cells. *Theoretical and Applied Genetics*, **81**, 581–588.

Ludwig, S. R., Somers, D. A., Peterson, W. L., Pohlman, R. F., Zarowitz, M. Z., Gengenbach, B. G. & Messing, J. (1985). High frequency callus formation from maize protoplasts. *Theoretical and Applied Genetics*, **71**, 344–351.

Lupotto, E. & Lusardi, M. C. (1990). A potential approach for transformation: DUT (desiccation-uptake technique). Evidence of transient expression in embryogenic structures and regenerated plants. *Maize Genetics Cooperative Newsletter*, **64**, 22–23.

Lyznik, L. A., Ryan, R., Ritchie, S. W. & Hodges, T. K. (1989). Stable co-transformation of maize protoplasts with *gusA* and *neo* genes. *Plant Molecular Biology*, **13**, 151–161.

Matzke, M. A., Primig, M., Trnovsky, J. & Matzke, A. J. M. (1989). Reversible methylation and inactivation of marker genes in sequentially transformed tobacco plants. *EMBO Journal*, **8**, 643–649.

Mazur, B. J. & Falco, S. C. (1989). The development of herbicide resistant crops. *Annual Review of Plant Physiology*, **40**, 441–470.

Mitchell, J. C. & Petolino, J. F. (1991). Plant regeneration from haploid suspension and protoplast cultures from isolated microspores of maize. *Journal of Plant Physiology*, **137**, 530–536.

Mol, J., VanBlokland, R. & Kooter, J. (1991). More about co-suppression. *Trends in Biotechnology*, **9**, 182–183.

Morocz, S., Donn, G., Nemeth, J. & Dudits, D. (1990). An improved system to obtain fertile regenerants via maize protoplasts isolated from a highly embryogenic suspension culture. *Theoretical and Applied Genetics*, **80**, 721–726.

Morota, H. & Uchimiya, H. (1988). Inheritance and structure of foreign DNA in progenies of transgenic tobacco obtained by direct gene transfer. *Theoretical and Applied Genetics*, **76**, 161–164.

Muller, A. J., Mendel, R. R., Schiemann, J., Simoens, C. & Inzé, D. (1987). High meiotic stability of a foreign gene introduced into tobacco by *Agrobacterium*-mediated transformation. *Molecular and General Genetics*, **207**, 171–175.

Napoli, C., Lemieux, C. & Jorgenson, R. (1990). Introduction of a chimeric chalcon synthase gene in petunia results in reversible co-suppression of homologous gene *in trans*. *Plant Cell*, **2**, 279–289.

Negrutiu, I., Shillito, R., Potrykus, I., Biasini, G. & Sala, F. (1987). Hybrid genes in the analysis of transformation conditions. *Plant Molecular Biology*, **8**, 363–373.

Neuhaus, G. & Spangenberg, G. (1990). Plant transformation by microinjection techniques. *Physiologia Plantarum*, **79**, 213–217.

Neuhaus, G., Spangenburg, G., Mittelsten Scheid, O. & Sweiger, H.-G. (1987). Transgenic rapeseed plants obtained by the microinjection of DNA into microspore-derived embryos. *Theoretical and Applied Genetics*, **75**, 30–36.

Ohta, Y. (1986). High-efficiency genetic transformation of maize by a mixture of pollen and exogenous DNA. *Proceedings of the National Academy of Sciences, USA*, **83**, 715–719.

Petersen, W. L., Sulc, S. & Armstrong, C. L. (1992). Effect of nurse cultures on the production of macro-calli and fertile plants from maize embryogenic suspension culture protoplasts. *Plant Cell Reports*, **10**, 591–594.

Potrykus, I. (1990). Gene transfer to plants: assessment and perspectives. *Physiologia Plantarum*, **79**, 125–134.

Potrykus, I., Harms, C. T. & Lörz, H. (1976). Problems in culturing cereal protoplasts. In *Cell Genetics in Higher Plants*, ed. D. Dudits, G. L. Farkas & G. Lazar, pp. 129–140. Akademiai Kiado, Budapest.

Potrykus, I., Paszkowski, J., Saul, M. W., Petruske, J. & Shillito, R. D. (1985). Molecular and general genetics of a hybrid foreign gene introduced into tobacco by direct gene transfer. *Molecular and General Genetics*, **199**, 169–177.

Prioli, L. M. & Sondahl, M. R. (1989). Plant regeneration and recovery of fertile plants from protoplasts of maize (*Zea mays* L.). *Bio/Technology*, **7**, 589–594.

Quayle, T. J. A. & Feix, G. (1992). Functional analysis of the −300 region of maize zein genes. *Molecular and General Genetics*, **231**, 369–374.

Ranieri, D. M., Bottino, P., Gordon, M. P. & Nester, E. W. (1990). *Agrobacterium*-mediated transformation of rice (*Oryza sativa* L.). *Bio/Technology*, **8**, 33–38.

Register III, J. C., Sillick, J. M., Peterson, D. J.,

Bullock, W. P. & Wilson, H. M. (1992). Molecular analysis of transgenic maize callus, plants and progeny. *Journal of Cellular Biochemistry*, **16F**, 210.

Rhodes, C. A., Lowe, K. S. & Ruby, K. L. (1988a). Plant regeneration from protoplasts isolated from embryogenic maize cell cultures. *Bio/Technology*, **6**, 56–60.

Rhodes, C. A., Pierce, D. A., Mettler, I. J., Mascarenhas, D. & Detmer, J. J. (1988b). Genetically transformed maize plants from protoplasts. *Science*, **240**, 204–207.

Riggs, C. D. & Bates, G. W. (1986). Stable transformation of tobacco by electroporation: evidence for plasmid concatenation. *Proceedings of the National Academy of Sciences, USA*, **83**, 5602–5606.

Roeckel, P., Heizmann, P., Dubois, M. & Dumas, C. (1988). Attempts to transform *Zea mays* via pollen grains. Effects of pollen and stigma nuclease activities. *Sexual Plant Reproduction*, **1**, 156–163.

Russell, J. A., Roy, M. K. & Sanford, J. C. (1992a). Physical trauma and tungsten toxicity reduce the efficiency of biolistic transformation. *Plant Physiology*, **98**, 1050–1056.

Russell, J. A., Roy, M. K. & Sanford, J. C. (1992b). Major improvements in biolistic transformation of suspension-cultured tobacco cells. *In Vitro Cellular and Developmental Biology*, **28P**, 97–105.

Sahi, S. V., Chilton, M. D. & Chilton, W. S. (1990). Corn metabolites affect growth and virulence of *Agrobacterium tumefaciens*. *Proceedings of the National Academy of Sciences, USA*, **87**, 3879–3883.

Sanford, J. C. (1990). Biolistic plant transformation. *Physiologia Plantarum*, **79**, 206–209.

Sanford, J. C., Klein, T. M., Wolf, E. D. & Allen, N. (1987). Delivery of substances into cells and tissues using a particle bombardment process. *Journal of Particle Science Technology*, **5**, 27–37.

Sanford, J. C., Skubik, K. A. & Reisch, B. I. (1985). Attempted pollen-mediated plant transformation employing genomic donor DNA. *Theoretical and Applied Genetics*, **69**, 571–574.

Sanford, J. C., Smith, F. D. & Russell, J. A. (1993). Optimising the biolistic process for different biological applications. *Methods in Enzymology*, **217**, 483–509

Saul, M. W. & Potrykus, I. (1990). Direct gene transfer to protoplasts: fate of the transferred genes. *Developmental Genetics*, **11**, 176–181.

Scheid, O. M., Paskowski, J. & Potrykus, I. (1991). Reversible inactivation of a transgene in *Arabidopsis thaliana*. *Molecular and General Genetics*, **228**, 104–112.

Schlappi, M. & Hohn, B. (1992). Competence of immature maize embryos for *Agrobacterium*-mediated gene transfer. *Plant Cell*, **4**, 7–16.

Schnorf, M., Neuhaus-Url, G., Galli, A., Lida, S.,

Potrykus, I. & Neuhaus, G. (1991). An improved approach for transformation of plant cells by microinjection: molecular and genetic analysis. *Transgenic Research*, **1**, 23–30.

Schocher, R. J., Shillito, R. J., Saul, M. W., Paszkowski, J. & Potrykus, I. (1986). Co-transformation of unlinked foreign genes into plants by direct gene transfer. *Bio/Technology*, **4**, 1093–1096.

Shillito, R. D., Carswell, G. K., Johnson, C. M., Dimaio, J. J. & Harms, C. T. (1989). Regeneration of fertile plants from protoplasts of elite inbred maize. *Bio/Technology*, **7**, 581–587.

Somers, D. A., Birnberg, P. R., Petersen, W. L. & Brenner, M. L. (1987). The effect of conditioned medium on colony formation from 'Black Mexican Sweet' corn protoplasts. *Plant Science*, **53**, 249–256.

Songstad, D. D., Armstrong, C. L. & Petersen, W. L. (1991). AgNO$_3$ increases type III callus production from immature embryos of maize inbred B73 and its derivatives. *Plant Cell Reports*, **9**, 699–702.

Spencer, T. M., Gordon-Kamm, W. J., Daines, R. J., Start, W. G. & Lemaux, P. G. (1990). Bialaphos selection of stable transformants from maize cell culture. *Theoretical and Applied Genetics*, **79**, 625–631.

Spencer, T. M., O'Brien, J. V., Start, W. G., Adams, T. R., Gordon-Kamm, W. J. & Lemaux, P. G. (1992). Segregation of transgenes in maize. *Plant Molecular Biology*, **18**, 201–210.

Sun, C. S., Prioli, L. M. & Sondahl, M. R. (1989). Regeneration of haploid and dihaploid plants from protoplasts of sweet and supersweet (sh2sh2) corn. *Plant Cell Reports*, **8**, 313–316.

Thompson, C. J., Movva, N. R., Tizard, R., Crameri, R., Davies, J. E., Lauwereys, M. & Botterman, J. (1987). Characterisation of the herbicide-resistance gene *bar* from *Streptomyces hygroscopicus*. *EMBO Journal*, **6**, 2519–2523.

Thompson, J. A., Abdullah, R. & Cocking, E. C. (1986). Protoplast culture of rice using media solidified with agarose. *Plant Science*, **47**, 123–133.

Tomes, D. T., Weissinger, A. K., Ross, M., Higgins, R., Drummond, B. J., Schaaf, S., Malone-Schoneberg, J., Staebel, M., Flynn, P., Anderson, J. & Howard, J. (1990). Transgenic tobacco plants and their progeny derived by microprojectile bombardment of tobacco leaves. *Plant Molecular Biology*, **14**, 261–268.

Töpfer, R., Gronenborn, B., Schaefer, S., Schell, J. & Steinbiss, H. (1990). Expression of engineered wheat dwarf virus in seed-derived embryos. *Physiologia Plantarum*, **79**, 158–162.

Töpfer, R., Gronenborn, B., Schell, J. & Steinbiss, H. (1989). Uptake and transient expression of chimeric genes in seed-derived embryos. *Plant Cell*, **1**, 133–139.

Vain, P., Flament, P. & Soudain, P. (1989a). Role of

ethylene in embryogenic callus initiation and regeneration in *Zea mays* L. *Journal of Plant Physiology*, **135**, 537–540.

Vain, P., Yean, H. & Flament, P. (1989*b*). Enhancement of production and regeneration of embryogenic type III callus in *Zea mays* L. by AgNO₃. *Plant Cell, Tissue and Organ Culture*, **18**, 143–151.

Van der Krol, A. R., Mur, L. A., Beld, M., Mol, J. N. M. & Stuitje, A. R. (1990). Flavenoid genes in petunia: addition of a limited number of gene copies may lead to a suppression of gene expression. *Plant Cell*, **2**, 291–299.

Wagner, V. T., Dumas, C. & Mogensen, H. L. (1990). Quantitative three-dimensional study on the position of the female gametophyte and its constituent cells as a prerequisite for corn (*Zea mays*) transformation. *Theoretical and Applied Genetics*, **79**, 72–76.

Waldron, J. C. (1987). Pollen transformation. *Maize Genetics Cooperative Newsletter*, **61**, 36–37.

Walters, D. A., Vetsch, C. S., Potts, D. E. & Lundquist, R. C. (1992). Transformation and inheritance of a hygromycin phosphotransferase gene in maize plants. *Plant Molecular Biology*, **18**, 189–200.

7

Barley, Wheat, Oat and Other Small-grain Cereal Crops

Ralf R. Mendel and Teemu H. Teeri

Introduction

The most important crop plants of the world belong to the monocotyledonous grasses. From the perspective of genetically transforming these successful outdoors plants with new genes, they have properties in common. The tissues of these grasses lack what is called the wound response – the characteristic of the cells at a site of physical damage to proliferate and form calloid protective tissue – the cereal tissue rather seals the wound site by lignification and sclerification of the cells.

This alternative strategy in reaction to wounding seems to be the basis of the recalcitrance of cereal plants to the manipulations of the experimentalist. The two most commonly used gene transfer techniques utilize either the capability of *Agrobacterium* to mate with plants or the capacity of a naked plant cell (protoplast), where DNA can be introduced by physical means, to regenerate its cell wall, start to divide and finally to differentiate into a mature plant. Neither of these methods functions easily for the cereal grasses. Agrobacteria deliver their transferred DNA into cells at a wound site – destined to deteriorate. Protoplasts of cereal plants have been very hard to regenerate and especially to retain their capacity to differentiate after regeneration. This situation has led to the development of alternative, often very imaginative, gene transfer techniques. The literature of cereal transformation is littered with reports of exotic methods that mostly have given ambiguous results and are very hard to repeat.

Still, the characteristics of the cereal life-style are not a fundamental obstacle to either plant regeneration from protoplasts or gene transfer itself. Investment of efforts has made it possible to generate stably transformed plants from both rice and maize (Rhodes *et al.*, 1988; Toriyama *et al.*, 1988). However, compared to the success with those plants, gene transfer to the small grain cereals of the moderate climatic regions is still in its beginning. Only a few papers are published and most information comes from congress abstracts and via personal communications. Still, remarkably, the first reports of obtaining transgenic, fertile wheat (Vasil *et al.*, 1992) and oat (Somers *et al.*, 1992) have recently been published.

In the context of this chapter, stable transformation means the introduction of one or more intact copies of one or more foreign genes into the genome of the plant, where the foreign genes are stably maintained through cell divisions and are expressed. Two forms of stably transformed plant tissue are possible to generate: (i) transgenic cell cultures, and (ii) transgenic mature plants. In the latter case, regular Mendelian inheritance of the foreign genes also is required.

For successful stable transformation, three conditions have to be met: (i) the foreign DNA must penetrate the cell, (ii) it has to stabilize by integration into the chromosomes, and (iii) the transformed cell must regenerate into either a proliferating tissue culture or a mature plant.

It is possible to monitor the first step of transformation, the penetration, independently from the steps that follow. Free DNA, once it has entered the plant cell, will persist there for several days, find its way to the nucleus, and have its genes expressed by the cell's machinery. Eventually the DNA is lost, but the phenomenon of transient expression is important in assessing penetration by using an established marker gene system. The most sensitive markers for penetration are complete genomes of plant viruses, which start to replicate and spread as a result of penetration of a single molecule. In this way it has been established that agrobacteria are capable of introducing DNA also to cereal cells (Grimsley *et al.*, 1987). The histochemically detectable reporter genes are, however, most useful as penetration markers. By staining the tissue after gene delivery, it is possible to count the number and evaluate the distribution of the cells that have been penetrated. Also, detection of the activity of histochemically detectable reporter genes gives most information for optimization, since the number of transformed foci is more important than the quantified average ability of the cells to express the particular construct used.

The next step of transformation, stabilization of the foreign genes, can be measured by following persisting marker gene expression. This is, however, more complicated because both of the histochemically detectable reporters that have been used for plants (*uid*A and *lacZ*) code for extremely persistent enzymes. Usually the assessment of stabilization is done together with regeneration. Active stabilization of foreign genes has not been used in direct gene transfer and the experimentalists have relied on natural stabilization through 'illegitimate recombination', a phenomenon with a fortunately high frequency of up to several per cent (Klein *et al.*, 1988).

The step from the penetration and subsequent stabilization to regeneration is not trivial. The logic is simple: DNA has to enter cells with full regeneration capacity. Penetration of foreign DNA to cereal cells has been easy to demonstrate with several techniques. The difficulty in the ability to penetrate regenerating cells is equal to the difficulty in obtaining transgenic cereal plants.

Stabilization of foreign genes in the cells of a differentiated plant is only the first step. Consistent inheritance of the transgenes in the next generation according to the laws of Mendel (1865) is necessary not only for the demonstration of true integration into chromosomes but also for the practical maintenance of the transgenic plants. The first transgenic dicotyledonous plants obtained with *Agrobacterium* vectors were shown to fulfil this condition by rule (Budar *et al.*, 1986). It is important to realize that this is not necessarily true for new plant species transformed with new

techniques (e.g. see Rogers & Rogers, 1992). It is clear that the assessment of gene transfer and transmission of the transgenes to further generations in these plants must be done by using the strongest and most clear-cut methods of molecular analysis (Potrykus, 1990).

Methods of transformation

The principal difficulty in transformation of cereal plants is not in the methods of DNA delivery, but in those of culturing the cells. It is, in fact, special that the Solanaceous species (tobacco, tomato, petunia, potato, etc.) react so well *in vitro* and enabled the whole concept of transgenic mature plants to be established ten years ago.

The development of cereal transformation methods has followed two lines of thought. First, much effort has been put into overcoming the technical problems in protoplast culture. The ability to regenerate protoplasts to mature plants practically means ability to transform. Second, attempts have been made to bypass the problems in the protoplast culture by developing alternative methods of DNA delivery into intact cells with more potential to regenerate into mature plants.

In the following, we discuss first the protoplast techniques and the one successful case of the alternative delivery method, particle bombardment. The required techniques of *in vitro* tissue culture are included. After a section for the marker and reporter gene constructions used, a collection of the other methods of DNA delivery are presented. At the moment, they have at the best a status of being 'promising', although some of them may prove to be 'successful' as well. Since the alternative methods represent a large amount of the work done towards transformation of cereal plants, we feel that they should be presented to the reader.

Protoplasts

The isolation of protoplasts from different tissues of the Gramineae is relatively straightforward (Morrish, Vasil & Vasil, 1987). Vital protoplasts can be obtained in large amounts, e.g. from young leaves. However, protoplasts isolated from differentiated cereal tissues have one thing in common: they are unable to divide. Nevertheless, mesophyll protoplasts, for example, are excellent material for physiological studies as well as for experiments of transient gene expression.

In order to test a given gene construct in transient expression, protoplasts have to be isolated in large amounts and should be uniform and vital. These requirements are met by protoplasts isolated either from leaves or from suspension cell cultures. The transformation scheme is principally the same as for other monocots and dicots, i.e. DNA-uptake is mediated by polyethylene glycol (PEG), electroporation, or a combination of both (see e.g. Lazzeri *et al.*, 1991). One to 3 days after DNA transfer, the cells are harvested, and expression of the reporter gene is measured usually by determining the enzymatic activity it codes for. Also immunodetection of the gene product or measurement of the reporter mRNA is possible. There are reports of use in cereals for most of the common reporter genes such as *npt*II, *uid*A, *cat*, *bar* and *luc*.

When using protoplasts to assay transient gene expression, it is important to realize that protoplasts are highly stressed cells and do not necessarily reflect all of the properties they had in the tissue from which they were derived. Still, a scheme for correct regulation of gene expression in cereal protoplasts by plant hormones has been published (see below).

The crucial point for stable transformation via the protoplast approach is the embryogenicity of the suspension culture as protoplast source. Two main sources of cells are successfully used for this purpose: immature embryos and anthers. Suspension cultures derived from these tissues have been shown to be embryogenic (Lührs & Lörz, 1987*a*,*b*; Jähne *et al.*, 1991*a*; Vasil, Redway & Vasil, 1990) and will frequently give protoplasts that are able to divide, depending on the skill of the experimenter, on the quality of the plant material, and on the genotype used. Also mesocotyl-derived suspension cultures have been shown to fulfill these criteria (Müller, Schulze & Wegner, 1989). Approximately 3–4 months after taking the immature embryos or the anthers into culture, the suspensions are fine enough to give large amounts of vital and dividing protoplasts that are able to form protoplast-derived colonies.

In principle, it should be possible to use these cell lines as starting material also for obtaining stably transformed protoplast-derived plants. However, the time window where the cell lines will retain their embryogenicity is rather narrow (1–3

months), and protoplasts isolated from later cultivation stages of the cultures will develop into calli that are no longer able to regenerate into plants. Thus, in order continuously to have embryogenic suspension cell cultures available, it is necessary to re-establish the embryogenic cell culture repeatedly, or to take the existing cell culture into cryopreservation. However, suspension cell cultures that have lost their embryogenic potential are not useless. They are often good sources of protoplasts for gene transfer experiments that aim to generate transgenic calli.

There are numerous basic and applied problems that can be studied at the level of transgenic cell cultures without the urgent necessity of having regenerated plants available. The study of mechanisms for the recognition and integration of foreign genes into cereal genomes, the physiological questions connected with housekeeping genes involved in primary metabolic steps, the recognition of targeting and processing signals of a foreign or endogenous gene by the cereal cell, and the ability of the cells to produce foreign protein in active form and its effects on the basic physiology of the cell can all be investigated in tissue culture. For all these areas of research, the protoplast approach is an excellent transformation system that gives stably transformed cell colonies in a short time.

Particle gun

Originally the particle gun was seen as one of the fantastic (but improbable) new ways to solve the problems in transformation of recalcitrant species. Remarkably, and unlike many of the other alternative methods, it could quickly be adopted by new laboratories, it functions very reproducibly, and it has proven to be a truly new approach to direct gene transfer. Thus, not only does it solve many existing problems, but also it penetrates into totally new areas such as transformation of organelle genomes (Svab, Hajdukiewicz & Maliga, 1990).

The principle of the particle gun is simple: DNA is precipitated on inert microcarriers of gold or tungsten, which can be accelerated to a speed where the particles have sufficient impact to penetrate intact plant cells. Once inside the cell, the DNA dissolves and what follows is common to direct transfer of DNA into wall-less protoplasts: transient expression of the genes in the transferred

DNA can be measured, and the sequences have a chance to integrate into the chromosomes by 'illegitimate recombination'.

The fundamental novelty of the method is that you end up with an intact cell with foreign DNA. Thus, this cell has not been subjected to the trauma of protoplasting and should have retained its capability of cell division and finally of differentiation to a whole new plant – if it ever had such potential.

The delivery of foreign DNA into cereal tissues by particle gun bombardment can be assessed by measuring transient expression of the transferred genes. Histochemically detectable reporter gene activities give most information for optimization, since they allow counting of the transformed foci. Transient assay of particle-bombarded tissue has also valuable applications of its own. It makes possible tissue-specific assays of gene constructions directly in intact tissue and without the need of stable transformation (see below).

In order to obtain transgenic plants, a natural target for bombardment would be a meristematic cell. However, these cells are small, compact and hard to reach. The *in vitro* induction of the regenerative potential in nonmeristematic cells is more practical. Tissue cultures initiated from immature embryos or microspores, and forming embryo-like structures regenerating to normal, fertile plants can be obtained from cultivars of barley (*Hordeum vulgare*; Ahloowalia, 1987; Jähne *et al.*, 1991a; Jähne, Lazzeri & Lörz, 1991b), wheat (*Triticum aestivum*; Vasil *et al.*, 1990) and oat (*Avena sativa*; Bregitzer *et al.*, 1991). It is also possible that the highly focussing bombardment device of Sautter *et al.* (1991) will prove useful in the transformation of cells that naturally have high regeneration capacity.

Promoters and marker genes

The 35S promoter of cauliflower mosaic virus is the most widely used promoter for plant transformation. Experiments of transient and stable transformation with barley and wheat protoplasts show that it is suitable also for driving gene expression in these small grain cereals. However, there seems to be a difference in the strength of this promoter between transient and stable transformation: in transient assays it is a rather weak promoter whereas its performance in stable transformants is comparable to those levels known from dicots

such as tobacco (Mendel *et al.*, 1990; J. Schulze, A. Nerlich & R. Mendel, unpublished data), allowing the selection of large numbers of transgenic colonies.

During the last years, numerous groups have tried to improve the performance of the 35S promoter in monocots by introducing monocot-specific sequences into the 5'-untranslated leader of reporter gene constructs, e.g. the exon 1 and intron 1 of the maize *sh1* gene (Maas *et al.*, 1991), the intron 1 of the maize *adh1* gene (Callis, Fromm & Walbot, 1987) and the intron 1 of the rice *act1* gene (McElroy *et al.*, 1990, 1991). These experiments were performed mainly in maize. It was shown that also in barley protoplasts the insertion of exon 1/intron 1 of the *sh1* gene into the nontranslated leader could increase the transient expression levels of *cat* up to 1000-fold and those of *luc* up to 250-fold (Maas *et al.*, 1991; Töpfer *et al.*, 1992). However, when exchanging *cat* for *uid*A as reporter gene, these effects were not any longer so pronounced (Lazzeri & Lörz, 1991).

Another way to boost transient expression levels in monocots is the systematic testing in transient assays of constitutive monocot promoters inserted in front of reporter genes. When the rice actin gene promoter was tested (McElroy *et al.*, 1990), fused in front of *uid*A, its performance was 1000-fold better than that of the 35S promoter in barley protoplasts (Spickernagel *et al.*, 1991) and about 50–100-fold better than that of 35S when using the particle bombardment approach with barley suspension cells (R. R. Mendel, unpublished data).

Last *et al.* (1991) found yet another possibility for obtaining high transient gene expression in monocots. They constructed an 'artificial' promoter consisting of six repeats of an anaerobic responsive element of maize plus four repeats of the *ocs* enhancer in front of a modified *adh1* promoter with the *adh1* intron 1. This promoter (pEmu), fused in front of *uid*A, showed a performance similar to that of the rice actin promoter when it was tested in the same transient assays (protoplasts and particle bombardment) that were used for the actin promoter (Spickernagel *et al.*, 1991; R. R. Mendel, unpublished data). With dicots, however, there are indications that too strong promoters are not optimal for selection of stable transformants: Hinchee *et al.* (1991) observed with petunia that the percentage of escapes increased with promoter strength due to cross-feeding protection so that a promoter of moderate strength is more suitable for driving the selected marker.

There seems to be a general problem in using *uid*A as an effective reporter gene in barley and wheat, and this is particularly obvious in stably transformed cells where in many cases the expression of *uid*A is completely repressed. This, however, is not due to the 35S promoter per se, or to the plasmid carrying the *uid*A gene as such, or to the lack of transcription factors etc., for the following reasons. (i) The 35S promoter is very active in combination with the *npt*II gene that is localized adjacent to 35S-*uid*A on the same piece of DNA that had been introduced into barley genome without rearrangements (Schulze, Nerlich & Mendel, 1991*b*). (ii) 35S-*uid*A can be effectively expressed in transient assays in barley cells when the microprojectile-bombardment approach is used (Mendel *et al.*, 1989, 1990; Schulze *et al.*, 1991*c*). (iii) Also the much stronger, monocot-specific promoter pEmu (Last *et al.*, 1991) driving the *uid*A gene is not able to increase β-glucuronidase (GUS) activity after stable integration into barley cells, although in transient assays after microprojectile bombardment there is a dramatic increase in *uid*A expression compared to the 35S-*uid*A construct (J. Schulze, A. Nerlich and R. Mendel, unpublished data). The integration of the *uid*A coding sequence into the genome seems to be the crucial point in repressing its activity. The reason for this phenomenon is not known. One could speculate that the base composition determining the methylation pattern of this bacterial gene is not compatible with the cereal gene expression machinery once the foreign gene has been integrated, whereas with the *npt*II gene sequence this would not be the case. Also in stably transformed wheat callus, *uid*A expression was observed to vary strongly (Mejza *et al.*, 1991; Vasil *et al.*, 1991).

When the performance of 35S-*uid*A or pEmu-*uid*A in barley and wheat on the one hand and in maize, rice and dicots on the other is compared with transient assays (bombardment of suspension culture cells), it is obvious that small-grain cereals are generally less effective expressing the introduced constructs (J. Schulze, A. Nerlich and R. Mendel, unpublished data; Brettschneider, Becker & Lörz, 1991). Taking into account the above-mentioned effects after stable integration one has to conclude that the *uid*A gene is a prob-lematic reporter gene for small-grain cereals. The gene *luc* coding for luciferase might be a better candidate, but information about its performance in these recipients is scarce. Maas *et al.* (1991) and Lanahan *et al.* (1992) have shown that it is principally possible to use *luc* as a reporter for transient gene expression in barley protoplasts and in our own experiments we confirmed these data (R. R. Mendel, unpublished data; R. Hannus, unpublished data).

For selecting stable transformants, the most useful marker genes are *npt*II and *bar*. For *npt*II, selection on kanamycin gives a considerable background growth of nontransformed colonies. Its analog, the compound G418 (also known as geneticin), is much more effective in selecting transgenic colonies in barley for example (Schulze *et al.*, 1991*b*). However, the selective agent may also have an effect on regeneration. We have experienced that G418 increases albinism in barley whereas kanamycin does not (A. Ritala, unpublished data). For rice, Battraw & Hall (1992) report a reverse effect of these chemicals.

The *npt*II gene under the control of the 35S promoter is completely sufficient for selecting large numbers of stably transformed cell colonies in barley (Lazzeri *et al.*, 1991; Schulze *et al.*, 1991*b*; Ritala *et al.*, 1993). The *bar* gene under control of the 35S promoter has also been successfully used to select stable transformants in barley (Louwerse *et al.*, 1991; Schulze & Mendel, 1991), wheat (Vasil *et al.*, 1992) and oat (Somers *et al.*, 1992). Genes mediating resistance to the antibiotics hygromycin (Kaneko *et al.*, 1991) and metothrexate (Chen, Dale & Mullineaux, 1991; Mejza *et al.*, 1991) can be used as well; however, there is not so much information available.

Other methods of DNA delivery

Macroinjection into floral tillers

Circumvention of *in vitro* regeneration would be highly advantageous and several approaches have been suggested in this direction. One of them is macroinjection of DNA directly into floral tillers. De la Peña, Lörz & Schell (1987) injected DNA containing the marker gene *npt*II into the tiller cavity of immature rye tillers (*Secale cereale*), thereby flooding the immature inflorescence, the florets and very likely also the enclosed sexual organs with DNA. It was thought that, once in the anther or ovule, the foreign DNA would penetrate the spore mother cell and would be incorporated into the micro- or megaspores. After fertilization, seed setting, harvesting and germination, the seedlings were screened for kanamycin resistance. From 3023 grains, three plants were obtained that grew in the presence of kanamycin and expressed *npt*II in their leaves. In genomic DNA, Southern blots revealed appropriately hybridizing restriction fragments. However, no transmission to the progeny was demonstrated.

Rogers & Rogers (1992) used the same approach with the genes for GUS and thaumatin, both under control of an aleurone-specific α-amylase promoter. Without any selection, the harvested barley grains were screened for the expression of the *uid*A gene by using a piece of aleurone tissue from each seed. From seeds positive in this test, the embryos were germinated and Southern blot analysis of DNA from the seedlings was performed. With a frequency of about 1%, positive plants were obtained that were fertile and could be selfed so that the next generation also could be analyzed. It turned out that some hybridizing bands were lost when different tillers from the same plant were compared. In addition, reduction of the intensity of hybridization signals, as well as the appearance of new hybridizing bands in different tissue samples of the same plants, were taken as an indication of ongoing rearrangements of the introduced foreign DNA within the recipient plant. Different stabilities of the two reporter genes coding for GUS and thaumatin were observed and conclusions about the importance of the methylation pattern for gene stability were drawn. It was observed that the very unstable *uid*A marker maintained the methylation pattern characteristic of prokaryotic DNA, while the monocot gene for thaumatin showed some loss of A-methylation and apparent acquisition of C-methylation in CpG dinucleotides. In further experiments with nonmethylated and with *in vitro* methylated plasmid DNA, it was shown that the presence of N^6-methyladenine in the transforming plasmid DNA was associated with rapid loss of the DNA from new tillers on the plants, while the absence of N^6-methyladenine coupled with methylation of C residues at CpG dinucleotides greatly increased the stability of the same plasmid and allowed transmission into the second generation. Rogers & Rogers speculated about the presence of a previously undefined system in barley that discriminates between foreign and genomic

DNA by identifying DNA that lacks the proper methylation pattern followed by removal of the latter in actively dividing cells. Integration into the genome was not shown, rather there were indications of an extrachromosomal location of the introduced DNA that persisted through many multiples of cell division and was inheritable by the F2 generation.

Mendel *et al.* (1990) used macroinjection into floral tillers for transferring the genes *npt*II and *uid*A. With a frequency of 1.6%, kanamycin-resistant plants were obtained that exhibited very low NPT II activities in leaves. No NPT II activity was detectable in F2 plants. DNA isolated from F1 and F2 plants was further analyzed. Polymerase chain reaction (PCR) primers specific for an internal 412 bp fragment of the *npt*II coding sequence allowed the amplification of the target sequence in selected F1 and F2 plants in those cases where NPT II activity was no longer detectable. Southern hybridization of genomic DNA demonstrated that a hybridizing 1.7 kb *npt*II-specific fragment was inherited from F1 to F2 plants; however, surprisingly, different plants showed the same fragment size even when *Pst*I was used to cut inside the *npt*II coding sequence in order to visualize the flanking DNA. It appeared that independent transformants showed an identical hybridization pattern that was inherited from F1 to F2. Further, the fragment size did not fit into the restriction map of the introduced plasmid DNA and, most unexpectedly, nondigested genomic DNA showed discrete *npt*II-hybridizing bands. Obviously the introduced plasmid DNA occurred in a nonintegrated, i.e. extrachromosomal, location. This phenomenon was also discussed by Rogers & Rogers (1992). It remains to be seen whether these plasmid molecules occur freely in the cytoplasm or are localized in organelles.

Du *et al.* (1989) utilized macroinjection into floral tillers of *Triticum monococcum* with *npt*II as the selected marker and the gene for soybean legumin as a cargo gene, and reported a frequency of 0.6% for the occurrence of kanamycin-resistant seedlings. Dot hybridizations of genomic DNA with legumin as probe were positive in 7 out of 20 tested green plants. No Southern blot data were presented.

In a critical review of cereal transformation, Potrykus (1990) argued that with *in vivo* techniques there is the chance that foreign DNA does not exist in barley cells but rather in contaminating endophytes that are intimately connected with the plant cell and very difficult to culture outside the plant. Rogers & Rogers (1992) exclude this possibility by showing that the α-amylase promoter-*uid*A construct was correctly transcribed in aleurone tissue from the barley promoter (as assessed by both S_1 nuclease protection assays and reverse transcriptase–PCR assays). Mendel et al. (1990, 1991) could not exclude this possibility and announced further investigations into this direction. Konstantinov et al. (1990), however, were able to recover kanamycin-resistant bacteria from the F5 generation of maize plants after they had macroinjected DNA with *npt*II as a marker into the plants before sporogenesis. In the F5 generation, *npt*II-specific hybridizing fragments were observed that were of uniform size in independent transformants, so that the endophyte theory was taken into consideration. The recovered bacteria contained plasmids that hybridized with the *npt*II probe, and were identified as *Pseudomonas* and *Acinetobacter* species.

The macroinjection approach is an interesting method for studying the fate of foreign DNA introduced into a full-size organism (i.e. an intact plant) rather than into a single cell. Presently, however, this procedure is of purely academic value and it is rather unlikely that it will develop into a reliable transformation method of general applicability for barley or other monocot or dicot plants.

DNA transfer via the pollen tube pathway
A long time before macroinjection was developed, other approaches were suggested to circumvent regeneration *in vitro*. Most of them proposed introduction of DNA into gametes, followed by fertilization and zygotic embryogenesis. This kind of approach would be simpler, faster and cheaper than the *in vitro* methods, and would also avoid the problem of somaclonal variation. It was a logical consequence to favor the use of pollen as a vector for DNA because ovules are difficult to isolate and the injection into the embryo sac *in situ* seemed to be too tedious and unpredictable. It was hoped that pollen was easily accessible for DNA transfer and that the pollen tube would deliver the DNA to the egg cell (for a review, see Hess, 1987).

De Wet et al. (1985) and Ohta (1986) suggested the use of DNA-treated pollen as a DNA vector for pollinating fertile plants of maize, and Hess (1987) described a similar transformation system for *Nicotiana glauca*, but without molecular-genetic evidence. Ahokas (1987) showed that DNA uptake into pea pollen was facilitated by liposomes, and Abdul-Baki et al. (1990) demonstrated the introduction of labeled DNA into pollen grains of *Nicotiana gossei* by electroporation. Twell et al. (1989) bombarded pollen with DNA using the particle gun approach and demonstrated transient gene expression of the marker gene *uid*A. However, although DNA could be taken up into pollen of plants of diverse phylogenetic origin, there is no case reported yet that decisively demonstrates a successful gene transfer using this kind of approach.

Another pollen tube pathway approach was described by Zhong-xun & Wu (1988) for rice where, after pollination, the stigmas were cut off and DNA solution was directly applied to the style using the pollen tube as a microcapillary. In DNA hybridizations of seed-derived marker-selected plants, positive signals were obtained. Our duplication of the experiment with barley was not successful (G. H. Patel & T. H. Teeri, unpublished data).

A third approach was proposed by Picard et al. (1988). They applied DNA with the *npt*II marker gene on to just-pollinated stigmas of wheat plants and found seedlings with kanamycin resistance and expression of the *npt*II gene among the progeny of the DNA-treated flowers. However, no molecular-genetic evidence was presented. Later, several groups tried to apply this method to other cereals, and failed. Only recently, by using the same approach, the group of Jacquemin reported the transfer of the maize zein gene into wheat plants (Delporte et al., 1991). *npt*II was used as the selectable marker for the nonselected cargo gene. DNA of kanamycin-resistant plants was shown to contain both *npt*II and zein-gene-specific hybridization signals, and after selfing, the tissue-specific expression of α-zein was immunologically detected in the endosperm. Further molecular-genetic work is in progress.

The pollen tube approach of Picard was also used by us in similar experiments with barley (Mendel et al., 1990). DNA with 35S-*npt*II as marker was directly applied to the stigma of barley plants that were pollinated 5–20 min before DNA application. DNA was applied in a buffer that maintained it stable for more than 1 h. In total

1058 plants were treated, 11 200 seeds were obtained and germinated on 150 mg kanamycin/l, and 305 green plants selected. Some of the selected plants showed very weak NPT II signals that, after selfing the plants, were lost in the progeny. Hybridization of genomic DNA (F1 and F2) with a *npt*II-specific probe revealed in some cases positive signals, the size of which, however, did not fit into the restriction map of the utilized gene construct. Moreover, in some selected plants an identical hybridization pattern occurred, although these plants were derived from independent mother plants. These observations are very similar to those of the macroinjection approach and can be interpreted in the same way (see above).

Kartel *et al.* (1990) applied the pollen tube approach for introducing the *npt*II gene into barley plants followed by selection for kanamycin resistance. In Southern blots, the selected plants showed *npt*II-specific fragments; however, no hybridizing fragments were detectable in the selfing progeny of kanamycin-resistant plants. NPT II assays were not performed.

Hess, Dressler & Nimmrichter (1990) presented a system named 'pollen-mediated indirect gene transfer' with *Agrobacterium* as the infective agent for the pollen. The authors pipetted *Agrobacterium tumefaciens* containing the marker gene *npt*II on to wheat spikelets. After selection of the progeny for kanamycin resistance, NPT II assays of the selected plants showed extremely low NPT II activities. The transmission of the transgenes was followed for two generations. Although transmission was observed, the ratios failed to follow Mendelian laws. Also, rearrangement and loss of signals in Southern analysis was detected, suggesting instability. The authors report that probing with Ti plasmid-specific *vir* probes was negative, thus demonstrating that the plants were not contaminated by *Agrobacterium*. A rigid molecular analysis of these plants with respect to the integrative state of the DNA will be necessary.

Recently, Heberle-Bors critically reviewed all kinds of pollen-mediated gene transfer approaches, and came to the conclusion that 'the elegant idea of using mature pollen as a super vector for gene transfer fell short of experimental reality' (Heberle-Bors *et al.*, 1990; Heberle-Bors, 1991). We do not exclude that some kind of DNA transfer is being brought about by these approaches. However, these experiments are far from being reproducible and predictable. From the present point of view, it is unlikely that the pollen tube approach will develop into a reliable transformation method of general applicability for monocot or dicot plants.

Electroporation of intact explants

With the most sensitive methods of detection (i.e. transformation with a virus genome), it was shown many years ago that nucleic acid can enter intact plant cells in electroporation conditions (Morikawa *et al.*, 1986). Lindsey & Jones (1987) were able to show low levels of transient expression from foreign DNA after electroporation of intact sugar beet tissue culture cells. Finally, Dekeyser *et al.* (1990) could develop the system to a level of practical use and to analyze tissue-specific expression of recombinant genes in cells of intact plant tissue. Significantly, this work was done with a monocot plant, rice, and an immediate application, as suggested by the authors, is stable transformation of cereal plants. They also tested for transient expression in leaf bases of several other plants, including barley. With a closely related method, stable transformation of maize was recently achieved (D'Halluin *et al.*, 1992). It remains to be experienced whether this method will be generally useful.

Imbibition of seeds

The establishment of genetic transformation techniques of plants during the last decade has also defined the most useful molecular-genetic tools in the assessment of gene transfer. This has made it possible to reevaluate claims of gene transfer to plants, presented at the time when only ambiguous markers for the event were available (e.g. Doy, Gresshoff & Rolfe, 1973; Johnson, Grierson & Smith, 1973). Töpfer *et al.* (1989) have tested whether in fact it is possible to enter plant cells by exogenous DNA during the imbibition of the dry embryo. They soaked dry embryos in DNA solution containing a specific reporter gene construct and, indeed, could show transient expression in both cereal and legume tissues. The natural continuation of these experiments is to select for stabilization of transferred resistance genes. No data from these kinds of experiment have so far been published, and the usefulness of dry seed imbibition as a transformation method is still open to debate.

Microinjection

Although plant cells have strong cell walls, it is possible to penetrate cells by techniques of microinjection (Toyoda *et al.*, 1990). As with other methods of direct gene transfer, the target cell must have the potential to regenerate into a whole plant. The advantage of microinjection is that the investigator can target the DNA into cells that he expects to regenerate. In spite of many trials, however, transgenic cereal plants have not yet been obtained (Potrykus, 1991).

Transformation with Agrobacterium

By far the most successful method of gene transfer for very many species of dicotyledonous plants is the use of *Agrobacterium*-based vectors. *Agrobacteria* have a unique ability to penetrate cells at a wound site and actively integrate the transferred DNA stably into plants chromosomes.

For penetration, *Agrobacterium* have utilized a variation of interbacterial mating, or conjugation. With a specific system, the bridge between the cells is generated and the transferred DNA enters the plant cell as a single-stranded molecule (Stachel & Zambryski, 1986). The integration of the transferred DNA is probably also specific, although the mechanism is as yet not understood.

The next step, regeneration and subsequent proliferation, is also controlled by the pathogenic *Agrobacterium*. The transferred DNA contains genes that are expressed in the transformed plant cell and lead to the synthesis of plant hormones, promoting further cell division and ensuring that the transformed cell, starting to produce unique nutrients to the benefit of the bacterium, will flourish and divide. When the *Agrobacterium* system is used for gene transfer by the investigator, the last step of control is removed and replaced by addition of exogenous growth factors, leading finally to the achieving of normally differentiated transgenic plants.

The first demonstration that *Agrobacterium* vectors can deliver DNA also into cereal (maize) cells was obtained by assaying the activity of infected viral sequences (Grimsley *et al.*, 1987). Since then, *Agrobacterium tumefaciens* has been shown to interact – very inefficiently – with the cereals maize (Gould *et al.*, 1991), wheat (Gould *et al.*, 1992) and rice (Raineri *et al.*, 1990; Li *et al.*, 1991). The main obstacle seems to be the growth habits of cereal cells, especially the fact that wounding is not followed by proliferation of the cells at the wound site. *Agrobacterium*-mediated gene transfer to monocot plants that, opposing to the rule, do have a wound response, was demonstrated early (Hernalsteens *et al.*, 1984; Hooyakaas-Van Slogteren, Hooyakaas & Schilperoort, 1984).

Although no breakthrough has yet been reported, there is certain promise in *Agrobacterium*-mediated gene transfer to cereals. In principle, agrobacteria could be used to transform any plant cells, provided that the transformed cell would have the capacity to regenerate. It has become clear that somehow bypassing the lack of wound response of cereals is critical.

Miscellaneous methods

Electrophoretic migration of DNA (35S-*uid*A) into barley caryopses, followed by transient expression of *uid*A, was described by Ahokas (1989). Using a laser microbeam, Kaneko *et al.* (1991) introduced the gene for hygromycin phosphotransferase (HPT) into barley microspore cells and selected for hygromycin-resistant callus.

In dicots, further ways for DNA transfer have been reported that are applicable also to cereal cells. For introducing DNA into tobacco protoplasts the use of PEG or electroporation could be replaced by a short pulse of mild sonication (Joersbo & Brunstedt, 1990). Sonication was also useful for stably introducing DNA into cells of whole leaf segments of tobacco (Zhang *et al.*, 1991). Another way was shown by Kaeppler *et al.* (1990): suspension culture cells of maize and tobacco showed transient gene expression after vortexing of the cells in the presence of silicon carbide fibers. Weber *et al.* (1990) used a laser microbeam for cutting holes of defined dimensions into cell walls of various dicot cells and tissues followed by DNA uptake and transient gene expression as well as stable transformation.

Transformation of small-grain cereals

Barley

Transient gene expression

Barley protoplasts are most easily isolated from young leaves or from an established tissue culture. For transient assays, the mesophyll protoplasts are equally suitable as suspension culture protoplasts, although the former are usually not able to divide

(Junker et al., 1987; Mendel et al., 1990). This also means that transient gene expression is not bound to cell division. In electroporated barley mesophyll protoplasts, it has been possible to demonstrate the chloroplast targeting of a chimeric nptII gene with a pea transit peptide (Teeri et al., 1989). Also protoplasts from the starchy endosperm (Lee et al., 1991) and from the aleurone layer (Lee et al., 1989) of developing barley grains were able to show transient expression of the reporter genes uidA and cat. In the case of endosperm protoplasts, two endosperm-specific promoters (of genes coding for wheat glutenin and barley chymotrypsin inhibitor 2), fused to the uidA coding sequence, could be shown to be functional.

While many dicot protoplasts quickly dedifferentiate and will not represent the tissue from which they were isolated, this does not always take place with cereal protoplasts. Several reports of correct gene regulation in barley protoplasts have been published. Jacobsen & Close (1991) could demonstrate the correct regulation of a barley high-pI α-amylase promoter, fused to uidA as a reporter gene, by gibberellic acid (GA$_3$) and abscisic acid (ABA) as tested by transient expression assays in protoplasts prepared from mature barley aleurone layers. Similar experiments were performed by Skriver et al. (1991) with barley aleurone protoplasts in order to delineate hormone-responsive elements of the ABA-responsive rice rab16A gene and of the barley α-amylase gene. Salmenkallio et al. (1990) compared the regulation of a low-pI α-amylase promoter of barley fused to nptII as a reporter. Correct regulation by GA$_3$ and ABA was observed in protoplasts isolated from aleurone layers. In protoplasts of the scutellar epithelium, GA$_3$ had no effect on the low level of expression, and in mesophyll protoplasts the gene construct was not expressed at all, both cases also reflecting the properties of the intact tissues.

By using particle bombardment, it is possible to assay transiently the expression of introduced genes in cells that still are integral to the plant tissue. Lanahan et al. (1992) mapped the GA$_3$ response complex in a barley α-amylase gene by performing transient assays in intact aleurone layers of a barley grain. Variation inherent in particle bombardment was reduced by coexpressing a second reporter (ubi-luc) and by comparing the expression of the tested uidA fusion with this standard control.

Stable transformation

Until now, there is no known case of genetically proven stably transformed, mature barley plants. However, transgenic protoplast derived or bombarded cell cultures containing and expressing one or more foreign genes have been obtained (Lazzeri & Lörz, 1990; Lazzeri et al., 1991; Mendel et al., 1990, 1991; Ritala et al., 1993).

Although protoplasts are easily prepared from various barley tissues, only those isolated from suspension cultures have been able to divide. If the suspension culture will form embryoids, there is a chance that the protoplasts will also regenerate into mature plants. Regeneration of protoplast-derived colonies has been reported by Yan et al. (1990) and Jähne et al. (1991b). In principle, transgenic mature barley should emerge from these experiments in the near future.

Two problems remain in stable transformation of protoplasts. First, the embryogenic suspension cultures are not stable, and must be reinitiated frequently. Second, they can all be initiated from only particular varieties. Therefore, the chances to develop the protoplast approach into a gene transfer system routinely applicable for a wide variety of genotypes are small.

Embryogenic tissue culture that does not go into suspension, and from which protoplast isolation is not possible, can be initiated from a much wider selection of barley cultivars (Jähne et al., 1991a,b). These tissue cultures are ideal for particle bombardment, and although no reports of transgenic barley by this method are published, it probably will be the most general method of barley transformation.

With barley tissue cultures that do not possess the capacity for differentiation, it has been easy to demonstrate, by selecting for the stabilization of an antibiotic resistance marker gene, that foreign DNA delivered by particle bombardment can be stably integrated into barley chromosomes (Mendel et al., 1991; Ritala et al., 1993). Similar results have been obtained by direct gene transfer to protoplasts (Lazzeri et al., 1991; Mendel et al., 1990; Schulze et al., 1991b). Stable integration of one or more copies of the foreign DNA into the host genome was demonstrated by hybridization of digested genomic DNA with probes derived from the coding regions of the transferred genes. It is important to realize that this is a major step in barley transformation and, as mentioned, scientists can already study many problems in transgenic barley tissue culture cells.

In order to study the ability of barley to produce foreign proteins, we have stably introduced the gene for the seed storage protein vicilin of *Vicia faba* (genomic sequence and complementary DNA (cDNA)) under control of the barley hordein promoter into barley cells in order to study its integration, stability and its possible expression (K. Greger *et al.*, unpublished data), as well as a fungal heat stable β-glucanase secreted in an active form from stably transgenic malting barley cells (Aspegren *et al.*, 1992).

Wheat

Transient gene expression
Back in 1986, transient expression of the *cat* gene (Ou-Lee, Turgeon & Wu, 1986) driven by the 35S promoter and of the *npt*II gene (Werr & Lörz, 1986) driven by the maize *sh1* promoter were reported for protoplasts of Einkorn wheat, *Triticum monococcum*. Oard, Paige & Dvorak (1989) showed for protoplasts of bread wheat, *Triticum aestivum*, transient expression of a 35S-*cat* construct. Inclusion of intron 6 of maize *adh1* gene between the 35S and *cat* sequence increased *cat* expression more than 100-fold. Also protoplasts from the aleurone layer of developing wheat grains were able to show transient expression of the reporter gene *cat* (Lee *et al.*, 1989).

Stable transformation
There is already vast experience of cell and tissue culture of wheat, and protoplast isolation from different explants is routine (e.g. see Morrish *et al.*, 1987). As early as 1985, Lörz, Baker & Schell reported the transfer of the *npt*II gene into protoplasts of *Triticum monococcum* suspension culture cells and the selection of stably transformed callus colonies that were not morphogenic. As mentioned above for barley, and for bread wheat, the main bottleneck for obtaining transgenic plants is the rapid loss of embryogenicity of morphogenic suspension cultures and the protoplasts derived from them. Again the regeneration is also strongly genotype dependent. There have been several attempts to improve the situation, e.g. by using different protocols for initiating the suspension cultures (Wang & Nguyen, 1990). Langridge, Lazzeri & Lörz (1991) used immature embryos of a wheat translocation line carrying the short arm

of rye chromosome 1RS instead of its own short arm of chromosome 1B and could define two distinct regions on 1RS that enhanced the embryogenicity of callus cultures derived from it.

Hayashi & Shimamoto (1988), isolating protoplasts directly from immature embryos, reported the formation of roots and albino shoots from protoplast-derived callus colonies (cv. Chinese Spring). Harris *et al.* (1988), using an antherderived suspension culture of the wheat cultivar Chris, obtained rare green plantlets from protoplasts. In 1990, Vasil's group reported the first successful case of regenerating green plants from protoplasts isolated from immature-embryoderived suspension cultures (Vasil *et al.*, 1990). After flowering, however, these otherwise healthy plants showed the phenomenon of precocious senescence so that no progeny could be obtained. Also Chang *et al.* (1991) were able to regenerate a few green plants from suspension-culture-derived protoplasts (cv. Mustang). So, in essence all prerequisites are given for generating the first transgenic protoplast-derived wheat plants. However, as yet there is no such case known.

As for barley, it is rather unlikely that the protoplast approach will develop into a gene transfer system routinely applicable for a wide variety of wheat genotypes. The particle bombardment method remains the most promising alternative. Tissue culture cells of wheat have been transformed to express foreign marker genes (Vasil *et al.*, 1991). Also, embryogenic tissue cultures, capable of regeneration into mature plants, have been available (Vasil *et al.*, 1990). These two methods were recently combined. Vasil *et al.* (1992) were able to obtain transgenic, fertile and (for the transgenes) true breeding wheat plants from two varieties of *Triticum aestivum*. They used as selectable marker the *bar* gene, coding for phosphinothricin resistance, in the control of the 35S promoter and enhanced with the maize *adh1* intron 1. Although the nonselected marker *uid*A controlled by the *adh1* promoter was expressed only weakly, this work, together with that done with oat described below, represent a major breakthrough in small-grain cereal transformation.

Oat

Using the particle gun approach, Kakuta *et al.* (1992) introduced 35S-*uid*A into embryo cells of imbibed oat and wheat seeds. Two to 7 days after bombardment, transient *uid*A expression was monitored in cells of the embryos, stems leaves and roots of the seedlings.

Most remarkably, fertile transgenic oat plants were reported recently by Somers *et al.* (1992). They used in their experiments an oat line that had been developed for the ability to produce regenerable, embryogenic tissue cultures. These cultures were bombarded with 35S-*bar* and *adh*1-*uid*A constructs, and selected for herbicide resistance. Thirty-eight transgenic plants could be regenerated and grown to maturity. One of them was fully fertile, and transferred both markers to its progeny, where they were expressed. Co-expression of the nonselected *uid*A marker (GUS activity) was observed in 75% of the transgenic tissue culture lines.

Rye

Not very much work towards transformation of rye has been published, except for that by De la Peña *et al.* (1987), where the macroinjection approach was first tried for a small-grain cereal. De la Peña *et al.* described injection of DNA (with the marker gene *npt*II) into the tiller cavity of immature rye tillers. After screening the progeny for kanamycin resistance, among 3023 resulting grains three plants were obtained that grew in the presence of kanamycin and expressed NPT II activity in the leaves. In genomic DNA, Southern blots revealed appropriately hybridizing restriction fragments. However, no transmission to the progeny was demonstrated. Later, preliminary experiments showed that the progeny plants contained *npt*II-specific hybridizing fragments; however, no correct segregation was observed (H. Lörz, personal communication).

Orchardgrass (*Dactylis glomerata*)

Horn *et al.* (1988) transferred the hygromycin resistance gene under control of the 35S promoter to protoplasts of orchardgrass and regenerated green, hygromycin-resistant plants that exhibited the correct hybridization pattern in a Southern analysis of genomic DNA.

Guinea grass (*Panicum maximum*)

Vasil *et al.* (1988) optimized DNA transfer into protoplasts of Guinea grass using a 35S-*cat* con-

struct and transient gene expression, and Hauptmann *et al.* (1988) obtained stably transformed cell lines with 35S-*hph* as a selectable marker.

Sorghum (*Sorghum vulgare*)

Hagio, Blowers & Earle (1991) transformed suspension cell cultures of sorghum using the particle gun approach and selected cell colonies stably expressing *hph* and *npt*II (driven by the 35S promoter). The cotransferred 35S-*uid*A was expressed at low levels. Also Blowers, Hagio & Earle (1991) successfully used the particle gun to obtain stably transformed cell cultures (*npt*II, *hph*, *uid*A). Battraw & Hall (1991) transformed protoplasts of *Sorghum bicolor* with the genes for NPT II and GUS and obtained stable transformants.

Turfgrass (*Festuca arundinacea*)

Ha, Wu & Thorne (1992) transformed tall fescue protoplasts with the gene for HPT and selected hygromycin-resistant cell lines that could be regenerated into plants. Using a similar approach and the marker genes *hph* and *bar*, Takamizo *et al.* (1992) obtained transgenic tall fescue plants expressing the transgenes.

Discussion

There are suitable promoters and effective selectable markers available for transforming cells of small-grain cereals and generating stably transformed cell cultures. Also the physical introduction of foreign DNA into a recipient cell does not seem to be problematic. However, the following steps – stable integration of the introduced DNA into the host genome and regeneration of the transformed cells to mature plants – seem to cause problems. Most importantly this is a problem of tissue culture technology. A successful transformation method requires that a large number of recipient cells can be induced to regenerate into mature plants. It might also be speculated that small-grain cereals could be more efficient than other species in recognizing the introduced DNA to be 'foreign' (e.g. by screening for the methylation pattern of the incoming gene) and activating mechanisms that ultimately lead either to the excision of the foreign DNA or to its irreversible repression. However, being successfully introduced into dividing cells, the foreign DNA is at least in some cases able to give high and stable expression (e.g. *npt*II driven by 35S in barley tissue culture). Yet we do not know whether or not the expression will change during the process of plant regeneration.

Furthermore, we do not know whether there are special requirements for efficient gene expression in cereal cells. It is well established that many highly expressed nuclear genes of cereals and other monocots are characterized by the occurrence of so called CpG islands within their promoter region and within the 5'-part of the coding sequence, whereas their intron regions do not exhibit these characteristics (Martinez, Martin & Cerff, 1989). However, all the marker and reporter genes commonly used are devoid of CpG islands. It remains to be established whether this structural characteristic will be of importance for the genetic stability of a transferred gene during sexual transmission and for the stability in the progeny. Other factors should also be taken into consideration when one is planning to transfer a given gene into a cereal species: splicing mechanisms of pre-mRNAs might be slightly different between cereals and dicots. This was indicated by a report of Keith & Chua (1986), who demonstrated that the wheat *rbc*S gene was not correctly spliced in tobacco. In another case, however, the gene for sorghum phosphoenolpyruvate carboxylase was correctly spliced in the dicot.

In summary, a large number of approaches has been tried to stably transform small-grain cereals, with the aim of obtaining green fertile plants. This goal has been reached for wheat and oat for the first time recently. Besides the protoplast approach and the particle bombardment method, there are numerous techniques belonging to the group of nonorthodox methods; however, the results are far from being reproducible and predictable. Most importantly, although there is evidence indicative of stabilization and correct expression of the foreign DNA transferred by these approaches, the DNA is rarely transmitted to the progeny. Meiotic divisions seem to eliminate the foreign genetic material. Thus, only the transfer of DNA into protoplasts and the microprojectile-mediated transfer of genes through intact cell walls can at the moment be recommended for approaching the problem of stably transforming small-grain cereals. But these two

methods have bottlenecks that have to be taken into account: with the protoplast approach it is the loss of the regenerative potential of the selected transgenic colonies, and with the particle bombardment it is the question of which is the appropriate target tissue and which is the most suitable system of selection. Nevertheless, both of these methods have a chance to develop into reliable methods of general applicability to small-grain cereals.

Recent developments

This chapter is based on literature that was available in the beginning of 1993. The field is developing rapidly and now, one year later, more breakthroughs have taken place. Barley, regenerating to fertile plants, has been transformed by one of us (Ritala *et al.*, 1994) and others (Wan & Lemaux, 1994), and sorghum by Casas *et al.* (1993). After the first reported success in a particular plant, it is necessary that transformation is repeated by others and made more efficient. This has taken place for wheat (Vasil *et al.*, 1993; Weeks, Anderson & Blechl, 1993; Becker, Brettschneider & Lörz, 1994; Nehra *et al.*, 1994). Common to the recent accomplishments has been the use of particle bombardment as method of DNA delivery, the *bar* gene as a selectable marker, and of immature embryos as targets. This gives a clear guideline for development of transformation systems for the remaining small-grain cereal species. A long and laborious phase of method development in cereal transformation has apparently come to an end. The exciting part of the work may begin.

References

Abdul-Baki, A. A., Saunders, J. A., Matthews, B. F. & Pittarelli, G. W. (1990). DNA uptake during electroporation of germinating pollen grains. *Plant Science*, **70**, 181–190.

Ahloowalia, B. S. (1987). Plant regeneration from embryo-callus culture in barley. *Euphytica*, **36**, 659–665.

Ahokas, H. (1987). Transfection by DNA-associated liposomes evidenced at pea pollination. *Hereditas*, **106**, 129–138.

Ahokas, H. (1989). Transfection of germinating barley seed electrophoretically with exogenous DNA. *Theoretical and Applied Genetics*, **77**, 469–472.

Aspegren, K., Ritala, A., Mannonen, L., Hannus, R., Kurtén, U., Salmenkallio, M., Kauppinen, V. & Teeri, T. H. (1992). Secretion of a fungal heat stable β-glucanase from tobacco and barley cells. *Progress in Molecular Biology (NOMBA Symposium)*, abstract 2, Helsinki.

Battraw, M. & Hall, T. C. (1991). Stable transformation of *Sorghum bicolor* protoplasts with chimeric neomycin phosphotransferase II and β-glucuronidase genes. *Theoretical and Applied Genetics*, **82**, 161–168.

Battraw, M. & Hall, T. C. (1992). Expression of a chimeric neomycin phosphotransferase II gene in first and second generation transgenic rice plants. *Plant Science*, **86**, 191–202.

Becker, D., Brettschneider, R. & Lörz, H. (1994). Fertile transgenic wheat from microprojectile bombardment of scutellar tissue. *Plant Journal*, **5**, 299–307.

Blowers, A. D., Hagio, T. & Earle, E. D. (1991). Isolation of autonomously-replicating plasmids from transformed Sorghum cells. In *Molecular Biology of Plant Growth and Development, Third International Congress of Plant Molecular Biology*, ed. R. B. Hallick, abstract 921. University of Arizona, Tucson.

Bregitzer, P., Bushnell, W. R., Rines, H. W. & Somers, D. A. (1991). Callus formation and plant regeneration from somatic embryos of oat (*Avena sativa* L.). *Plant Cell Reports*, **10**, 243–246.

Brettschneider, R., Becker, D. & Lörz, H. (1991). Transient gene expression in cereal cells after microprojectile bombardment. In *International Association of Plant Cell and Tissue Culture, Meeting of the German Section*, abstract v15. University of Hamburg, Hamburg.

Budar, F., Thia-Toong, L., Van Montagu, M. & Hernalsteens, J.-P. (1986). *Agrobacterium*-mediated gene transfer results mainly in transgenic plants transmitting T-DNA as a single mendelian factor. *Genetics*, **114**, 303–313.

Callis, J., Fromm, M. & Walbot, V. (1987). Introns increase gene expression in cultured maize cells. *Genes and Development*, **1**, 1183–1200.

Casas, A. M., Kononowicz, A. K., Zehr, U. B., Tomes, D. T., Axtell, J. D., Butler, L. G., Bressan, R. A. & Hasegawa, P. M. (1993). Transgenic sorghum plants via microprojectile bombardment. *Proceedings of the National Academy of Sciences, USA*, **90**, 11212–11216.

Chang, Y.-F., Wang, W. C., Warfield, C. Y., Nguyen, H. T. & Wong, J. R. (1991). Plant regeneration from protoplasts isolated from long-term cell cultures of wheat (*Triticum aestivum* L.). *Plant Cell Reports*, **9**, 611–614.

Chen, D. F., Dale, P. J. & Mullineaux, P. M. (1991). Transformation of *Triticum monococcum* L. cells and analysis of long term stability of foreign genes. In *Molecular Biology of Plant Growth and Development, Third International Congress of Plant Molecular Biology*, ed. R. B. Hallick, abstract 418. University of Arizona, Tucson.

De la Peña, A., Lörz, H. & Schell, J. (1987). Transgenic rye plants obtained by injecting DNA into young floral tillers. *Nature*, **325**, 274–276.

De Wet, J. M. J., Bergquist, R. R., Harlan, J. R., Brink, D. E., Cohen, C. E., Newell, C. A. & De Wet, A. E. (1985). Exogenous gene transfer in maize (*Zea mays*) using DNA-treated pollen. In *Experimental Manipulation of Ovule Tissue*, ed. G. P. Chapman, S. H. Mantell & W. Daniels, pp. 197–209. Longman, London.

Dekeyser, R. A., Claes, B., De Rycke, R. M. U., Habets, M. E., Van Montagu, M. C. & Caplan, A. B. (1990). Transient gene expression in intact and organized rice tissues. *Plant Cell*, **2**, 591–602.

Delporte, F., Mingeo, C., Salomon, F. & Jacquemin, J.-M. (1991). Expression of zein family protein in wheat endosperm (*Triticum aestivum* L.) after transformation by copollination with plasmidic DNA. In *Eucarpia Symposium on Genetic Manipulation in Plant Breeding*, abstract I/P4. Tarragona.

D'Halluin, K., Bonne, E., Bossut, M., De Beuckeleer, M. & Leemans, J. (1992). Production of fertile transgenic maize plants by electroporation of immature embryos and type I callus. *Plant Cell*, **4**, 1495–1505.

Doy, C. H., Gresshoff, P. M. & Rolfe, B. G. (1973). Biological and molecular evidence for the transgenesis of genes from bacteria to plant cells. *Proceedings of the National Academy of Sciences, USA*, **70**, 723–726.

Du, J., Liu, H., Xie, D. X., Hua, X. J. & Fan, Y. L. (1989). Injection of exogenous DNA into young floral tillers of wheat. *Genetic Manipulation in Plants*, **5**, 8–12.

Gould, J., Devey, M., Hasegawa, O., Ulian, E. C., Peterson, G. & Smith, R. H. (1991). Transformation of *Zea mays* L. using *Agrobacterium tumefaciens* and the shoot apex. *Plant Physiology*, **95**, 426–434.

Gould, J. H., Devey, M. E., Ko, T. S., Peterson, G., Hasegawa, O. & Smith, R. H. (1992). Transformation of Gramineae using *Agrobacterium tumefaciens*. *Journal of Cellular Biochemistry*, **16F**, 207.

Grimsley, N., Hohn, T., Davies, J. W. & Hohn, B. (1987). *Agrobacterium*-mediated delivery of infectious maize streak virus into maize plants. *Nature (London)*, **325**, 177–179.

Ha, S. B., Wu, F.-S. & Thorne, T. K. (1992). Transgenic tall fescue (*Festuca arundinacea* Schreb.) plants regenerated from protoplasts. *Journal of Cellular Biochemistry*, **16F**, 207.

Hagio, T., Blowers, A. D. & Earle, E. D. (1991). Stable transformation of sorghum cell cultures after bombardment with DNA-coated microprojectiles. *Plant Cell Reports*, 10, 260–264.

Harris, R., Wright, M., Byrne, M., Varnum, J., Brightwell, B. & Schubert, K. (1988). Callus formation and plantlet regeneration from protoplasts derived from suspension cultures of wheat (*Triticum aestivum* L.). *Plant Cell Reports*, 7, 337–340.

Hauptmann, R. M., Vasil, V., Ozias-Akins, P., Tabaeizadeh, Z., Rogers, S. G., Fraley, R. T., Horsch, R. B. & Vasil, I. K. (1988). Evaluation of selectable markers for obtaining stable transformants in the Gramineae. *Plant Physiology*, 86, 602–606.

Hayashi, Y. & Shimamoto, K. (1988). Wheat protoplast culture: embryogenic colony formation from protoplasts. *Plant Cell Reports*, 7, 414–417.

Heberle-Bors, E. (1991). Germ line transformation in higher plants. *Newsletter of the International Association of Plant Cell and Tissue Culture*, 64, 2–10.

Heberle-Bors, E., Benito Moreno, R. M., Alwen, A., Stöger, E. & Vincente, O. (1990). Transformation of pollen. In *Progress in Plant Cellular and Molecular Biology*, ed. H. J. J. Nijkamp, L. H. W. van der Plas & J. van Aartrijk, pp. 244–251. Kluwer Academic Publishers, Dordrecht.

Hernalsteens, J.-P., Thia-Toong, L., Schell, J. & Van Montagu, M. (1984). An *Agrobacterium*-transformed cell culture from the monocot *Asparagus officinalis*. *EMBO Journal*, 3, 3039–3041.

Hess, D. (1987). Pollen-based techniques in genetic manipulation. *International Review of Cytology*, 107, 367–395.

Hess, D., Dressler, K. & Nimmrichter, R. (1990). Transformation experiments by pipetting *Agrobacterium* into the spikelets of wheat (*Triticum aestivum* L.). *Plant Science*, 72, 233–244.

Hinchee, M. A. W., Fry, J. E., Sanders, P. R., Armstrong, C. L., Petersen, W. L., Songstad, D. D., Mathis, N. L., Muskopf, Y. M., Barnason, A. R., Laytom, J. G., Corbin, D. R., Connor-Ward, D. V. & Horsch, R. B. (1991). Experimental gene transfer systems for plants. In *Molecular Biology of Plant Growth and Development, Third International Congress of Plant Molecular Biology*, ed. R. B. Hallick, abstract 63. University of Arizona, Tucson.

Hooyakaas-Van Slogteren, G. M. S., Hooyakaas, P. J. J. & Schilperoort, R. A. (1984). Expression of Ti plasmid genes in monocotyledonous plants infected with *Agrobacterium tumefaciens*. *Nature (London)*, 311, 763–764.

Horn, M. E., Shillito, R. D., Conger, B. V. & Harms, C. T. (1988). Transgenic plants of orchardgrass (*Dactylis glomerata* L.) from protoplasts. *Plant Cell Reports*, 7, 469–472.

Jacobsen, J. V. & Close, T. J. (1991). Control of transient expression of chimaeric genes by gibberellic acid and abscisic acid in protoplasts prepared from mature barley aleurone layers. *Plant Molecular Biology*, 16, 713–724.

Jähne, A., Lazzeri, P. A., Jäger-Gussen, M. & Lörz, H. (1991*a*). Plant regeneration from embryogenic cell suspensions derived from anther cultures of barley (*Hordeum vulgare* L.). *Theoretical and Applied Genetics*, 82, 74–80.

Jähne, A., Lazzeri, P. A. & Lörz, H. (1991*b*). Regeneration of fertile plants from protoplasts derived from embryogenic suspensions of barley (*Hordeum vulgare* L.). *Plant Cell Reports*, 10, 1–6.

Joersbo, M. & Brunstedt, J. (1990). Direct gene transfer to plant protoplasts by mild sonication. *Plant Cell Reports*, 9, 207–210.

Johnson, C. B., Grierson, D. & Smith, H. (1973). Expression of λplac5 DNA in cultured cells of a higher plant. *Nature New Biology*, 244, 105–107.

Junker, B., Zimny, J., Lührs, R. & Lörz, H. (1987) Transient expression of chimaeric genes in dividing and non-dividing cereal protoplasts after PEG-induced DNA uptake. *Plant Cell Reports*, 6, 329–332.

Kaeppler, H. F., Gu, W., Somers, D. A., Rines, H. W. & Cockburn, A. F. (1990). Silicon carbide fiber-mediated DNA delivery into plant cells. *Plant Cell Reports*, 9, 415–418.

Kakuta, H., Hashidoko, Y., Seki, T., Matsui, M., Anai, T., Hasegawa, K. & Mizutani, J. (1992). In situ foreign gene expression in seedlings of some monocotyledonous crops using the improved particle gun driven by compressed N_2 gas. *Journal of Cellular Biochemistry*, 16F, 237.

Kaneko, T., Kuroda, H., Hirota, N., Kishinami, I., Kimura, N. & Ito, K. (1991). Two transformation systems of barley – electroporation and laser perforation. In *Barley Genetics VI* Vol. 1, ed. L. Munck, K. Kirkegaard & B. Jensen, pp. 231–233. Munksgaard International Publishers, Copenhagen.

Kartel, N. A., Zabenkova, K. I., Maneshina, T. V. & Ablov, S. E. (1990). Barley plants with introduced gene for kanamycin resistance. *Doklady Akademii Nauk BSSR*, 34 (N3), 261–263.

Keith, B. & Chua, N.-H. (1986). Monocot and dicot pre-mRNAs are processed with different efficiencies in transgenic tobacco cells. *EMBO Journal*, 5, 2419–2425.

Klein, T. M., Harper, E. C., Svab, Z., Sanford, J. C., Fromm, M. E. & Maliga, P. (1988). Stable genetic transformation of intact *Nicotiana* cells by the particle bombardment process. *Proceedings of the National Academy of Sciences, USA*, 85, 8502–8505.

Konstantinov, K. L., Mladenovic, S. D., Denic, M. P. & Plecas, M. (1990). Maize genome and genome of bacteria living in maize – transformation with a bacterial gene. In *VI NATO Advanced Studies – Plant Molecular Biology*, p. 99. Schloss Elmau, Bavaria.

Lanahan, M. B., Ho, T.-H. D., Rogers, S. W. & Rogers, J. C. (1992) A gibberellin response complex in cereal α-amylase gene promoters. *Plant Cell*, **4**, 203–211.

Langridge, P., Lazzeri, P. & Lörz, H. (1991). A segment of rye chromosome 1 enhances growth and embryogenesis of calli derived from immature embryos of wheat. *Plant Cell Reports*, **10**, 148–151.

Last, D. I., Brettell, R. I. S., Chamberlain, D. A., Chaudhury, A. M., Larkin, P. J., Marsh, E. L., Peacock, W. J. & Dennis, E. S. (1991). pEmu: an improved promoter for gene expression in cereal cells. *Theoretical and Applied Genetics*, **81**, 581–588.

Lazzeri, P. A., Brettschneider, R., Lührs & Lörz, H. (1991). Stable transformation of barley via PEG-induced direct DNA uptake into protoplasts. *Theoretical and Applied Genetics*, **81**, 437–444.

Lazzeri, P. A. & Lörz, H. (1990). Regenerable suspension and protoplast cultures of barley and stable transformation via DNA uptake into protoplasts. In *Genetic Engineering in Crop Plants*, ed. G. W. Lycett & D. Grierson, pp. 231–238. Butterworths, London.

Lazzeri, P. A. & Lörz, H. (1991). Protoplast culture and transformation in barley. In *Barley Genetics VI*, Vol. 1, ed. L. Munck, K. Kirkegaard & B. Jensen, pp. 240–244. Munksgaard International Publishers, Copenhagen.

Lee, B. T., Murdoch, K., Topping, J., Jones, M. G. K. & Kreis, M. (1991). Transient expression of foreign genes introduced into barley endosperm protoplasts by PEG-mediated transfer or into intact endosperm tissue by microprojectile bombardment. *Plant Science*, **78**, 237–246.

Lee, B., Murdoch, K., Topping, J., Kreis, M. & Jones, M. G. K. (1989). Transient gene expression in aleurone protoplasts isolated from developing caryopses of barley and wheat. *Plant Molecular Biology*, **13**, 21–39.

Li, X.-Q., Liu, C., Peng, J., Gelvin, S. B. & Hodges, T. K. (1991). GUS expression in rice tissues using *Agrobacterium*-mediated transformation. In *Molecular Biology of Plant Growth and Development, Third International Congress of Plant Molecular Biology*, ed. R. B. Hallick, abstract 385. University of Arizona, Tucson.

Lindsey, K. & Jones, M. G. K. (1987). Transient gene expression in electroporated protoplasts and intact cells of sugar beet. *Plant Molecular Biology*, **10**, 43–52.

Lörz, H., Baker, B. & Schell, J. (1985). Gene transfer to cereal cells mediated by protoplast transformation. *Molecular and General Genetics*, **199**, 178–182.

Louwerse, J. D., Gersman, M., van Dam, K. J., Hensgens, L. A. M. & van der Mark, F. (1991). Towards the stable transformation of barley using particle bombardment. In *Molecular Biology of Plant Growth and Development, Third International Congress of Plant Molecular Biology*, ed. R. B. Hallick, abstract 910. University of Arizona, Tucson.

Lührs, R. & Lörz, H. (1987a). Initiation of morphogenic cell-suspension and protoplast cultures of barley. *Planta*, **175**, 71–81.

Lührs, R. & Lörz, H. (1987b). Plant regeneration in vitro from embryogenic cultures of spring- and winter-type barley (*Hordeum vulgare* L.) varieties. *Theoretical and Applied Genetics*, **75**, 16–25.

Maas, C., Laufs, J., Grant, S., Korfhage, C. & Werr, W. (1991). The combination of a novel stimulatory element in the first exon of the maize Shrunken–1 gene with the following intron 1 enhances reporter gene expression up to 1000-fold. *Plant Molecular Biology*, **16**, 199–207.

Martinez, P., Martin, W. & Cerff, R. (1989). Structure, evolution and anaerobic regulation of a nuclear gene encoding cytosolic glyceraldehyde-3-phosphate dehydrogenase from maize. *Journal of Molecular Biology*, **208**, 551–565.

McElroy, D., Blowers, A. D., Jenes, B. & Wu, R. (1991). Construction of expression vectors based on the rice actin 1 (Act1) 5′ region for use in monocot transformation. *Molecular and General Genetics*, **231**, 150–160.

McElroy, D., Zhang, W., Cao, J. & Wu, R. (1990). Isolation of an efficient actin promoter for use in rice transformation. *Plant Cell*, **2**, 163–171.

Mejza, S. J., Nguyen, T., Driller, J. P., Walker, L. S., Ulrich, T. & Wong, J. R. (1991). Stable co-transformation of genes in *Triticum aestivum* cells after particle bombardment. In *Molecular Biology of Plant Growth and Development, Third International Congress of Plant Molecular Biology*, ed. R. B. Hallick, abstract 411. University of Arizona, Tucson.

Mendel, G. (1865). Versuche über Pflanzenhybriden. *Verhandlungen des naturforschenden Vereines in Brünn*, **4**, 3–47.

Mendel, R. R., Clauss, E., Hellmund, R., Schulze, J., Steinbiss, H. H. & Tewes, A. (1990). Gene transfer to barley. In *Progress in Plant Cellular and Molecular Biology*, ed. H. J. J. Nijkamp, L. H. W. van der Plas & J. van Aartrijk, pp. 73–78. Kluwer Academic Publishers, Dordrecht.

Mendel, R. R., Müller, B., Schulze, J., Kolesnikov, V. & Zelenin, A. (1989). Delivery of foreign DNA to intact barley cells by high-velocity microprojectiles. *Theoretical and Applied Genetics*, **78**, 31–34.

Mendel, R. R., Schulze, J., Clauss, E., Nerlich A., Cerff, R. & Steinbiss, H. H. (1991). Gene transfer to barley. In *Barley Genetics VI*, Vol. 1, ed. L. Munck, K. Kirkegaard & B. Jensen, pp. 240–244. Munksgaard International Publishers, Copenhagen.

Morikawa, H., Iida, A., Matsui, C., Ikegami, M. & Yamada, Y. (1986). Gene transfer into intact plant cells by electroinjection through cell walls and membranes. *Gene*, **41**, 121–124.

Morrish, F. M., Vasil, V. & Vasil, I. K. (1987). Developmental morphogenesis and genetic manipulation in tissue and cell cultures of the Gramineae. *Advances in Genetics*, **24**, 431–499.

Müller, B., Schulze, J. & Wegner, U. (1989). Establishment of barley cell suspension cultures of mesocotyl origin suitable for isolation of dividing protoplasts. *Biochemie und Physiologie der Pflanzen*, **185**, 123–130.

Nehra, N. S., Chibbar, R. N., Leung, N., Caswell, K., Mallard, C., Steinhauer, L., Baga, M. & Kartha, K. K. (1994). Self-fertile transgenic wheat plants regenerated from isolated scutellar tissues following microprojectile bombardment with two distinct gene constructs *Plant Journal*, **5**, 284–297.

Oard, J. H., Paige, D. & Dvorak, J. (1989). Chimeric gene expression using maize intron in cultured cells of breadwheat. *Plant Cell Reports*, **8**, 156–160.

Ohta, U. (1986). High efficiency genetic transformation of maize by a mixture of pollen and exogenous DNA. *Proceedings of the National Academy of Sciences, USA*, **83**, 715–719.

Ou-lee, T. M., Turgeon, R. & Wu, R. (1986). Expression of a foreign gene linked to either a plant-virus or a *Drosophila* promoter, after electroporation of protoplasts of rice, wheat and sorghum. *Proceedings of the National Academy of Sciences, USA*, **83**, 6815–6819.

Picard, E., Jacquemin, J. M., Granier, F., Bobin, M. & Forgeois, P. (1988). Genetic transformation of wheat (*Triticum aestivum*) by plasmid DNA uptake during pollentube germination. In *VIIth International Wheat Genetics Symposium*, pp. 779–781. Cambridge University Press, Cambridge.

Potrykus, I. (1990). Gene transfer to cereals: an assessment. *Bio/Technology*, **8**, 535–542.

Potrykus, I. (1991). Gene transfer to plants: assessment of published approaches and results. *Annual Review of Plant Physiology and Plant Molecular Biology*, **42**, 205–225.

Raineri, D. M., Bottino, P., Gordon, M. P. & Nester, E. W. (1990) *Agrobacterium*-mediated transformation of rice (*Oryza sativa* L.). *Bio/Technology*, **8**, 33–38.

Rhodes, C. A., Pierce, D. A., Mettler, I. J., Mascarenhas, D. & Detmer, J. J. (1988). Genetically transformed maize plants from protoplasts. *Science*, **240**, 204–207.

Ritala A., Aspegren K., Kurtén U., Salmenkallio-Marttila M., Mannonen L., Hannus, R., Kauppinen, V., Teeri, T. H. & Enari, T.-M. (1994). Fertile transgenic barley by particle bombardment of immature embryos. *Plant Molecular Biology*, **24**, 317–325.

Ritala, A., Mannonen, L., Salmenkallio, M., Kurtén, U., Hannus, R., Aspegren, K., Mendez-Lozano, J.,

Teeri, T. H. & Kauppinen, V. (1993). Stable transformation of barley tissue culture by particle bombardment. *Plant Cell Reports*, **12**, 435–440.

Rogers, S. W. & Rogers, J. C. (1992). The importance of DNA methylation for stability of foreign DNA in barley. *Plant Molecular Biology*, **18**, 945–962.

Salmenkallio, M., Hannus, R., Teeri, T. H. & Kauppinen, V. (1990). Regulation of α-amylase promoter by gibberellic acid and abscisic acid in barley protoplasts transformed by electroporation. *Plant Cell Reports*, **9**, 352–355.

Sautter, C., Waldner, H., Neuhaus-Url, G., Galli, A., Neuhaus, G. & Potrykus, I. (1991). Micro-targeting: high efficiency gene transfer using a novel approach for the acceleration of micro-projectiles. *Bio/Technology*, **9**, 1080–1085.

Schulze, J. & Mendel, R. R. (1991). Effect of 5-azacytidine on the expression of foreign genes in stably transformed barley cell lines. In *International Association of Plant Cell and Tissue Culture, Meeting of the German Section*, abstract v12. University of Hamburg, Hamburg.

Schulze, J., Nerlich, A. & Mendel, R. R. (1991*a*). Effect of 5-azacytidine on the expression of foreign genes in stably transformed barley. In *Molecular Biology of Plant Growth and Development, Third International Congress of Plant Molecular Biology*, ed. R. B. Hallick, abstract 360. University of Arizona, Tucson.

Schulze, J., Nerlich, A., Ryschka, S., Steinbiss, H. H. & Mendel, R. R. (1991*b*). Stable transformation of barley using a highly-efficient protoplast system. *Physiologia Plantarum*, **82**, A32.

Schulze, J., Nerlich, A., Ryschka, S., Steinbiss, H. H. & Mendel, R. R. (1991*c*). Expression of foreign genes in stably transformed barley. In *Eucarpia Symposium on Genetic Manipulation in Plant Breeding*, abstract I/P16. Tarragona.

Skriver, K., Olsen, F. L., Rogers, J. C. & Mundy, J. (1991). *Cis*-acting DNA elements responsive to gibberellin and its antagonist abscisic acid. *Proceedings of the National Academy of Sciences, USA*, **88**, 7266–7270.

Somers, D. A., Rines, H. W., Gu, W., Kaeppler, H. F. & Bushnell, W. R. (1992). Fertile, transgenic oat plants. *Bio/Technology*, **10**, 1589–1594.

Spickernagel, R., Lazzeri, P. A., Lütticke, S. & Lörz, H. (1991). Transient marker gene expression from different promoter constructs after PEG mediated DNA uptake into barley protoplasts. In *International Association of Plant Cell and Tissue Culture, Meeting of the German Section*, abstract 39. University of Hamburg, Hamburg.

Stachel, S. E. & Zambryski, P. C. (1986). *Agrobacterium tumefaciens* and the susceptible plant cell: a novel adaptation of extracellular recognition and DNA conjugation. *Cell*, **47**, 155–157.

Svab, Z., Hajdukiewicz, P. & Maliga, P. (1990). Stable transformation of plastids in higher plants. *Proceedings of the National Academy of Sciences, USA*, **87**, 8526–8530.

Takamizo, T., Wang, Z., Suginobu, K., Perez-Vincente, R., Potrykus, I. & Spangenberg, G. (1992). Biotechnological approaches to forage grass improvement: *Festuca* and *Lolium*. *Journal of Cellular Biochemistry*, **16F**, 212.

Teeri, T. H., Patel, G. H., Aspegren, A. & Kauppinen, V. (1989). Chloroplast targeting of neomycin phosphotransferase II with a pea transit peptide in electroporated barley mesophyll protoplasts. *Plant Cell Reports*, **8**, 187–190.

Töpfer, R., Gronenborn, B., Schell, J. & Steinbiss, H.-H. (1989). Uptake and transient expression of chimeric genes in seed-derived embryos. *Plant Cell*, **1**, 133–139.

Töpfer, R., Maas, C., Höricke-Grandpierre, C., Schell, J. & Steinbiss, H. H. (1992). Expression vectors for high-level gene expression in dicotyledonous and monocotyledonous plants. In *Methods in Enzymology*, vol. 217, ed. R. Wu, pp. 66–78. Academic Press, New York.

Toriyama, K., Arimoto, Y., Uchimiya, H. & Hinata, K. (1988). Transgenic rice plants after direct gene transfer into protoplasts. *Bio/Technology*, **6**, 1072–1074.

Toyoda, H., Yamaga, T., Matsuda, Y. & Ouchi, S. (1990). Transient expression of the β-glucuronidase gene introduced into barley coleoptile cells by microinjection. *Plant Cell Reports*, **9**, 299–302.

Twell, D., Klein, T. M., Fromm, M. E. & McCormick, S. (1989). Transient expression of chimeric genes delivered into pollen by microprojectile bombardment. *Plant Physiology*, **91**, 1270–1274.

Vasil, V., Brown, S. M., Re, D., Fromm, M. E. & Vasil, I. K. (1991). Stably transformed callus lines from microprojectile bombardment of cell suspension cultures of wheat. *Bio/Technology*, **9**, 743–747.

Vasil, V., Castillo, A. M., Fromm, M. E. & Vasil, I. K. (1992). Herbicide resistant fertile transgenic wheat plants obtained by microprojectile bombardment of regenerable embryogenic callus. *Bio/Technology*, **10**, 667–674.

Vasil, V., Hauptmann, R., Morrish, F. M. & Vasil, I. K. (1988). Comparative analysis of free DNA delivery and expression into protoplasts of *Panicum maximum* Jacq. (Guinea grass) by electroporation and polyethylene glycol. *Plant Cell Reports*, **7**, 499–503.

Vasil, V., Redway, F. & Vasil, I. K. (1990). Regeneration of plants from embryogenic suspension culture protoplasts of wheat (*Triticum aestivum* L.). *Bio/Technology*, **8**, 429–434.

Vasil, V., Srivastava, V., Castillo, A. M., Fromm, M. E. & Vasil, I. K. (1993). Rapid production of transgenic wheat plants by direct bombardment of cultured immature embryos. *Bio/Technology*, **11**, 1553–1558.

Wan, Y. & Lemaux, P. G. (1994). Generation of large numbers of independently transformed fertile barley plants. *Plant Physiology*, **104**, 37–48.

Wang, W. C. & Nguyen, H. T. (1990). A novel approach for efficient regeneration from long-term suspension culture of wheat. *Plant Cell Reports*, **8**, 639–642.

Weber, G., Monejambashi, S., Wolfrum, J. & Greulich, K.-O. (1990). Genetic changes induced in higher plant cells by a laser microbeam. *Physiologia Plantarum*, **79**, 190–193.

Weeks, J. T., Anderson, O. D. & Blechl, A. E. (1993). Rapid production of multiple independent lines of fertile transgenic wheat (*Triticum aestivum*). *Plant Physiology*, **102**, 1077–1084.

Werr, W. & Lörz, H. (1986). Transient gene expression in a Gramineae cell line: a rapid procedure to study plant promoters. *Molecular and General Genetics*, **202**, 368–373.

Yan, Q., Zhang, X., Shi, J. & Li, J. (1990). Long-term callus cultures of diploid barley (*Hordeum vulgare* L.). *Kexue tonqbao*, **35**, 1581–1583.

Zhang, L.-J., Cheng, L.-M., Xu, N., Zhao, N.-M., Li, C.-G., Yuan, J. & Jia S.-R. (1991). Efficient transformation of tobacco by ultrasonication. *Bio/Technology*, **9**, 996–997.

Zhong-xun, L. & Wu, R. (1988). A simple method for the transformation of rice via the pollen-tube pathway. *Plant Molecular Biology Reporter*, **6**, 165–174.

PART III TRANSFORMATION OF INDUSTRIALLY IMPORTANT CROPS

8

Leguminous Plants

Jack M. Widholm

Introduction

Many of the most important grain and forage crops of the world are legumes so there has been an interest in their tissue culture and transformation for many years. Since legumes are dicotyledons (dicots), they are in general very susceptible to infection by both *Agrobacterium tumefaciens* and *Agrobacterium rhizogenes,* so these systems have been used extensively.

Legumes in general and grain legumes in particular have been recalcitrant as far as plant regeneration from tissue culture is concerned. This has made many of the transformation techniques difficult, and so has delayed the overall progress. However, it is possible to transform the two most important crop legumes, soybean and alfalfa, and regenerate transformed fertile plants. Thus, we are now ready to attempt to improve leguminous crops using transformation. A review of legume transformation has been presented by Nisbet & Webb (1990).

In the following sections, we summarize many of the transformation studies that have been carried out with leguminous crops species and draw conclusions about the shortcomings and prospects for each. The species are divided into sections on Grain and Forage, based upon primary crop use.

Grain legumes

Soybean (*Glycine max*)

Soybean is a very important oil- and protein-producing crop, so many studies have been carried out with the goal of obtaining transformed plants.

Transformation by Agrobacterium tumefaciens
While most dicots, including legumes as described in other sections of this review, are very susceptible to *A. tumefaciens*, soybean plants do not form tumors very well when inoculated with virulent strains and there are large differences in responses found with different genotypes (e.g. see Owens & Cress, 1985; Byrne *et al.*, 1987). Similar genotype effects were also found in the formation of kanamycin-resistant callus on aseptic hypocotyl segments following inoculation with *A. tumefaciens* carrying the *npt*II gene (Hinchee *et al.*, 1988). These reports identified Peking as one of the most responsive genotypes, and this black-seeded noncommercial line has been used in many subsequent studies.

In early soybean transformation work, cotyledons, nodes and internodes of 2–3-week-old aseptic Forrest soybean seedlings were injected with *A. tumefaciens* strain A722 by Facciotti *et al.* (1985). This strain harbored a cointegrate, virulent plasmid containing the *npt*II gene driven by a small subunit promoter from soybean ribulose-bisphosphate carboxylase, with an octopine synthase terminator. Hormone-independent, octopine-positive and kanamycin-resistant callus and suspension cultures could be recovered from tissues near the wound sites after growth on MS basal medium with 500 mg carbenicillin/l. When the transformed callus was grown in the light the *npt*II poly(A$^+$)RNA was 5–10 times higher than when in the dark. Light also increased the NPT II enzyme activity and the kanamycin resistance of the callus. Thus, the light-inducible promoter operates properly in transformed callus even when driving an *npt*II gene.

A shoot regeneration system from detached cotyledons of 4–10-day-old, aseptically germinated soybean seedlings was used to obtain transgenic plants with the cultivars Peking and Maple Presto, which were identified as being the most responsive to *A. tumefaciens* out of 100 genotypes (Hinchee *et al.*, 1988). The cotyledons were dipped in *A. tumefaciens* strain A208 carrying *nos*, *npt*II, *nos* and cauliflower mosaic virus (CaMV) 35S, *uid*A, *nos* (pMON 9749) and were placed adaxial side down on B5 medium (Gamborg, Miller & Ojima, 1968) salts with 1.15 mg 6-benzyladenine (BA)/l, 500 mg carbenicillin/l and 100 mg cefotaxime/l to inhibit bacterial growth, and 200–300 mg kanamycin/l to select for transformed cells. After 3 weeks some of the tissues expressed β-glucuronidase (GUS) enzyme activity in tissue sections when measured histochemically with X-Gluc as substrate.

This procedure was used with 1400 cotyledons, mostly from Peking plants, using cointegrate plasmids pMON9749 or pMON894, which carries *npt*II and a modified petunia EPSPS (enolpyruvylshikimate-3-phosphate synthase) gene which confers glyphosate resistance, driven by an enhanced CaMV 35S promoter. A total of 128 plants developed via adventitious shoot formation in the presence of kanamycin. Eight of these plants were actually transformed when NPT II or GUS activity was measured in leaves or resistance to kanamycin or glyphosate was measured in leaf callus. No transgenic plants were recovered without kanamycin selection when 100 shoots were analyzed.

Two transgenic plants produced seed that showed a 3:1 segregation of the transforming DNA and the encoded traits (NPT II and GUS enzyme activity in one case and kanamycin and glyphosate resistance in the other). The genetic and molecular biological data are consistent with a single site of integration of one or a few gene copies.

Zhou & Atherly (1990) used a modification of the Hinchee *et al.* (1988) method with cotyledons from 5-day-old Peking seedlings. The cotyledons were removed, air-dried for 30 min and inoculated by wounding the proximal end with a scalpel dipped in *A. tumefaciens* strain A281 with the virulent plasmid pTiBo542 and a binary with *npt*II and *uid*A. The cotyledons were incubated on a medium similar to that used before for 48 h in the dark and for 24 h in the light. The cotyledons

were then transferred to the same medium with 500 mg carbenicillin/l, 100–200 mg cefotaxime/l and 250 mg kanamycin/l where callus developed on up to 94% of the cotyledons and shoot primordia on 8% within 2–3 weeks. Fourteen shoots that formed were maintained on the same medium with the antibiotics but without BA. Six plants were recovered and three out of five plants that contained the maize *Ac* active transposable element inserted inside the *uid*A gene had GUS-positive sectors when assayed with X-Gluc. These results show that this cotyledon system can produce transformed plants and that the *Ac* element is active in soybean. This activity was studied further with transformed callus where kanamycin-selected callus with the uninterrupted *uid*A gene showed GUS activity in 100% of the cases, while 45% of the calli with *Ac* in the gene showed GUS activity. There was no GUS activity detected in selected callus containing the *uid*A gene with the inactive *Ac* element *Ds* inserted into it.

Transgenic soybean plants were also produced by injecting about 30 μl of *A. tumefaciens* C58Z707 carrying a binary plasmid with *npt*II into the plumule, cotyledonary node and adjacent regions of 18–24 h aseptically germinated A0949 seedlings with one of the cotyledons removed (Chee, Fober & Slightom, 1989). The seedlings were incubated for 4 h on moistened paper towels and were then transplanted to soil for normal plant growth. A total of about 2200 plants were recovered from about 4000 inoculated seedlings and leaves of these were analyzed for NPT II activity by the *in situ* gel assay. Sixteen plants contained NPT II activity and this was not due to contaminating bacteria since no *A. tumefaciens* were recovered when the leaf extracts were plated on a selective medium. Ten of the 16 NPT II positive plants clearly contained the *npt*II gene and this was confirmed by the polymerase chain reaction (PCR). However, progeny from only one of these plants carried *npt*II sequences and then only three out of 36 were positive. This low inheritance ratio may be caused by chimerism of the R0 plant.

This method of injecting germinating seedlings did produce some transformed progeny but the method is labor intensive and the frequency very low (0.07% of the originally injected seedlings produced transgenic plants). Also the method may not show genotype specificity.

Another transformation system utilized immature cotyledons that were wounded by being pushed into 500 μm stainless steel or nylon mesh with a spatula and then were incubated overnight with *A. tumefaciens* strains LBA4404 or EHA101 with a binary plasmid containing *npt*II and the maize zein gene (Parrott *et al.*, 1989). The tissue was then incubated in an embryogenesis induction medium with high naphthaleneacetic acid (NAA) (10 mg/l) and 500 mg cefotaxime/l and in some cases with 10 mg G418/l to select transformed cells. Embryos were removed at 30 days and then placed on a maturation and germination medium with the cefoxitin. There were great differences in the numbers of plants produced by the different genotypes used and LBA4404 seemed to decrease the regeneration frequency. From a total of 9800 cotyledons, 14 plants were recovered, three of which were established to have been transformed by Southern hybridization. Two of these three plants came from embryos formed without G418 selection. These plants produced seed but none of the progeny were transformed, indicating that the R0 plants were apparently chimeric and none of the seed was produced from transformed cells.

Transformation by Agrobacterium rhizogenes
Hairy roots have been produced by inoculating wounds on 2–3-week-old greenhouse grown plants made with a multiple needle device between nodes 2 and 3 with *A. rhizogenes* strain R1000 (pRiA46) (Owens & Cress, 1985). When 23 soybean cultivars and three *Glycine soja* plant introductions were inoculated, hairy roots formed on five of the soybean and two of the *G. soja* genotypes. In this case, Peking did respond but Biloxi produced a larger number of roots. Inoculation of aseptic cotyledons from 1- or 3-day-old seedlings in cuts on the adaxial side also resulted in hairy root formation with Peking and Biloxi.

Savka *et al.* (1990) tested four *A. rhizogenes* strains with aseptic 6–15-day-old seedlings of 10 soybean cultivars by inoculating with a scalpel dipped in bacteria by cutting the cotyledon abaxial surface several times and by making 2 cm long longitudinal cuts on the hypocotyls. The seedlings were incubated in large culture tubes where roots formed at both inoculation sites. The *A. rhizogenes* strain K599 (cucumopine) produced roots on cotyledons of all cultivars, whereas the other three strains (8196, mannopine; 1855, agropine; A4, agropine) were much less effective. The K599-

induced roots were all opine positive, but only about a third of those produced by the other strains were. None of the roots produced on hypocotyls contained opines, so these were not considered to be transformed.

The hairy roots grew rapidly in liquid or on solidified media and were used in soybean cyst nematode (*Heterodera glycines*) propagation studies. The nematodes formed mature cysts on the cultured roots, followed by formation of second-generation cysts, which indicates that an entire life cycle can be completed with this system.

So far there has been no indication that the soybean hairy root cultures can produce shoots by direct shoot formation or by embryogenesis or organogenesis following callus induction.

Hairy root cultures have also been produced with the wild perennial *Glycine* species *canescens*, *clandestina* and *argyrea* by inoculating hypocotyls of 6–12-day-old aseptic seedlings with their roots removed by puncturing with a needle on a syringe containing *A. rhizogenes* (Rech *et al.*, 1988, 1989; Kumar, Jones & Davey, 1991). Transgenic plants were produced when the hairy roots from *G. canescens* and *G. argyrea* were cultured on a medium with 1.1–10 mg BA/l and 0.05 mg indolebutyric acid (IBA)/l. The roots formed callus and then shoots formed on the callus.

Transformation by microprojectile bombardment
In the first work with a particle acceleration device with soybean, Christou, McCabe & Swain (1988) bombarded immature embryos with 1–5 μm gold spheres coated with plasmid DNA containing the *npt*II gene with the CaMV 35S promoter and *nos* polyadenylation sequences. The acceleration device utilizes a force generated by the volatilization of a water droplet by an electrical discharge to accelerate an aluminum foil sheet that hits a stopping screen, thus allowing the gold particles to continue on into the target tissue. The system is operated under a partial vacuum. Protoplasts prepared from the bombarded tissue were cultured and 3 weeks after the bombardment were selected with 50 or 100 mg kanamycin/l. Selected colonies, which were recovered at a frequency of about 10^{-5}, were grown on 100 mg kanamycin/l. These cell clones all contained NPT II activity and *npt*II gene sequences. However, plants could not be regenerated from these transformed cultures.

Meristems from immature soybean seed embryonic axes have also been bombarded with gold particles coated with DNA containing the CaMV 35S, *npt*II, *nos* or CaMV 35S, *uid*A, *nos* gene constructs (McCabe *et al.*, 1988). The tissues were then cultured on a high cytokinin (13.3 μM BA) medium to induce multiple shoot formation in the dark for 1–2 weeks. Shoots formed on a medium with 1.7 μM BA and these could be rooted on hormone-free medium or be grafted onto soybean seedlings to obtain whole plants. Three to eight shoots could be recovered from each axis. Histochemical assays for GUS activity showed that expression could be observed in sectors of the tissues 2 days after bombardment and NPT II activity could also be measured in tissue homogenates. Overall about 2% of the shoots contained some transformed tissue. In one case, one plant resulting from 389 grafted shoots showed NPT II enzyme activity and three out of ten progeny were also positive. One of these progeny plants contained transforming DNA.

Two of the transformed plants, obtained following bombardment of meristems with DNA-coated gold particles with both *uid*A and *npt*II (plant 4615) or with *uid*A (plant 3993) genes as described above, were analyzed for gene expression and inheritance (Christou *et al.*, 1989). These two plants expressed the genes in all leaves so were not sectorial chimeras. Plant 4615 produced transformed progeny in the ratio 2.3:1 and the R1 transformed plants produced progeny that either were all transformed or in a 3:1 ratio, which fits a homozygous or heterozygous R1 state, respectively. Plant 3993 produced transformed R1 progeny at a 1:1 ratio and this was also 1:1 in the R2 indicating some problem with transmission of the transformed trait through the pollen. This was confirmed by staining pollen for GUS activity. In all cases the enzyme activities and gene sequences were inherited together. In plant 4615 progeny, the *npt*II and *uid*A genes were inherited as a block, indicating integration of both genes at a single site.

Since the soybean, particle bombardment, transformation system does not employ a resistance marker for selection, the transformed plants are identified by histochemical assay of the expression of the *uid*A gene (Christou, 1990; Christou & McCabe, 1992). The first step is to find expression in 2–4 mm thick sections taken from the basal end of elongating shoots. GUS-positive shoots are grown further, and primary leaves and petioles and midribs of the second and third trifoliate leaves are assayed. A wide variety of GUS expression patterns are found, which clearly supports the chimeric nature of the transformation events. Plants with patterns expected for transformation of the L_2 or germ layer were found to produce transformed progeny. In about 450 experiments where about 180 000 shoots were regenerated, 21 600 (12%) showed GUS expression but apparently only about 2% of those that expressed GUS were germline transformants and produced transformed progeny.

Transgenic R1 plants obtained by particle bombardment with the CaMV 35S, *uid*A, *nos* gene were assayed for GUS activity histochemically using X-Gluc (Yang & Christou, 1990). The activity staining of thin sections showed that certain tissues expressed GUS activity at high levels (root pericycle, xylem parenchyma and stem and leaf phloem), others at intermediate levels (leaf mesophyll, certain ground tissues of stem and leaf midrib and trichome and guard cells) and some at low or zero levels (root procambium, phloem and cortex, stem vascular cambium and the majority of leaf midrib cortex cells).

An embryogenic suspension culture of the cultivar Fayette was initiated by Finer & Nagasawa (1988); it grew relatively slowly but consisted of small clumps of globular stage embryos. These cultures continuously increased in mass by adventitious embryo formation from cells at or near the surface of the preexisting embryos. The cultures were initiated and maintained on a high 2,4-dichlorophenoxyacetic acid (2,4-D) medium by manually selecting the embryogenic-appearing clumps at each transfer. Embryo formation can be induced within a month on a regeneration medium without hormones and these embryos can be matured within another month and, following desiccation, can germinate and form complete plants.

These embryogenic suspension cultures were used in particle bombardment studies with the Biolistics Particle Delivery System from DuPont with 1.1 μm tungsten particles coated with DNA carrying *hph* and/or *uid*A genes driven by the CaMV 35S promoter (Finer & McMullen, 1991). Selection for hygromycin (50 μg/ml) resistance was imposed 1–2 weeks after bombardment and within 4–6 weeks live growing regions were visible on the cell clumps, which were selectively transferred to fresh medium for another 2–3 months at which point plant regeneration procedures were initiated. When GUS activity was measured histochemically, 3 days after bombardment, there were an average of 709 positive foci, and each bombardment produced about three stably transformed clones. These results give a stable transformation versus transient expression frequency of about 0.4%. An average of 25 embryos were produced per clump and 20% of these germinated to form plants from a total of 12 different transgenic clones. The regenerated plants expressed GUS activity and contained transforming DNA. Some transformed progeny were also produced.

Transformation of protoplasts by direct uptake of DNA

The first report of the regeneration of soybean plants from protoplasts was that of Wei & Xu (1988), who obtained plants from four out of six Chinese cultivars used. The protoplasts were isolated from immature cotyledons and nodular callus formed on a medium with 2,4-D and BA and shoots subsequently formed on a medium with NAA, BA, kinetin, zeatin and casein hydrolysate. Up to 35% of the calli formed shoots with one cultivar (Heilong 26) and the regenerated plants were fertile. Plant regeneration from soybean immature embryo protoplasts has also been reported by Luo, Zhao & Tian (1990).

We have been able to repeat this work using several more cultivars (7 out of 15 cultivars produced plants) including some current commercially important ones (Dhir, Dhir & Widholm 1991*c*, 1992*b*). So far the regenerated plants have been fertile and usually produce more than 20 seeds. To date little somaclonal variation has been noted in hundreds of R2 Clark 63 progeny.

Transient expression studies were carried out with Clark 63 protoplasts using the *cat* and *uid*A genes to optimize the DNA uptake methods (Dhir *et al.*, 1991*b*). The optimal conditions included a single electroporation pulse (500 V/cm field strength and 1000 μF capacitance) with about 4% polyethylene glycol (PEG) 6000. To obtain stable transformation, Clark 63 protoplasts were electroporated with two different plasmids, one with *hph* and *uid*A genes controlled by CaMV 35S promoters and *nos* terminators and another with *npt*II with a CaMV 35S promoter and *nos* terminator and the mannityl opine biosynthesis region (Dhir *et al.*, 1991*b*, 1992*a*), using similar electroporation methods, selection for the antibiotic resistance markers carried by the plasmids and regeneration

of shoots and plants. However, further study of the progeny does not confirm the presence of the transformed genes or opine synthesis so one cannot say that transformed progeny can be obtained using this system at present. Because of these inconsistencies these papers have been withdrawn. Other studies in this laboratory have shown that transformed calli can be clearly produced but the regeneration of plants is difficult.

Transformation by microinjection

An ovary microinjection method was used by Liu *et al.* (1990) to transfer the *Solanum nigrum psb*A atrazine resistance gene into soybean. Two microliters of 100 μg plasmid DNA/ml, carrying the *psb*A gene, was injected into 507 ovaries 1 day after pollination and 1220 seeds were harvested from these. Seed set was only slightly decreased in comparison with uninjected controls. Plants grown from the treated seeds in the field were sprayed with the herbicide atrazine and seven of the 866 plants that grew from the seed showed resistance. Fluorescence induction curves confirmed the atrazine tolerance of some plants. Dot-blot hybridization with isolated chloroplast DNA from two of the resistant plants showed that there were pBR322 plasmid sequences present. The plasmid sequences, rather than the *psb*A gene, were used for probing, since the resistance is caused by a single base change and the gene sequences are highly conserved between species. The authors feel that the *psb*A gene was in the chloroplast because the pBR322 sequences were found in purified chloroplast DNA and the *psb*A gene used had only the native promoter, which should be expressed only in the chloroplast.

This microinjection method looks promising but it must require careful and precise injection techniques to be successful so may not be possible in many laboratories. Likewise additional genetic and molecular analyses are needed to clearly substantiate the claims.

Conclusions about soybean transformation

Soybean has been transformed by *A. tumefaciens* and microprojectile bombardment to obtain transformed progeny. None of the methods used is ideal at present so the one chosen would depend upon the laboratory skills, funding levels and the genotypes desired. The *A. tumefaciens* method is low frequency, is very genotype specific and produces chimeric plants. The microprojectile method is also low frequency, produces chimeric plants, requires much labor and costly equipment (including royalty payments), but apparently is not genotype specific. This method, when applied to embryogenic suspensions, needs first of all the suspension cultures, which require time and skill for initiation, but the selection is straightforward and nonchimeral plants can be obtained in about 8 months.

Pea (*Pisum sativum*)

There have been a number of studies carried out with *P. sativum* (pea) because this is an important food crop. In one study, seven wild-type *A. tumefaciens* strains were used to induce tumors on greenhouse-grown plants, shoot cultures and aseptically germinated seedlings of five cultivars (Puonti-Kaerlas, Stabel & Eriksson, 1989). Three of the strains were very effective on all cultivars, with A281 being most effective, followed by C58 and GV3101 (pGV3304). These results show that the bacterial strain was more important than pea genotype as far as tumor induction is concerned.

Kanamycin-resistant (75 μg/ml) callus was obtained in cocultivation experiments with epicotyls, stem and leaf pieces, with a frequency usually near 30%. Cocultivation with microcolonies formed from protoplasts also produced kanamycin-resistant colonies with two cultivars (frequencies of 5×10^{-3} for Filby and 10^{-4} for Bello). Nopaline synthase activity was expressed in 91% of the kanamycin-resistant calli and all 12 lines analyzed contained *npt*II sequences.

Pieces of pea shoot cultures and epicotyls from axenic pea seedlings were cocultivated for 2–3 days with *A. tumefaciens* carrying *hph* or *npt*II genes, in liquid medium containing 0.5 mg BA/l and 2,4-D (Puonti-Kaerlas, Eriksson & Engström, 1990). The explants were washed and were transferred to callus initiation medium with 500 μg claforan/ml to inhibit bacterial growth and 15 μg hygromycin B/ml or 75 μg kanamycin/ml to select transformed cells. The best medium for transformed callus formation contained 0.5 mg BA/1 and 2,4-D, where from 135–392 hygromycin-resistant calli formed on 200 explants from three cultivars within 1 month. Other trials with all five cultivars with both kanamycin or hygromycin produced 30–240 resistant calli per 200 explants. Shoots could be regenerated from hygromycin-selected callus of one cultivar, whereas none could

be obtained from any of the kanamycin-selected calli. The transformed shoots contained *hph* sequences. The regenerated plants had produced flowers but no seed at the time of publication.

Recently the pea cvs. Puget and Stivo have been transformed using *A. tumefaciens* strain GV3101 (pGV2260::pGV1503) carrying *hph* using the methods of Puonti-Kaerlas *et al.* (1990) as described above (Puonti-Kaerlas, Eriksson & Engström, 1992*a*). Shoot induction occurred with 0.2%–15% of the hygromycin selected calli in 4–9 months and about half of these could be rooted. Plants regenerated from different Stivo-transformed calli were fertile and the R1 and R2 progeny inherited the transformed gene as a Mendelian, dominant trait. Progeny were analyzed by Southern and Northern blots and a leaf spot test with concentrated hygromycin solutions (1%) where lesions formed on untransformed leaves. Hygromycin resistance was confirmed by callus formation from leaf pieces on a medium with 15 mg hygromycin/ml. The transformed plants and their progeny were tetraploids. Whether the production of tetraploids is associated with the transformation or the tissue culture process is not known. However, ploidy increases are common in plants regenerated from tissue cultures.

Tumor induction frequencies induced by five different *A. tumefaciens* strains were measured with epicotyl segments from 7-day-old axenic, etiolated, Madria pea seedlings (Kathen & Jacobsen, 1990). The segments were soaked for 1 h with the bacterial suspensions, were blotted dry and then were incubated on MS medium with 0.1 mg BA/l and 0.1 mg picloram/l. Tumor induction frequencies of from 26% to 36% were obtained with three strains, C58C1 being most effective. No tumors could be induced on immature zygotic embryos or apical domes, both of which are known to produce somatic embryos (Kysely *et al.*, 1987). Low frequencies were found with immature leaflet (5%) and leaf discs (11%), while epicotyl segments, nodal explants and whole apices gave frequencies from 47% to 68%. Cocultivation with 50 μM acetosyringone was inhibitory and increasing the cocultivation time to 5 days stimulated tumor formation. Transformation frequency was also affected by the pea genotype.

Following transformation of nodus explants with *A. tumefaciens* carrying either *hph* or *npt*II selectable marker genes, selection either during shoot formation or after shoot formation at the rooting stage produced from 2.5% to 4.9% and 0.9% to 1.1% transformed shoots from 1075 nodus explants. Analysis of eight kanamycin-resistant plants detected NPT II activity in five. Twelve of these plants flowered but no seed was set.

It is of interest to note that the frequency of recovery of transformed shoots was similar for the *npt*II and *hph* containing strains even though the latter was seven-fold more efficient with hypocotyl segments at producing transformed callus. The authors feel that, owing to the greater toxicity of hygromycin, the selection of resistant shoots was too stringent with this antibiotic in comparison with kanamycin.

A plant regeneration system from nodal thin cell layer segments obtained from 10–12 day-old axenic pea seedlings was developed by Nauerby *et al.* (1991). After 14 days in liquid medium, up to eight 1–2 cm long shoots could be removed as they formed over a 2–3 month period. The original explant contained one preformed bud. The four pea cultivars tested produced an average of 3.3–4.1 shoots per explant with a shoot-forming frequency of between 88% and 97%. The rooting frequency varied from 14% to 49% and the pod-forming frequencies of the regenerated plants varied from 12% to 57%. This method can produce rooted plants within 7 weeks, but the number of shoots, the rooting frequency and fertility are all relatively low.

Preliminary transformation experiments with the nodal, thin cell layer segments have shown GUS expression in shoots regenerated from eight out of 148 explants following a 3-day cocultivation with *A. tumefaciens* carrying the *uid*A gene on a binary vector. Following cocultivation the explants were incubated in a medium containing 750 μg vancomycin/ml to stop bacterial growth but apparently not with any antibiotic or other agent to select for transformants.

There have been several studies, in addition to those described above, that have optimized *A. tumefaciens* transformation of pea tissues but have not yet produced transformed plants. Hobbs, Jackson & Mahon (1989) inoculated 17-day-old plants grown in growth chambers of 13 different genotypes with wild-type *A. tumefaciens* strains C58, A281 and ACH5. The A281 strain induced more and larger tumors. Similar results were found with immature leaves that were cocultivated

for 2 days and then incubated on medium with no growth regulators but with 500 μg carbenicillin/ml. In all cases there were positive responses but there were differences between pea genotypes. The tumor tissue grew without growth regulators, contained the specific opines for each strain and had both right (TR) and left (TL) transferred DNA (T-DNA) sequences.

This same laboratory has also transformed 5-day-old seedling stem segments with several disarmed *A. tumefaciens* strains (Lulsdorf *et al.*, 1991). Following cocultivation, transformed callus was induced from 2–3 mm end-pieces on medium with 1 mg BA/l and 1 mg 2,4-D/l and 500 mg carbenicillin/l and 100 mg cefotaxime/l to inhibit bacterial growth, and 50 mg kanamycin/l or 25 mg hygromycin/l to select the transformed callus. Optimal conditions included a 4-day cocultivation period with a tobacco, nurse, suspension culture and the *A. tumefaciens* strain EHA101 (pBI1042), where about 75% of the sections formed transformed callus using kanamycin (35S, 35S, alfalfa mosaic virus enhancer, *npt*II) or hygromycin (*nos*, *hph*) selection. The cointegrate vector was more efficient than the binary vector system used.

Hussey, Johnson & Warren (1989) studied the induction of tumors or hairy roots when various tissues of cultured pea shoots were inoculated with wild-type strains. Hairy roots could be induced by the *A. rhizogenes* strain 9402 and tumors by *A. tumefaciens* strain C58 but not by ACH5. Inoculation of 0.5 mm slices of the shoot, beginning at the tip by a 48 h cocultivation and then incubation on a medium with 0.5 mg BA/l, 0.25 mg IBA/l and 500 mg ampicillin/l, produced tumors and hairy roots first on the youngest tissue and then progressively down the stem. Tumors and hairy roots could also be induced by stabbing the apical dome, following dissection of the leaves and primordia, with a needle dipped in *A. tumefaciens* or *A. rhizogenes*, respectively. These studies indicate that meristematic cells can be transformed by the *Agrobacterium* systems.

Hairy roots had also been induced on axenic, 3-day-old, pea epicotyls from five cultivars either following decapitation or on wounded regions along the epicotyl (Bercetche *et al.*, 1987). The inoculated plants were incubated in medium in tubes in the dark. Transformed roots were obtained with *A. rhizogenes* strain 1855 but not with strains 2659 and 8196. The cortical cells from which the hairy roots were ultimately derived were partly polyploid but the resulting roots were diploid.

Pea protoplasts should also provide a system for transformation by direct DNA uptake if plant regeneration can be obtained. Puonti-Kaerlas & Eriksson (1988) were able to obtain callus formation from protoplasts isolated from epicotyls of 12-day-old aseptic seedlings of 10-pea cultivars. Plating in agarose slabs 2 days after preparation was beneficial for increasing the plating efficiencies. High plating efficiencies were obtained with protoplasts from leaves, shoot cultures and epicotyls of the best cultivar, Filby. Shoot-like structures developed in about 1% of the calli from two of the genotypes but normal plants with roots did not form.

Protoplasts prepared from shoot apices of young, in vitro-grown Filby and Belman pea seedlings were electroporated with plasmids containing the nptII or hph genes (Puonti-Kaerlas, Ottoson & Eriksson, 1992b). After two days the protoplasts were embedded in 0.6% (w/v) Sea Plaque agarose and selection began immediately or after 12 days with 75 mg kanamycin/l or 15 mg hygromycin/l. Resistant transformed colonies were selected with hygromycin and none was recovered with kanamycin. No plants could be regenerated from the transformed colonies.

Protoplasts were also prepared from shoot apices or lateral shoot buds from aseptically grown embryonic axes (Lehminger-Mertens & Jacobsen, 1989). When the protoplast derived callus was grown on media with 5–10 μM picloram or 2,4-D, somatic embryos formed on 10% –30% of the calli from two cultivars, whereas 1% or less of the calli from three other cultivars formed embryos. Embryo maturation was stimulated by 1.5 μM gibberellic acid (GA$_3$) and 0.12 μM NAA and germination occurred with 2.9 μM GA$_3$. Complete plants formed and could be potted in soil. These plants had normal morphology and produced seed.

Leaf mesophyll protoplasts were used in transient expression experiments following electroporation with DNA carrying the uidA gene (Hobbs et al., 1990). The nos and CaMV 35S promoters gave similar expression levels but the duplicated 35S promoter had higher activity. One of the genotypes gave 5–20-fold higher levels of GUS activity than did the other genotype.

These studies indicate that it should be possible to transform protoplasts from certain genotypes using direct DNA uptake methods and following the selection of the transformed colonies plant regeneration can be accomplished. However, the successful use of such a system has not been reported. Likewise the A. tumefaciens transformation system with pea explants that can regenerate plants seems to be relatively efficient and the passage of the transformed genes to progeny has been demonstrated by Puonti-Kaerlas et al. (1992a).

Moth bean (*Vigna aconitifolia*)

Protoplasts isolated from primary leaves of 9–10-day-old axenic moth bean seedlings were incubated with plasmids carrying the nptII gene driven by a pair of tandemly arranged nos promoters or by the CaMV 35S promoter (Köhler et al., 1987b). DNA uptake was induced by PEG and in some cases by electroporation. Selection was applied after 7 days, when the protoplasts were embedded in agarose slabs and 75 μg kanamycin/ml was added. Transformation frequencies of about 10^{-5} were obtained with cultivar 560, but the frequency was about 10-fold lower with 909. Electroporation did not increase the frequency when applied in addition to PEG.

Electroporation or PEG were used with the protoplast system described above, with the same two cultivars to stimulate the uptake of a plasmid carrying the nptII gene alone and one with nptII and cat (Kohler et al., 1987a). The transformation rate was higher when kanamycin selection began 7 days after DNA uptake rather than after 28 days. Again the transformation frequency was much higher with cultivar 560 than with 909. The culture medium was also very important in determining transformation frequencies. Transient expression of the cat gene could be observed 48 h after electroporation. No stable transformation or transient expression was seen with suspension cultured cells without cell wall removal. Southern hybridization analysis showed that nptII DNA sequences were present in the selected callus lines. When the selected callus lines were placed on regeneration medium without growth regulators, about 10% of the calli regenerated plants.

Moth bean protoplasts isolated and cultured as described above were cocultivated 3 days after preparation for 48 h with A. tumefaciens containing a cointegrate plasmid containing nptII driven by the nos promoter (Eapen et al., 1987). After being washed, the cells were cultured in a medium with 500 μg cefotaxime/ml to inhibit bacterial growth. After 21 days the cells were plated in agarose and 75 μg kanamycin/ml added to select transformants. Colonies growing after 3–4 weeks were grown with 100 μg kanamycin/ml for an additional 2–4 weeks. Again the transformation frequency of the cultivar 560 (4×10^{-5}) was much higher than with 909 (5×10^{-7}). The selected calli contained *npt*II DNA sequences and 23% of them contained nopaline synthase activity. Shoot buds and plants were then regenerated.

The same moth bean protoplast system was used in another study of PEG-mediated transformation with a plasmid carrying the *npt*II gene driven by the CaMV 35S promoter (Köhler *et al.*, 1989). Again the cultivar 560 was transformed at a much higher rate than another cultivar, in this case 88. The transformation frequency was increased by 1–2-fold when the protoplasts were X-irradiated (10 Gy) 1 h after incubation with DNA.

These studies with moth bean protoplasts show that transformation by direct DNA uptake can be relatively efficient and shoots can be regenerated within about 3 months from about 10% of the selected calli (Kohler *et al.*, 1987*a*; Eapen *et al.*, 1987). Even though transformed plants were regenerated, no information concerning the characteristics of the plants including fertility and passage of the transforming DNA to progeny have been reported.

Broad bean (*Vicia faba*)

Hairy roots were induced on axenic 5–6-day-old *V. faba* seedling epicotyls by wounding with a syringe filled with *A. rhizogenes* strains A4 and 15834 containing the binary plasmid pGSGLUC1, which carries *npt*II and *uid*A genes driven by the 1′ and 2′ promoters, respectively, of the TR-DNA of *A. tumefaciens* (Schiemann & Eisenreich, 1989). Hairy roots formed within 14–21 days and 45% expressed GUS activity. These roots could be cultured on hormone-free solid, but not on liquid medium. No antibiotics were needed to suppress bacterial growth.

Hairy roots were also induced on axenic *V. faba* cotyledons and stems of 10-day-old seedlings using *A. rhizogenes* strain LBA9402 with the Ri plasmid and the binary vector pBIN19, which carries the *npt*II gene (Ramsey & Kumar, 1990). The

stems were inoculated by placing bacteria in a 1 cm long cut, half-way through the stem above the cotyledonary node. The excised cotyledons were cut with a scalpel to make five slits on the adaxial surface and bacteria were spread over the surface. After 2 days of cocultivation the tissues were transferred to a medium with 250 μg cefotaxime/ml to inhibit bacterial growth. Roots formed after 14 days on all eight genotypes but there were large differences in the frequencies. Sixteen of 25 roots analyzed contained NPT II activity. Chromosome counts showed the roots to be 51% diploid, 45% tetraploid, with the rest having a higher ploidy level. Four of 65 root clones examined had chromosome rearrangements. Stems produced a higher proportion of diploids than did cotyledons. Apparently plants could not be regenerated from the transformed roots.

These two studies show that transformed hairy root cultures can be easily initiated from many faba bean genotypes but with no apparent plant regeneration.

Vicia narbonensis

A system for regenerating plants has been developed for *V. narbonensis*, a close relative of *V. faba*, via embryogenesis from callus obtained from excised young shoot tips (Pickardt, Huancaruna Perales & Schieder, 1989). This regeneration protocol was coupled with *A. tumefaciens* transformation to produce transformed plants by Pickardt *et al.* (1991). Shoot tips and epicotyls from 2–3-day-old aseptic seedlings were precultured for 1 day in liquid MS medium containing 4 mg picloram/l with shaking. A small amount of *A. tumefaciens* strain C58C1 containing the cointegrate vector pGV3850HPT (CaMV 35S, *hph* and *nos*, *nos*) was added to the liquid medium and cocultivation was carried out for 48 h with shaking. The segments were rinsed and incubated for 48 h on the same medium, solidified with 0.25% (w/v) Gelrite, with 500 μg claforan/ml to inhibit bacterial growth and then were selected on this medium with 30 μg hygromycin/ml. After 28 days the cultures were transferred to the same medium except that 60 μg hygromycin/ml and 200 μg claforan/ml were used. After 3 months on this medium with transfers every 28 days, the cultures were placed on regeneration medium with lowered auxin content and transferred every 28 days until somatic embryos formed. These were matured and germinated to form complete plants. Hygromycin concentrations sufficient to prevent growth of controls were present in each medium used.

A total of 438 of the original 2500 treated explants had callus still growing on the regeneration medium and 41 of the 44 tested contained nopaline. From a total of 142 embryos formed from these calli, 26 showed hygromycin resistance and 7 of 12 shoots formed from these were nopaline positive. Southern hybridization of five nopaline positive shoots showed the presence of T-DNA. The rooted shoots failed to develop further.

Recently, Pickardt et al. (1992) have reported the regeneration of two fertile *V. narbonensis* plants transformed with the Brazil nut 2S albumin gene using methods similar to those described above. Some of the progeny expressed this high sulfur amino acid protein in leaves but the levels in seed were very low, perhaps because the CAMV 35S promoter was used. This work demonstrates that *V. narbonensis* can be transformed to produce transformed progeny.

Lentil (*Lens culinaris*)

Two-week-old greenhouse-grown lentil seedlings were inoculated with four wild-type *A. tumefaciens* strains (C58, ACH5, GV3111, A281) by stabbing with a needle at the first and second internodes (Warkentin & McHughen, 1991). Shoot apices excised from 4-day-old seedlings were also inoculated by immersing in a bacterial suspension. The explants were blotted on filter paper and were cocultivated for 3 days on MS basal medium followed by rinsing with 500 µg carbenicillin/ml and incubation on the same medium with 100 µg cefotaxime/ml. All four *A. tumefaciens* strains formed tumors at all of the wound sites on the lentil stems. The tumors were larger when induced by strains GV3111 and ACH5. From 81% to 100% of the shoot apices developed tumors, the largest being formed by strain A281. The tumors produced the appropriate opine or opines and the one line analyzed contained T-DNA sequences. These results indicate that lentil transformation can be carried out efficiently. If this can be coupled with the apex regeneration system described by Williams & McHughen (1986) then transformed plants should be obtainable.

Cowpea (*Vigna unguiculata*)

Tumors could be induced on stems of 7-day-old cowpea seedlings when virulent *A. tumefaciens* strains LBA1010 (with the octopine type plasmid pTiB6) and LBA958 (with the nopaline type plasmid pTiC58) were inoculated by puncturing the stem with a toothpick soaked in the bacteria (Garcia, Hille & Goldbach, 1986).

Leaf discs from 6-day-old cowpea plants were also submerged for 30 s with diluted *A. tumefaciens* C58C1 carrying an *npt*II (*nos*, *npt*II, *ocs*) gene on the nononcogenic Ti plasmid pGV3850::1103 neo (dim) (Garcia et al., 1986). The discs were washed briefly, blotted dry and incubated on callus-inducing medium with a petunia cell feeder layer and a filter paper for separation. After 2 days the discs were transferred to medium with 200 µg cefotaxime/ml and 200 µg vancomycin/ml to inhibit bacterial growth and with 50 µg G418/ml to select transformed cells. After six 20-day transfers, on a medium with the cefotaxime, vancomycin and 100 µg kanamycin/ml, the selected calli were grown on a medium without any antibiotics. The selected calli contained nopaline and transforming DNA sequences (Southern hybridization). Plants could not be regenerated from the transformed calli.

The same cowpea leaf disc method, with somewhat altered selection protocols, was used to insert a full-length copy of the cowpea mosaic virus (CPMV) mRNA carried on a binary vector with *npt*II (Garcia et al., 1987). The selected transformed callus was used to study the expression of the CPMV cDNA and showed that the CaMV 35S promoter was 10-fold more active than the *nos* promoter.

Cowpea transformation was also carried out with surface-sterilized mature seeds of five genotypes, which were incubated in water for 12 h and then the embryos with the cotyledons removed were sliced longitudinally (Penza, Lurquin & Fillippone, 1991). These pieces were soaked for 2 min with *A. tumefaciens* virulent strain A281 carrying the *npt*II gene and were then transferred to a culture medium containing 5 mg BA/l and 1 mg NAA/l for 2 days. The explants were rinsed in a medium containing 500 mg claforan/l and after blotting were grown on a medium containing 50 mg G418/l to select transformed cells. Antibiotic-resistant callus formed on 10% or fewer of the treated embryos but no plants could be regenerated. The growing cells did contain *npt*II sequences, as shown by dot blots.

The embryo sections were also cocultivated with a disarmed *A. tumefaciens* strain C58 (pGV2260) with p35SGUSINT as the binary containing both *uid*A (containing an intron) and *npt*II genes. In this case the cocultivation and regeneration media both lacked growth regulators but the latter contained 500 mg claforan/l to inhibit bacterial growth. The shoots that formed after 21 days from axillary buds were sectioned and stained for GUS activity. About 50% of the shoots contained GUS-expressing cells in the stem but not in leaf tissue. While some cells were apparently transformed, none of the shoots survived when placed on a medium containing 50 mg G418/l indicating that too few of the cells were expressing NPT II activity.

This cowpea system appears to have promise for producing transformed shoots, since chimeral tissues were readily obtained. The use of selection from the beginning to eliminate escapes and to channel only the transformed cells into shoot formation would seem to be one possible way to improve the results.

Dry bean (*Phaseolus vulgaris*)

Hairy root cultures have been initiated by Aird, Hamill & Rhodes (1988) by infecting aseptic dry bean plants at puncture sites on the stems with *A. rhizogenes* LBA9402. The hairy roots were excised and were cultured in B5 liquid medium with 500 mg ampicillin/l. The cultures contained TL and TR T-DNA sequences, as shown by Southern hybridization. The chromosome number was the euploid number of 22 after 7 months in culture. No additional information was presented.

Hairy root cultures were also initiated by McClean *et al.* (1991) from 1 cm long hypocotyl segments from aseptic seedlings of one genotype by inoculating *A. rhizogenes* strain A4RS (pRi B278b) carrying the binary plasmid pGA482 on the cut end of the segments set upright. After 2 days of cocultivation, the segments were transferred two times at 2-day intervals on MS medium with 800 μg cefotaxime/ml and 500 μg carbenicillin/ml to inhibit bacteria and then to the same medium with 75 μg kanamycin/ml to select for transformed roots. The roots could be removed and be cultured in liquid medium with 75 μg kanamycin/ml. The roots contained NPT II enzyme activity and *npt*II gene sequences.

Neither of the reports of hairy root induction on dry bean plants described plant regeneration so this must not have occurred.

Tumor induction assays were performed with 1-day germinated, surface-sterilized dry bean seedlings of 19 diverse genotypes that were inoculated with the virulent *A. tumefaciens* strains A208 (nopaline), A281 (agropine) and LBA 4001 (octopine) by puncturing the cotyledonary node region three times after removal of the cotyledons and then placing bacteria on the wound (McClean *et al.*, 1991). Following growth for 7 days on moistened paper towels in a culture room, the seedlings were transplanted in soil and gall formation determined after 14 days. All genotypes showed a high susceptibility to *A. tumefaciens*, with little difference between strains.

One genotype was also inoculated with the avirulent strain C58Z707(C58Z707/pGA482) with a binary plasmid containing *nos*, *npt*II, *nos*. Cotyledonary nodal regions from aseptic 1–7-day-old seedlings were inoculated with bacteria after puncturing with a needle. Following a 3-day cocultivation on MS medium, the explants were transferred three times at 3-day intervals on MS medium with 500 μg carbenicillin/ml and 800 μg cefotaxime/ml to eliminate bacteria. The explants were then cultured on the same medium with the carbenicillin and 200 μg kanamycin/ml to select for transformed cells. The kanamycin concentration was gradually increased over a 2-month period to 500 μg/ml. The selected growing callus did contain NPT II enzyme activity.

While it is possible to induce hairy roots and tumors readily with dry bean, no transformed plants have apparently been regenerated using these systems.

The navy bean cv. Seafarer has been transformed using electric-discharge particle acceleration to deliver gold particles coated with plasmid DNA into apical meristems of germinating seeds (Russell *et al.*, 1993). Multiple shoots that formed on a high cytokinin medium were screened for GUS activity histochemically or for bialaphos resistance by placing in medium with the herbicide. About two shoots formed on each explant and about 0.5% of these expressed GUS activity in at least a small portion. About 0.03% of the shoots produced seed carrying the GUS activity. Some of the GUS-negative shoots showed bialaphos resistance and these produced progeny that were not damaged when sprayed with Basta, showing that the *bar* gene driven by the CAMV 35S promoter can impart clearcut herbicide resistance. Thus, stably transformed, fertile, dry bean plants can be produced by the electric-discharge particle acceleration technique.

Peanut (*Arachis hypogaea*)

Peanut, or groundnut, is an important food crop that, like other large-seeded legumes, has been difficult to manipulate in culture to obtain regenerated plants. There has been one report describing tumor induction with peanut tissues (Lacorte *et al.*, 1991). Four-week-old, greenhouse-grown plants of five cultivars were inoculated with four different virulent *A. tumefaciens* strains (A281, Bo542, A208, T37) by applying the bacteria to needle wounds on stem internodes. All of the strains except T37 induced tumors on all genotypes, with A281 producing tumors that appeared earlier and were larger. Tumors could also be produced by all four strains on wounded stems of *in vitro*-grown aseptic plants. These tumors would grow as friable callus on MS medium with no growth regulators. The tumor tissues contained the correct opine.

The *A. tumefaciens* strain A281 carrying pTDO2 with the *npt*II and *uid*A genes was used to infect cotyledon segments and embryonic axes from surface-sterilized seed and leaf and petiole explants from 7-day-old aseptic, *in vitro*-grown plantlets. The tissues were inoculated with bacteria, blotted and cultured on MS medium for 24 h. After being rinsed and blotted, the tissues were cultured on MS medium with 500 μg cefotaxime/ml to inhibit bacterial growth. Tumors formed with all five cultivars on all explant sources except cotyledons which formed roots. Bacteria-free tumor tissues would grow without growth regulators and contained agropine. About half of the tissues were kanamycin resistant and contained NPT II enzyme activity.

This study shows that peanut can be readily transformed by *A. tumefaciens*, but there is a need for plant regeneration protocols so that transformed plants can be recovered.

Forage legumes

Alfalfa (*Medicago sativa*), lucerne

Alfalfa is the most widely grown forage legume and has been one of the easiest legumes to regenerate from callus (Bingham *et al.*, 1975), suspension cultures (Atanassov & Brown, 1984) and protoplasts (Kao & Michayluk, 1980). Genotype specificity for plant regeneration is very evident and there has even been a recurrent selection program for plant regeneration ability that has resulted in the release of genotypes for general use (Bingham, 1989, 1991). The first line released, Regen-S, was selected mostly from the cultivar Saranac, which showed 12% regeneration in cycle 0, 50% in cycle 1 and 67% in cycle 2 (Bingham *et al.*, 1975). Regen-S produces about 90% as much herbage as does Saranac and so performs in a manner close to that of a commercial cultivar. The line Regen-SY is especially susceptible to *A. tumefaciens* transformation. Ideally, however, transformation should be done with the best current commercial varieties to produce rapidly acceptable materials that do not need to be backcrossed. However, Bingham & McCoy (1986) believed that alfalfa plants regenerated from tissue culture contain enough somaclonal variation that backcrossing would always be needed anyway.

There are several examples of alfalfa transformation using *A. tumefaciens*, including a report by Shahin *et al.* (1986). Surface-sterilized 5 cm long stem segments from mature plants (cultivar CUF101) grown in the field were cut into 2–3 mm slices, submerged in a suspension of *A. tumefaciens* (LBA4404 containing a binary vector with *nos*, *npt*II), blotted and then incubated on a filter paper on medium containing 2 mg 2,4-D/l and 1 mg BA/l. After 48 h the slices were placed on a similar medium with lower BA (0.25 mg/l) and with 500 mg cefatoxime/l to eliminate bacteria and 50 mg kanamycin/l to select transformed calli. Kanamycin-resistant calli formed on 12%–15% of the slices and many plants were regenerated following somatic embryo induction on media with 2 mg (or higher) 2,4-D/l. All of the selected calli tested contained NPT II enzyme activity, but not all of the regenerated plants did. The presence of transforming DNA in the regenerated plants was confirmed by Southern hybridization. The regenerated plants were morphologically normal and were fertile.

An autotetraploid *Medicago varia* Hungarian genotype, which was selected by A. Atanassov from a *M. sativa* × *Medicago falcata* hybrid, was used in transformation experiments by Chabaud *et al.* (1988). Wounded leaflet and petiole segments from aseptic seedlings were dipped into an *A. tumefaciens* (two strains both with pVW130 carrying *npt*II) suspension, blotted with filter papers and cocultivated for 2–4 days on a complex medium (B5h), which is B5 with 1 mg 2,4-D/l, 0.2 mg kinetin/l, 800 mg glutamine/l, 100 mg serine/l, 10 mg glutathione/l and 1 mg adenine/l using a feeder layer of alfalfa suspension cultured cells or 50 μM acetosyringone under a filter paper. The explants were rinsed with water and were blotted and then transferred to the same medium with 300 μg carbenicillin/ml to inhibit bacterial growth and 50 or 100 μg kanamycin/ml to select for transformed calli and subsequently somatic embryos that begin to form after 6–8 weeks. Embryos were matured on a medium lacking growth regulators but containing the original antibiotics. The mature embryos were rooted on a medium that also contained the antibiotics.

In these experiments the leaflets produced more transformed callus (36%) than petioles (9%) and *A. tumefaciens* strain A281 was more efficient (34%) than strain LBA4404 (14%). A 4-day cocultivation was better (43%) than 2 (10%) or 3 days (19%). There was not much difference in transformation frequency between 50 and 100 μg kanamycin/ml for selection. The regenerated plants produced callus with NPT II enzyme activity and *npt*II sequences. Maximum transformation frequencies near 70% were obtained when all parameters were optimized.

D'Halluin, Botterman & De Greef (1990) have transformed *M. sativa* line RA-3, which is a highly regenerable genotype selected from Regen-S, using *A. tumefaciens* strain C58C1Rif[R] carrying the disarmed T-DNA vector pMP90. The two binary vectors used contained *npt*II driven by the *nos* promoter and the *bar* gene driven by the CaMV 35S promoter or the *npt*II and *bar* genes driven by the 1' and 2' promoters, respectively, of the divergent TR T-DNA promoter as well as *hph* driven by CaMV 35S. Stem and petiole segments 1 cm long from aseptically grown shoot cultures were dipped in a late log-phase *A. tumefaciens* culture and then were cocultivated on top of a filter paper covering a tobacco feeder layer in a medium containing 25 μM NAA and 10 μM kinetin. After

2–3 days the sections were rinsed and were then grown on the same medium with 500 or 1000 μg carbenicillin/ml to eliminate bacteria, and kanamycin, hygromycin or phosphinothricin to select transformed cells. After about two 3–4-week subcultures on the selective medium, somatic embryogenesis was induced with a 3–4 day pulse of 50 μM 2,4-D in liquid selective medium. The embryos were allowed to germinate to form complete plants on selective media over a period of one to several months.

In this system petioles produced transformed callus more readily than they did stems. Selection immediately after cocultivation with 50 or 100 μg/ml kanamycin, 75 μg hygromycin/ml or 5 or 10 μg phosphinothricin/ml produced transformed plants, with kanamycin and hygromycin giving a higher frequency. A large number of plants were regenerated and most of these were resistant to the herbicide Basta when sprayed. Plants with the *bar* gene driven by the CaMV 35S promoter tended to be more resistant than those with the TR2'-*bar* gene and the levels of phosphinothricin acetyltransferase activity generally correlated with the herbicide resistance. Most of the transformed regenerated plants also expressed NPT II.

This study shows that alfalfa can be transformed with a useful gene to produce plants that express the gene at varying levels. In this case the herbicide resistance gene is expressed at a high enough level in some plants to impart resistance to the usual field application rates.

In another attempt to place a useful gene into alfalfa, Schroeder *et al.* (1991) used *A. tumefaciens* strain LBA4404 with a binary plasmid containing a chicken ovalbumin cDNA clone driven by the CaMV 35S promoter with a *nos* terminator. The plasmid also contained *npt*II. The ovalbumin gene was used because the ovalbumin protein contains a relatively high level of cysteine plus methionine (6.5%). Leaves from greenhouse-grown plants of three cultivars, two of which are of commercial importance, were surface sterilized; segments were submerged in a bacterial suspension for 10 min, blotted and cocultivated for 3 days on the complex B5h medium described above. The segments were then washed and placed on B5h medium with 50 μg kanamycin/ml to select transformed callus and 100 μg cefotaxime/ml to eliminate the bacteria. Callus was transferred after 4 weeks to the same medium or to the same medium with higher 2,4-D (11 mg/l) and kinetin (1.1 mg/l), where somatic embryos formed in three more weeks. The embryos were allowed to mature and form plants on media without kanamycin.

The noncommercial cultivar Rangelander produced transformed calli on 10%–15% of the explants, whereas the success rate with the two commercially important ones was only 0.1%–0.5%, even with an altered protocol. The production of somatic embryos from each callus was also much lower with the latter two cultivars. However, transformed plants were obtained from all three genotypes; they did express the ovalbumin cDNA at varying levels to accumulate from about 0.001% to 0.01% of the total leaf soluble protein as measured by immunoblots. The protein appeared to be stable, as shown by pulse-chase labeling experiments and by levels found in leaves of different ages. Of 33 Rangelander plants analyzed, 31 contained the ovalbumin cDNA but only 23 contained the protein.

These results show that it is possible to obtain expression of a useful gene in commercially important alfalfa varieties, but the levels of the protein accumulated seems to be too low to be significant for increasing the leaf sulfur amino acid content. Whether another promoter or altered gene can increase the production is open to debate.

There are also a number of reports of the production of hairy roots using *A. rhizogenes*, followed by regeneration of transformed plants. An example is the work of Sukhapinda, Spivey & Shahin (1987), who used stem segments from mature plants of two cultivars grown in the greenhouse. The sections were surface sterilized, rinsed three times, cut into 1–2 cm sections and placed in an inverted position on a medium containing 250 μg cefatoxime/ml. A drop of a bacterial suspension was spread on the top of each section. Hairy roots, which formed profusely on most sections in 2–3 weeks, were excised and placed on a growth-regulator-free medium with 100 μg cefatoxime/ml to inhibit bacterial growth. From 43% to 60% of the roots produced nopaline and they also expressed the typical hairy root phenotype, lack of geotropism and extensive branching. Embryogenesis was induced by transfer of the roots to a medium with 50 μM 2,4-D to induce callus formation and then to a complex medium containing asparagine, glutamine, adenine sulfate, 5 μM 2,4-D and 0.5 μM BA. After a week the

medium was modified by substituting 2% (w/v) sucrose for 2% glucose. Plants were regenerated, from the embryos formed, by continuous subculturing on a medium lacking growth regulators.

A total of 382 transformed plants were produced using *A. rhizogenes* strain A4 with binary vectors containing two soybean genes or the maize transposable element *Mu* 1 as well as *nos*. There were multiple copies of the TL- and TR-DNA as well as the Ri plasmid. The plants contained nopaline as did their progeny. The transformed plants were phenotypically normal except for extensive and shallow roots.

In a report by Spano *et al.* (1987), stem segments from 1–2-month-old aseptically grown plants of three cultivars were inoculated with *A. rhizogenes* strain NCPPB1855 harboring the pRi1855 (agropine type) plasmid by injection with a syringe or by placement on the cut ends. Hairy roots were removed after 1 month and were placed in liquid MS medium with 1000 mg carbenicillin/l to inhibit bacterial growth. About 50% of the hairy roots contained agropine. Bacteria-free callus was formed within 2 months on solid medium with 2–5 mg 2,4-D/l and 0.25 mg kinetin/l with decreasing levels of the antibiotic. Somatic embryos then formed on solid medium without any growth regulators and these then formed complete plants on the same medium. Only one of the cultivars, which had previously been shown to be highly regenerative, formed plants. One of the plants that contained agropine also contained three to five copies of the TL-DNA and two copies of the TR-DNA in truncated form. The regenerated transformed plants had larger, shortened roots with more highly developed lateral roots, unlike the usual tap root system of normal alfalfa plants. The transformed plants had shortened internodes and more and smaller leaves. The plants were fertile.

These two examples of hairy root transformation did produce transformed, fertile plants after callus formation and embryogenesis and not directly by shoot formation from the roots. Similar results were obtained by Golds *et al.* (1991), where transformed hairy roots were produced by three cultivars, but plants were obtained from roots of only one of the cultivars and only through a callus phase. The hairy root production was efficient from stem sections with cotransformation frequencies of about 50% using both cointegrate and binary vector systems. The callus formation and embryogenesis system appears to require several months and is genotype specific. The plants produced do have abnormal, compact root systems that are likely to be detrimental under most field conditions.

Alfalfa mesophyll protoplasts have been transformed via electroporation or PEG-mediated direct DNA uptake using plasmids with *nos*, *hph*, *nos* or CaMV 35S, *npt*II (Larkin *et al.*, 1990). Transformed colonies were selected with hygromycin or kanamycin, respectively, at frequencies near 10^{-4} to 10^{-5}. Plants were regenerated from the kanamycin-selected colonies via somatic embryogenesis. Samples of selected calli and regenerated plants contained NPT II enzyme activity.

Protoplasts from leaves of *M. borealis* line 94 shoot cultures were electroporated with pGA472, which carries the *npt*II gene (Kuchuk *et al.*, 1990). The protoplast viability was about 50%–60% after electroporation with 10 μg plasmid DNA and 40 μg of calf thymus DNA using 5×10^5 protoplasts at 560 V/cm and 1 μF. Colonies which formed after 2–3 weeks were transferred to a solidified medium with 50 μg kanamycin sulfate/ml and 1 mg BA/l. A total of 56 green colonies formed, of which 15 expressed stable kanamycin resistance. Shoot regeneration was obtained from five of these clones and roots did form on these shoots in the presence of kanamycin. The two plants tested contained NPT II enzyme activity.

Another possible method for transforming alfalfa was investigated by Senaratna *et al.* (1991), who prepared desiccated somatic embryos (about 10%–15% water) and then these were incubated for 2 h with pBI221.2 with the *uid*A gene driven by the CaMV 35S promoter. GUS activity was seen after histochemical staining on the surface of germinating embryos 3 days after imbibition and on up to 80% of the plantlets formed 14 days after imbibition. PCR analysis also showed the presence of a 250 bp segment spanning the promoter and *uid* A gene in embryos imbibed with the plasmid DNA.

These results indicate that some cells can take up plasmid DNA upon imbibition but few of the apical meristematic cells that form the plant were transformed. This method has been used previously with imbibing seeds (Ledoux & Huart, 1968) but in general the overall results obtained in the past work did not clearly demonstrate transformation of the resulting plants. Thus, the usefulness of this general method needs further confirmation.

Certain alfalfa genotypes can be readily transformed using *A. tumefaciens* or by direct DNA uptake with protoplasts to produce transgenic fertile plants that express the integrated genes. A limitation on the process is the genotype specificity of plant regeneration. *Agrobacterium rhizogenes* also can transform efficiently, but here again there is plant regeneration genotype specificity and the transformed plants have the generally undesirable hairy root phenotype of shortened and more highly branched roots.

The genotype specificity of transformation of alfalfa apparently may not be an important limitation because there is enough variation induced during tissue culture so that several backcrosses will be needed to eliminate this and produce commercially acceptable lines, according to Bingham & McCoy (1986).

Birdsfoot trefoil (*Lotus corniculatus*)

A number of forage legumes other than alfalfa have been transformed using *A. rhizogenes* and *A. tumefaciens*. There have been several reports with *L. corniculatus*, including that by Jensen *et al.* (1986), who infected wound sites of aseptic seedlings with *A. rhizogenes* strain 15834 carrying a *cat* gene driven by a soybean leghemoglobin promoter inserted into the T-region of the Ri plasmid. The roots that formed at the wound sites were cultured in the presence of 500 μg claforan/ml to eliminate the bacteria. Plants were regenerated apparently after callus induction with 2,4-D through somatic embryogenesis or organogenesis. The methods used to accomplish this were not explained. The regenerated plants were studied at the molecular level as far as integrated DNA structure and expression of the *cat* gene under the control of the soybean leghemoglobin promoter. The gene control was root nodule specific just as in the soybean plant.

Birdsfoot trefoil was also transformed by Petit *et al.* (1987): hypocotyls of aseptic seedlings were wounded with a needle or scalpel dipped in *A. rhizogenes* strains 15834 (agropine-type C58) or 8196 (mannopine-type). After 14 days, stem pieces with developing roots (up to 90% formed roots) were transferred to medium with 500 μg claforan/ml and then the roots were excised and cultured further on the same medium for 20–30 days in the dark. Shoots formed within 20–30 days in the light and could be excised and rooted to form complete plants. The process from inoculation to potting of plants took about 132 days. About 90% of the 30 transformed plants examined were transformed, since they synthesized opines and contained Ri DNA sequences. Root nodules appeared to have normal nitrogenase activities and transcript levels for nodule-specific leghemoglobin and constitutive ubiquitin. These plants apparently had the typical hairy root phenotype and have been used in basic studies of this phenotype (Shen *et al.*, 1988). These studies showed that, unlike *A. tumefaciens* where auxin and cytokinin synthesis carried by the Ti DNA caused the tumorous phenotypes, the Ri genes somehow make the transformed cells more sensitive to auxin.

In another study, seedling hypocotyl and stem segments from 1–2-month-old *L. corniculatus* plants were incubated with *A. rhizogenes* for 1–2 min, and after 2–3 days were then incubated with 300 μg cefotaxime/ml to eliminate the bacteria (Tabaeizadeh, 1989). The wild-type *A. rhizogenes* contained the binary pBin19 with *nos*, *npt*II and CaMV 35S, *cat*, *nos* gene constructs. Roots formed only on inoculated segments (15% and 35% of the stem and hypocotyl segments, respectively). Plant regeneration was obtained on regeneration medium containing up to 30 μg kanamycin/ml from about 50% of the hairy roots. Shoots could be regenerated from wild-type roots in up to 10 μg kanamycin/ml. Enzyme activities (CAT, NPT II) could be detected in transformed shoots and roots, and *cat* gene sequences were also present in the transformed plants. The process from inoculation to transformed plant production required only 2 months. These regenerated transformed *L. corniculatus* plants 'did not show any morphological differences with respect to leaf shape, plant height, etc. compared to seed grown plants' according to the author Tabaeizadeh (1989).

Hairy root cultures were also initiated by Webb *et al.* (1990) from stolon segments or seedlings of *L. corniculatus*, *T. repens* and *T. pratense*, following incubation with a wild-type *A. rhizogenes* strain C58C1 using a needle. Following decontamination by growth with several antibiotics during several subcultures, the cultures were maintained in liquid or on solid medium. The integrated copy number of the TL-DNA varied from one to eight, while the chromosome number was near normal. Plants could not be regenerated from the two *Trifolium* species but shoots regenerated spontaneously from the *L. corniculatus* hairy roots after 8 weeks in culture. Many of these plants had normal characteristics (height, stem thickness, tannin content, flower color, plant dry weight) but root morphology, chlorophyll content, leaf size and flower and seed production were altered. Nitrogen fixation was not affected.

Transgenic *L. corniculatus* plants containing a full-length cDNA clone of the soybean cytosolic glutamine synthetase, which is expressed in roots and root nodules, were obtained, using the methods of Petit *et al.* (1987), by Miao *et al.* (1991). By using a glutamine synthetase–*uid*A gene fusion construct the expression was followed by measuring GUS activity histochemically. The gene was expressed with the correct tissue specificity in *L. corniculatus* and tobacco, which was also transformed.

Rapidly growing birdsfoot trefoil hairy root cultures were obtained following transformation of 8–10-day-old aseptic seedling hypocotyls with wild-type *A. rhizogenes* (C58C1-pRi15834) (Morris & Robbins, 1992). Shoots formed readily upon illumination in liquid medium. These shoots could form fertile regenerated plants or could be maintained in liquid medium as shoot-organ cultures. Condensed tannin formation was studied in these cultures and in callus initiated from them.

Transformation of *L. corniculatus* was also accomplished using *A. tumefaciens* by Yu & Shao (1991). Cotyledons excised from axenic 5–10-day-old seedlings were cut in half, incubated for 10 min with *A. tumefaciens* strain C58C1 containing a selectable marker gene (*nos*, *npt*II) and a reporter gene (*nos*), blotted dry with filter paper and then cocultivated for 2 days on a medium with 0.1 mg BA/l. Selection was then carried out on the same medium with 100 mg kanamycin/l and bacterial growth inhibited by 300 μg cefotaxime/ml. Within 21 days, buds had formed on growing calli on 80 of the 200 treated cotyledon explants. Shoots longer than 3 cm were rooted in growth-regulator-free medium and were then transplanted to pots where seed was produced. All selected calli, shoots and plants contained nopaline. The one plant analyzed contained NPT II activity and T-DNA sequences as determined by dot blot. The progeny contained nopaline, indicating that the transforming DNA was passed to the next generation.

Birdsfoot trefoil thus can be readily transformed with both *A. rhizogenes* or *A. tumefaciens* to obtain fertile transformed plants. The success is due to the rapid regeneration of shoots directly from hairy roots and from callus. The plants regenerated from hairy roots generally do have the typical altered root phenotype, which is likely to affect field performance.

White clover (*Trifolium repens*)

A highly regenerable white clover genotype was transformed by placing *A. tumefaciens* on the cut ends of stolon internode segments followed by incubation for 1–3 days in a moist Petri dish (White & Greenwood, 1987). The ends were cut off and placed on callusing medium with 500 µg cefotaxime/ml to stop bacterial growth and 100 µg kanamycin/ml to select transformed plant cells. Kanamycin-resistant growth was obtained, with the four different binary vectors used, on 24%–44% of the treated segments with a 3-day cocultivation period. The callus was cultured on the antibiotic medium with 21-day subculture intervals for about 3 months, when plants were regenerated on regeneration medium with or without 100 µg kanamycin/ml. The regenerated shoots produced nopaline with certain constructs, had NPT II enzyme activity and contained one to two copies of the *npt*II gene sequences. One reason for the success of this system is the high regeneration capacity of the callus of the genotype used.

Hairy roots were also produced by Webb *et al.* (1990) from stolon segments or seedlings of *T. repens* and *T. pratense*, as described in the birdsfoot trefoil section, above. Whole plants could not be regenerated from these roots.

Stylosanthes species

Tumors were induced on leaf sections of the *Stylosanthes* species *humilis, hamata, guianensis* and *scabra* by *A. tumefaciens* strains A281 and C58 at relatively low frequencies, with no tumors being formed on some cultivars of the latter two species (Manners, 1987).

Stems and leaf sections from 4–5-week-old *S. humilis* seedlings grown under aseptic conditions were treated with an *A. rhizogenes* suspension for 2 min, blotted dry and incubated for up to 4 days

before washing and transfer to medium with 500 µg cefotaxime/ml to inhibit bacterial growth and 50 µg kanamycin/ml to select transformed cells (Manners & Way, 1989). The *A. rhizogenes* contained the binary plasmid pGA492 (*nos, nos–npt*II gene fusion). The highest frequency of transformed root production was 86% with stems, following a 3-day coincubation period in the dark. After 6 weeks the bacteria-free kanamycin-resistant transformed roots were placed on shoot regeneration medium with different BA concentrations, where shoots formed on about 23% of the roots with 2 mg BA/l. Transformed plants could be produced within about 18–24 weeks after inoculation.

Opines were found in some of the transformed roots but not in any of the regenerated plants, whereas NPT II activity was found in all roots and regenerated plants. Most of the plants had morphological abnormalities, with about half being dwarf and multistemmed. All plants were fertile and the dwarfness could be separated from kanamycin resistance in the next generation. Only one of the 25 regenerated plants had completely normal morphology and this plant was kanamycin resistant but lacked the TL-DNA sequences, indicating that the Ri virulence plasmid was not transferred in this case.

Lotononis bainesii

Lotononis bainesii is a subtropical forage legume that can be easily regenerated from tissue cultures (Bovo, Mroginski & Rey, 1986). Leaf discs were transformed (Wier *et al.*, 1988) by *A. tumefaciens* carrying a gene fusion of the *cat* gene and a synthetic DNA sequence that codes for a peptide containing relatively high concentrations of the essential amino acids lysine (16.7%, w/v), tryptophan (11.5%) and methionine (8.3%); methods used were similar to those developed for tobacco by Horsch *et al.* (1985). The chimeric fused gene was driven by the CaMV 19S promoter with the *nos* terminator. Transformed callus was selected when the discs were incubated after 48 h on tobacco nurse cultures on medium with 30 µg kanamycin sulfate/ml (the plasmid also carried the *npt*II gene) and 500 µg carbenicillin/ml to eliminate the bacteria. The callus was placed on fresh selective medium, where shoots grew and formed roots on 300 µg kanamycin/ml but untransformed shoots did not grow or form roots and had lowered chlorophyll levels. Seeds obtained from three of the transformed plants showed kanamycin resistance with segregation ratios fitting a dominant, single gene model. Amino acid analysis of whole plants showed some increases in certain of the expected amino acids in some plants.

Sainfoin (*Onobrychis viciifolia*)

Hypocotyls and cotyledons of aseptic 10–14-day-old sainfoin plants were inoculated by injecting *A. rhizogenes* into wounds and then were incubated in an upright position for 21–28 days in solid medium (Golds *et al.*, 1991). From 50% to 78% of the hypocotyls or cotyledons, respectively, formed hairy roots with strain A4T, which carried pRiA46. Inoculated leaves were unresponsive. Most (70% to 80%) of the roots would grow rapidly on a medium with 500 μg cefotaxime/ml. Shoots regenerated spontaneously from some of the root cultures to form plants that were shorter and had smaller leaves and prolific, irregularly branched masses of fine negatively geotropic roots. The transformed roots contained agropine and mannopine but the regenerated plants did not. Southern hybridization confirmed the presence of TL- and TR-DNA sequences in the regenerated plants.

Transformed plants can be readily obtained with birdsfoot trefoil, white clover, *S. humilus*, *L. bainesii* and sainfoin as described in this section using *Agrobacterium* systems and either shoot regeneration from hairy roots or plant regeneration from callus. The transformation success depends greatly on the relative ease of obtaining regeneration from these systems.

Summary and conclusions

A summary of the published results for each species discussed here is given in Table 8.1. A number of the spaces are blank because it is difficult to determine the category to use on the basis of the results presented. This summary is given to show what is reported at this time and many changes would be expected in the future as more progress is made.

The grain legumes as a group have produced transformed progeny in the cases of soybean, dry bean, pea and *V. narbonensis* (Table 8.1). Trans-

formed plants have also been produced by moth bean but there are no reports of progeny. The other grain legumes listed, broad bean, lentil, cowpea and peanut, have not produced transformed plants. The deficiency with the grain legumes lies predominantly with the recalcitrance of these species in plant regeneration. One would anticipate that further effort will surmount this problem, just as it has with soybean where about 10 years ago some workers felt that this species could not regenerate plants from tissue culture. Since then, callus, suspension and protoplast regeneration systems have been developed.

The soybean transformation is not easy and all of the successful techniques, *A. tumefaciens* infection, microprojectile bombardment and protoplast electroporation, have shortcomings. The *A. tumefaciens* system is low frequency, produces chimeric plants with many escapes and is very genotype specific. The microprojectile system is not genotype specific but requires large numbers of immature embryos, is low frequency, produces chimeric plants and uses expensive equipment and supplies, and royalties also apply if a commercial product is produced. If microprojectiles are used to transform suspension cultures then the embryogenic culture must first be initiated, which is difficult. The use of suspensions improves the selection so that nonchimeric plants can be recovered, but again the equipment and royalty costs can be high. The protoplast system should be applicable to about half the genotypes, is slow (8 months or more) but should produce nonchimeric, fertile plants, although this has not been confirmed as yet. Certainly improvements may be made in all of the systems to make them more ideal.

There has been better success with the forage legumes where alfalfa and birdsfoot trefoil can be readily transformed to produce regenerated, fertile plants. The one problem with alfalfa is that plant regeneration shows genotype specificity so that the best commercial genotypes cannot be transformed easily.

The other four forage legumes listed, white clover, *Stylosanthes*, *L. bainesii* and sainfoin, have been successfully transformed to produce plants and those of *Stylosanthes* and *L. bainesii* have produced seed. The most apparent reason for the better success with the forage, in contrast to that with the grain legumes, is the superior plant regeneration capabilities of the former.

Table 8.1. *Summary of transformation results with leguminous plants*

Species	Method[a]	Starting material	Genotype specific?	Regenerate plants?	Plants fertile?	Remarks
Soybean	*A.t.*	Seedling	Yes	Yes	Yes	
	A.r.	Seedling	Intermed.	No		
	Proj.	Immature seed	No	Yes	Yes	
	Proj.	Suspension culture	No	Yes	Yes	
	Proto.	Immature seed	Intermed.	Yes	Yes	Plants not transformed
Pea	*A.t.*	Plant	Intermed.	Yes	No	
	A.r.	Plant	Yes	No		
	Proto.	Seedling	Yes	No		
Moth bean	Proto.	Seedling	Yes	Yes	No	
Faba bean	*A.r.*	Seedling		No		
Vicia narbonensis	*A.t.*	Seedling		Yes	Yes	
Lentil	*A.t.*	Seedling	No			
Cowpea	*A.t.*	Seedling		No		
Dry bean	*A.t.*	Seedling	No	No		
	A.r.	Seedling		No		
	Proj.	Mature seed	No	Yes	Yes	
Peanut	*A.t.*	Plant		No		
Alfalfa	*A.t.*	Seedling	Yes	Yes	Yes	
	A.r.	Plants	Yes	Yes	Yes	
	Proto.	Leaves		Yes		
Birdsfoot trefoil	*A.t.*	Seedling		Yes	Yes	Rapid
	A.r.	Seedling		Yes	Yes	Rapid, shoots directly from root
White clover	*A.t.*	Stolon	Yes	Yes		
	A.r.	Stolon Seedling		No		
Stylosanthes	*A.t.*	Leaves	Yes	No		
	A.r.	Plant		Yes	Yes	
Lotononis bainesii	*A.t.*	Leaves		Yes	Yes	
Sainfoin	*A.t.*	Seedling	Yes			

Notes:
[a] *A.t., Agrobacterium tumefaciens; A.r., Agrobacterium rhizogenes*; Proj., microprojectile bombardment; Proto., protoplast direct DNA uptake.

While there have been no reports of the production of transformed forage legume plants using protoplast systems, there have been many reports describing the regeneration of plants from protoplasts, so the technique should be possible. Plants have been regenerated from protoplasts of *S. guianensis, L. corniculatus* and *T. repens* (Webb, Woodcock & Chamberlain, 1987) among others, as reviewed by Vieira *et al.* (1990).

As with many other species, some legumes can form shoots from hairy roots, although only birdsfoot trefoil seems to form shoots directly without an intervening callus phase. In a review by Tepfer (1990), a total of 116 dicotyledonous species are listed that have been transformed with *A. rhizo-*

genes; transformed plants have been regenerated from 37 of them. However, not all of the plant regeneration was direct formation from the root.

In the transformation systems that utilize *A. rhizogenes* for gene insertion, almost all of the regenerated plants carry the Ri T-DNA, which imparts certain abnormal characteristics to the plants, including shortened and enlarged roots. While this type of root, which permeates a smaller soil mass than normal, might be advantageous under certain conditions, in general this phenotype would be detrimental.

It should be possible with binary Ri plasmid systems to separate the integrated Ri plasmid from the integrated binary vector DNA by genetic segregation if these sequences are not closely linked on the chromosome. This has been accomplished with one Ti system (de Framond *et al.*, 1986). This separation could eliminate the detrimental hairy root phenotype but retain the useful gene.

Most legumes, being dicots, are sensitive to both *A. tumefaciens* and *A. rhizogenes*, although there can be some host genotype specificity, as well as bacterial strain specificity. Overall, however, relatively good infection rates can be attained with the right combinations of plant genotype and *Agrobacterium* strain.

In several places in this chapter the production of chimeric plants was described. This conclusion was reached because only certain sectors of the transformed plants expressed the transformed gene and in a few documented cases only seeds from these sectors produced plants that also expressed the transformed genes. This information does not allow a determination of whether or not the nonexpressing tissue actually contains the transformed gene in an inactive state. Christou (1990) showed, however, that, when some of the GUS nonexpressing tissues in chimeric soybean plants were analyzed, the *uid*A gene was absent. Thus, only in rare cases has the absence of the transformed gene been documented in chimeric plants. The most correct description of these plants would be that they are chimeric for expression of the transformed genes unless further molecular studies have been carried out.

While somaclonal variation does occur with most tissue culture systems, the results presented here and our results with soybean protoplasts and organogenic and embryogenic cultures (Barwale & Widholm, 1987) indicate that somaclonal variation occurs at a low enough frequency that the few variants can be discarded with little harm. The remaining normal-appearing soybean families do not show any detrimental agronomic characteristics in field trials (Stephens, Nickell & Widholm, 1991). One should therefore always regenerate a number of plants (10–20?) so that the undesirable ones can be discarded. This is especially important from the standpoint of gene expression, since this is known to often be variable in different individuals. Enough plants need to be available to give a choice in order to find the desired expression level. In the case of alfalfa, however, Bingham & McCoy (1986) felt that the somaclonal variation is so great in any regenerated plants that backcrossing is required in all cases.

There can be some debate about the effect of how the genes are integrated because the *Agrobacterium* vectors integrate a more or less set piece of DNA, in contrast to the direct DNA uptake and microprojectile systems where random pieces and concatemers can be integrated. This discussion can best be carried out with other systems, since the data with legumes are rather limited at the moment, although successful gene expression in progeny has certainly been observed with both systems.

There are many general traits that could be improved in any crop plant including disease, drought, heat, cold, salt and insect resistance, depending upon the growing area. Some of these can already be targeted since there are antifungal (chitinase, β-glucanase) and insecticidal (*Bacillus thuringiensis* endotoxins) genes available. Most of the other traits cannot be approached presently, since the genes controlling these traits are not available.

Nutritional quality is also very important to both grain and forage legumes. Legume seeds are generally deficient in sulfur amino acids so the addition of proteins high in methionine is being attempted (Parrott *et al.*, 1989).

There have already been a number of transgenic plants produced with genes that could improve the crop. These would include soybean with glyphosate (Hinchee *et al.*, 1988) or atrazine resistance (Liu *et al.*, 1990), alfalfa with Basta resistance (D'Halluin *et al.*, 1990) or chicken ovalbumin expression (Schroeder *et al.*, 1991), dry bean with Basta resistance (Russell *et al.*, 1993) and *L. bainesii* with a synthetic gene encoding a protein with a high lysine, tryptophan and methionine content (Bovo *et al.*, 1986). Many of the transgenic plants also synthesize opines so might be useful in testing the 'opine concept' (Petit *et al.*, 1983), which postulates that opines produced by plants act as chemical mediators of parasitism between *Agrobacterium* and the plants. A symbiosis might be encouraged if beneficial bacteria contained the opine utilization genes.

Basic studies include the integration of transposable elements into soybean (Zhou & Atherly, 1990) and alfalfa (Sukhapinda *et al.*, 1987) and studies of tissue-specific expression of the soybean leghemoglobin gene (Jensen *et al.*, 1986) and the glutamine synthetase gene (Miao *et al.*, 1991), both in birdsfoot trefoil. Basic studies on the auxin sensitivity of hairy roots were also done with birdsfoot trefoil (Shen *et al.*, 1988). Facciotti *et al.* (1985) studied light induction of a chimeric soybean small subunit gene in soybean callus transformed by *A. tumefaciens*.

While the transformation progress with legumes has been slow, we now have some systems that are successful and more progress is expected. Thus, there should be clear evidence of crop improvement in the next few years. Then questions about which new genes to use, degree of stability of gene expression and the effect of these genes on plant performance will have to be addressed.

Acknowledgements

I thank E. T. Bingham and Guangbin Luo for reading the manuscript and for making useful suggestions, and T. Eriksson, J. Puonti-Kaerlas, O. Schieder and D. R. Russell for supplying manuscripts in press or unpublished information. The unpublished data were obtained with the support of the Illinois Agricultural Experiment Station and the Illinois Soybean Program Operating Board.

References

Aird, E. L. H., Hamill, J. D. & Rhodes, M. J. C. (1988). Cytogenetic analysis of hairy root cultures from a number of plant species transformed by *Agrobacterium rhizogenes*. *Plant Cell, Tissue and Organ Culture*, 15, 47–57.

Atanassov, A. & Brown, D. C. W. (1984). Plant regeneration from suspension culture and mesophyll protoplasts of *Medicago sativa* L. *Plant Cell, Tissue and Organ Culture*, 3, 149–162.

Barwale, U. B. & Widholm, J. M. (1987). Somaclonal variation in plants regenerated from cultures of soybean. *Plant Cell Reports*, 6, 365–368.

Bercetche, J., Chriqui, D., Adam, S. & David, C. (1987). Morphogenesis and cellular reorientations induced by *Agrobacterium rhizogenes* (strains 1855, 2659 and 8196) on carrot, pea and tobacco. *Plant Science*, 52, 195–210.

Bingham, E. T. (1989). Registration of Regen-S alfalfa germplasm useful in tissue culture and transformation research. *Crop Science*, 29, 1095–1096.

Bingham, E. T. (1991). Registration of alfalfa hybrid Regen-SY germplasm for tissue culture and transformation research. *Crop Science*, 31, 1098.

Bingham, E. T., Hurley, L. V., Kaatz, D. M. & Saunders, J. W. (1975). Breeding alfalfa which

regenerates from callus tissue in culture. *Crop Science*, 15, 719–721.

Bingham, E. T. & McCoy, T. J. (1986). Somaclonal variation in Alfalfa. *Plant Breeding Reviews*, 4, 123–152.

Bovo, O. A., Mroginski, L. A. & Rey, H. Y. (1986). Regeneration of plants from callus tissue of the pasture legume *Lotononis bainesii*. *Plant Cell Reports*, 5, 295–297.

Byrne, M. C., McDonnell, R. E., Wright, M. S. & Carnes, M. G. (1987). Strain and cultivar specificity in the *Agrobacterium* – soybean interaction. *Plant Cell, Tissue and Organ Culture*, 8, 3–15.

Chabaud, M., Passiatore, J. E., Cannon, F. & Buchanan-Wollaston, V. (1988). Parameters affecting the frequency of kanamycin resistant alfalfa obtained by *Agrobacterium tumefaciens* mediated transformation. *Plant Cell Reports*, 7, 512–516.

Chee, P. O., Fober, K. A. & Slightom, J. L. (1989). Transformation of soybean (*Glycine max*) by infecting germinating seeds with *Agrobacterium tumefaciens*. *Plant Physiology*, 91, 1212–1218.

Christou, P. (1990). Morphological description of transgenic soybean chimeras created by the delivery, integration and expression of foreign DNA using electric discharge particle acceleration. *Annals of Botany*, 66, 379–386.

Christou, P. & McCabe, D. E. (1992). Prediction of germ-line transformation events in chimeric R0 transgenic soybean plantlets using tissue-specific expression patterns. *Plant Journal*, 2, 283–290.

Christou, P., McCabe, D. E. & Swain, W. F. (1988). Stable transformation of soybean callus by DNA-coated gold particles. *Plant Physiology*, 87, 671–674.

Christou, P., Swain, W. F., Yang, N. S. & McCabe, D. E. (1989). Inheritance and expression of foreign genes in transgenic soybean plants. *Proceedings of the National Academy of Sciences, USA*, 86, 7500–7504.

de Framond, A. J., Back, E. W., Chilton, W. S., Kayes, L. & Chilton, M. D. (1986). Two unlinked T-DNAs can transform the same tobacco plant cell and segregate in the F1 generation. *Molecular and General Genetics*, 202, 125–131.

D'Halluin, K., Botterman, J. & De Greef, W. (1990). Engineering of herbicide-resistant alfalfa and evaluation under field conditions. *Crop Science*, 30, 866–871.

Dhir, S. K., Dhir, S., Hepburn, A. & Widholm, J. M. (1991a). Factors affecting transient gene expression in electroporated *Glycine max* protoplasts. *Plant Cell Reports*, 10, 106–110.

Dhir, S. K., Dhir, S., Savka, M. A., Belanger, F., Kriz, A. L., Farrand, S. K. & Widholm, J. M. (1992a). Regeneration of transgenic soybean (*Glycine max*) plants from electroporated protoplasts. *Plant Physiology*, 99, 81–88.

Dhir, S. K., Dhir, S., Sturtevant, A. P. & Widholm,

J. M. (1991b). Regeneration of transformed shoots from electroporated soybean (*Glycine max* (L.) Merr.) protoplasts. *Plant Cell Reports*, **10**, 97–101.

Dhir, S. K., Dhir, S. & Widholm, J. M. (1991c). Plantlet regeneration from cotyledonary protoplasts of soybean (*Glycine max* (L.) Merr.). *Plant Cell Reports*, **10**, 39–43.

Dhir, S. K., Dhir, S. & Widholm, J. M. (1992b). Regeneration of fertile plants from protoplasts of soybean (*Glycine max* L.): genotypic differences in culture response. *Plant Cell Reports*, **11**, 285–289.

Eapen, S., Kohler, F., Gerdemann, M. & Schieder, O. (1987). Cultivar dependence of transformation rates in moth bean after co-cultivation of protoplasts with *Agrobacterium tumefaciens*. *Theoretical and Applied Genetics*, **75**, 207–210.

Facciotti, D., O'Neal, J. K., Lee, S. & Shewmaker, C. K. (1985). Light-inducible expression of a chimeric gene in soybean tissue transformed with *Agrobacterium*. *Bio/Technology*, **3**, 241–246.

Finer, J. J. & McMullen, M. D. (1991). Transformation of soybean via particle bombardment of embryogenic suspension culture tissue. *In Vitro Cellular and Developmental Biology*, **27P**, 175–182.

Finer, J. J. & Nagasawa. A. (1988). Development of an embryogenic suspension culture of soybean [*Glycine max* (L.) Merrill]. *Plant Cell, Tissue and Organ Culture*, **15**, 125–136.

Gamborg, O. L., Miller, R. A. & Ojima, K. (1968). Nutrient requirements of suspension cultures of soybean root cells. *Experimental Cell Research*, **50**, 148–151.

Garcia, J. A., Hille J. & Goldbach, R. (1986). Transformation of cowpea (*Vigna unguiculata*) cells with an antibiotic resistance gene using a Ti plasmid-derived vector. *Plant Science*, **44**, 37–46.

Garcia, J. A., Hille, J., Vos, P. & Goldbach, R. (1987). Transformation of cowpea *Vigna unguiculata* with a full-length DNA copy of cowpea mosaic virus m-RNA. *Plant Science*, **48**, 89–98.

Golds, T. J., Lee, J. Y., Husnain, T., Ghose, T. K. & Davey, M. R. (1991). *Agrobacterium rhizogenes* mediated transformation of the forage legumes *Medicago sativa* and *Onobrychis viciifolia*. *Journal of Experimental Botany*, **42**, 1147–1157.

Hinchee, M. A. W., Conner-Ward, D. V., Newell, C. A., McDonnell, R. E., Sato, S. J., Gasser, C. S., Fischhoff, D. A., Re, D. B., Fraley, R. T. & Horsch, R. B. (1988). Production of transgenic soybean plants using *Agrobacterium*-mediated DNA transfer. *Bio/Technology*, **6**, 915–922.

Hobbs, S. L. A., Jackson, J. A., Baliski, D. S., DeLong, C. M. O. & Mahon, J. D. (1990). Genotype- and promoter-induced variability in transient β-glucuronidase expression in pea protoplasts. *Plant Cell Reports*, **9**, 17–20.

Hobbs, S. L. A., Jackson, J. A. & Mahon, J. D. (1989).

Specificity of strain and genotype in the susceptibility of pea to *Agrobacterium tumefaciens*. *Plant Cell Reports*, **8**, 274–277.

Horsch, R. B., Fry, J. E., Hoffmann, N. L., Eichholtz, D., Rogers, S. G. & Fraley, R. J. (1985). A simple and general method for transferring genes into plants. *Science*, **227**, 1229–1231.

Hussey, G., Johnson, R. D. & Warren, S. (1989). Transformation of meristematic cells on the shoot apex of cultured pea shoots by *Agrobacterium tumefaciens* and *A. rhizogenes*. *Protoplasma*, **148**, 101–105.

Jensen, J. S., Marcker, K. A., Otten, L. & Schell, J. (1986). Nodule-specific expression of a chimaeric soybean leghaemoglobin gene in transgenic *Lotus corniculatus*. *Nature (London)*, **321**, 669–674.

Kao, K. N. & Michayluk, M. R. (1980). Plant regeneration from mesophyll protoplasts of alfalfa. *Zeitschrift für Pflanzenphysiologie*, **96**, 135–141.

Kathen, A. D. & Jacobsen, H. J. (1990). *Agrobacterium tumefaciens*-mediated transformation of *Pisum sativum* L. using binary and cointegrate vectors. *Plant Cell Reports*, **9**, 276–279.

Köhler, F., Cardon, G., Pohlman, M., Gill, R. & Schieder, O. (1989). Enhancement of transformation rates in higher plants by low-dose irradiation: are DNA repair systems involved in the incorporation of exogenous DNA into the plant genome? *Plant Molecular Biology*, **12**, 189–199.

Köhler, F., Golz, C., Eapen, S., Kohn, H. & Schieder, O. (1987a). Stable transformation of moth bean *Vigna aconitifolia* via direct gene transfer. *Plant Cell Reports*, **6**, 313–317.

Köhler, F., Golz, C., Eapen, S. & Schieder O. (1987b). Influence of plant cultivar and plasmid-DNA on transformation rates in tobacco and moth bean. *Plant Science*, **53**, 87–91.

Kuchuk, N., Komarnitski, I., Shakhovsky, A. & Gleba, Y. (1990). Genetic transformation of *Medicago* species by *Agrobacterium tumefaciens* and electroporation of protoplasts. *Plant Cell Reports*, **8**, 660–663.

Kumar, V., Jones, B. & Davey, M. R. (1991). Transformation by *Agrobacterium rhizogenes* and regeneration of transgenic shoots of the wild soybean *Glycine argyrea*. *Plant Cell Reports*, **10**, 135–138.

Kysely, W., Myers, J. R., Lazzeri, P. A., Collins, G. B. & Jacobsen, H. J. (1987). Plant regeneration via somatic embryogenesis in pea (*Pisum sativum* L.). *Plant Cell Reports*, **6**, 305–308.

Lacorte, C., Mansur, E., Timmerman, B. & Cordeiro, A. R. (1991). Gene transfer into peanut (*Arachis hypogaea* L.) by *Agrobacterium tumefaciens*. *Plant Cell Reports*, **10**, 354–357.

Larkin, P. J., Taylor, B. H., Gersmann, M. & Brettell, R. I. S. (1990). Direct gene transfer to protoplasts. *Australian Journal of Plant Physiology*, **17**, 291–302.

Ledoux, L. & Huart, R. (1968). Integration and replication of DNA of *M. lysodeikticus* in DNA of germinating barley. *Nature (London)*, **218**, 1256–1259.

Lehminger-Mertens, R. & Jacobsen, H. J. (1989). Plant regeneration from pea protoplasts via somatic embryogenesis. *Plant Cell Reports*, **8**, 379–382.

Liu, B., Yue, S., Hu, N., Li, X., Zhai, W., Li, N., Zhu, R., Zhu, L., Mao, D. & Zhou, P. (1990). Transfer of the atrazine-resistance gene of black nightshade to soybean chloroplast genome and its expression in transgenic plants. *Science in China B*, **33**, 444–452.

Lulsdorf, M. M., Rempel, H., Jackson, J. A., Baliski, D. S. & Hobbs, S. L. A. (1991). Optimizing the production of transformed pea (*Pisum sativum* L.) callus using disarmed *Agrobacterium tumefaciens*. *Plant Cell Reports*, **9**, 479–483.

Luo, X., Zhao, G. & Tian, Y. (1990) Plant regeneration from protoplasts of soybean (*Glycine max* L.). *Acta Botanica Sinica*, **32**, 616–621.

Manners, J. M. (1987). Transformation of *Stylosanthes* spp. using *Agrobacterium tumefaciens*. *Plant Cell Reports*, **6**, 204–207.

Manners, J. M. & Way, H. (1989). Efficient transformation with regeneration of the tropical pasture legume *Stylosanthes humilis* using *Agrobacterium rhizogenes* and a Ti plasmid-binary vector system. *Plant Cell Reports*, **8**, 341–345.

McCabe, D. E., Swain, W. F., Martinell, B. J. & Christou, P. (1988). Stable transformation of soybean (*Glycine max*) by particle acceleration. *Bio/Technology*, **6**, 923–926.

McClean, P., Chee, P., Held, B., Simental, J., Drong R. F. & Slightom, J. (1991). Susceptibility of dry bean (*Phaseolus vulgaris* L.) to *Agrobacterium* infection: transformation of cotyledonary and hypocotyl tissues. *Plant Cell, Tissue and Organ Culture*, **24**, 131–138.

Miao, G. H., Hirel, B., Marsolier, M. C., Ridge, R. W. & Verma, D. P. S. (1991). Ammonia-regulated expression of a soybean gene encoding cytosolic glutamine synthetase in transgenic *Lotus corniculatus*. *Plant Cell*, **3**, 11–22.

Morris, P. & Robbins, M. P. (1992). Condensed tannin formation by *Agrobacterium rhizogenes* transformed root and shoot organ cultures of *Lotus corniculatus*. *Journal of Experimental Botany*, **43**, 221–231.

Murashige, T. & Skoog, F. (1962). A revised medium for rapid growth and bioassays with tobacco tissue cultures. *Physiologia Plantarum*, **15**, 473–497.

Nauerby, B., Madsen, M., Christiansen, J. & Wyndaele R. (1991). A rapid and efficient regeneration system for pea (*Pisum sativum*), suitable for transformation. *Plant Cell Reports*, **9**, 676–679.

Nisbet, G. S. & Webb, K. J. (1990). Transformation in legumes. In *Biotechnology in Agriculture and Forestry*, Vol. 10, *Legumes and Oilseed Crops* I, ed. Y. P. S. Bajaj, pp. 38–48. Springer-Verlag, Berlin.

Owens, L. D. & Cress, D. E. (1985). Genotypic variability of soybean response to *Agrobacterium* strains harboring the Ti or Ri plasmids. *Plant Physiology*, **77**, 87–94.

Parrott, W. A., Hoffman, L. M., Hildebrand, D. F., Williams, E. G. & Collins, G. B. (1989). Recovery of primary transformants of soybean. *Plant Cell Reports*, **7**, 615–617.

Penza, R., Lurquin P. F. & Fillippone, E. (1991). Gene transfer by cocultivation of mature embryos with *Agrobacterium tumefaciens*: applications to cowpea (*Vigna unguiculata* Walp). *Journal of Plant Physiology*, **138**, 39–43.

Petit, A., David, C., Dahl, G., Ellis, J., Guyon, P., Casse Delbart, F. & Tempé, J. (1983). Further extension of the opine concept: plasmids in *Agrobacterium rhizogenes* cooperate for opine degradation. *Molecular and General Genetics*, **190**, 204–214.

Petit, A., Stougaard, J., Kiihle, A., Marcker, K. A. & Tempé, J. (1987). Transformation and regeneration of the legume *Lotus corniculatus*: a system for molecular studies of symbiotic nitrogen fixation. *Molecular and General Genetics*, **207**, 245–250.

Pickardt, T., Huancaruna Perales, E. & Schieder, O. (1989). Plant regeneration via somatic embryogenesis in *Vicia narbonensis*. *Protoplasma*, **149**, 5–10.

Pickardt, T., Meixner, M., Schade, V. & Schieder, O. (1991). Transformation of *Vicia narbonensis* via *Agrobacterium*-mediated gene transfer. *Plant Cell Reports*, **9**, 535–538.

Pickardt, T., Saalbach, I., Machemehl, F., Saalbach, G. Schieder, O. & Muntz, K. (1992). Expression of the sulfur-rich Brazil nut 2S albumin in transgenic *Vicia narbonensis* plants. *In Vitro Cellular Developmental Biology*, Suppl. **28**, 92A.

Puonti-Kaerlas, J. & Eriksson, T. (1988). Improved protoplast culture and regeneration of shoots in pea (*Pisum sativum* L.). *Plant Cell Reports*, **7**, 242–245.

Puonti-Kaerlas, J., Eriksson, T. & Engström, P. (1990). Production of transgenic pea (*Pisum sativum* L.) plants by *Agrobacterium tumefaciens*-mediated gene transfer. *Theoretical and Applied Genetics*, **80**, 246–252.

Puonti-Kaerlas, J., Eriksson, T. & Engström, P. (1992a). Inheritance of a bacterial hygromycin phosphotransferase gene in the progeny of primary transgenic pea plants. *Theoretical and Applied Genetics*, **84**, 443–450.

Puonti-Kaerlas, J., Ottosson, A. & Eriksson, T. (1992b). Survival and growth of pea protoplasts after transformation by electroporation. *Plant Cell, Tissue and Organ Culture*, **30**, 141–148.

Puonti-Kaerlas, J., Stabel, P. & Eriksson, T. (1989).

Transformation of pea (*Pisum sativum* L.) by *Agrobacterium tumefaciens*. *Plant Cell Reports*, **8**, 321–324.

Ramsey, G. & Kumar, A. (1990). Transformation of *Vicia faba* cotyledon and stem tissues by *Agrobacterium rhizogenes*: infectivity and cytological studies. *Journal of Experimental Botany*, **41**, 841–847.

Rech, E. L., Golds, T. J., Hammatt, N., Mulligan, B. J. & Davey, M. R. (1988). *Agrobacterium rhizogenes* mediated transformation of the wild soybeans *Glycine canescens* and *G. clandestina*: production of transgenic plants of *G. canescens*. *Journal of Experimental Botany*, **39**, 1225–1285.

Rech, E. L., Golds, T. J., Husnain, T., Vainstein, M. H., Jones, B., Hammatt, N., Mulligan, B. J. & Davey, M. R. (1989). Expression of a chimaeric kanamycin resistance gene introduced into the wild soybean *Glycine canescens* using a cointegrate Ri plasmid vector. *Plant Cell Reports*, **8**, 33–36.

Russell, D. R., Wallace, K. M., Bathe, J. H., Martinell, B. J. & McCabe, D. E. (1993). Stable transformation of *Phaseolus vulgaris* via electric-discharge mediated particle acceleration. *Plant Cell Reports*, **12**, 165–169.

Savka, M. A., Ravillion, B., Noel, G. R. & Farrand, S. K. (1990). Induction of hairy roots on cultivated soybean genotypes and their use to propagate the soybean cyst nematode. *Phytopathology*, **80**, 503–508.

Schiemann, J. & Eisenreich, G. (1989). Transformation of field bean (*Vicia faba* L.) cells: expression of a chimaeric gene in cultured hairy roots and root-derived callus. *Biochemie und Physiologie der Pflanzen*, **185**, 135–140.

Schroeder, H. E., Khan, M. R. I., Knibb, W. R., Spencer, D. & Higgins, T. J. V. (1991). Expression of a chicken ovalbumin gene in three lucerne cultivars. *Australian Journal of Plant Physiology*, **18**, 495–505.

Senaratna, T., MacKensie, B. D., Kasha, K. J. & Procunier, J. D. (1991). Direct DNA uptake during the imbibition of dry cells. *Plant Science*, **79**, 223–228.

Shahin, E. A., Spielmann, A., Sukhapinda, K., Simpson, R. B. & Yashar, M. (1986). Transformation of cultivated alfalfa using disarmed *Agrobacterium tumefaciens*. *Crop Science*, **26**, 1235–1239.

Shen, W. H., Petit, A., Guern, J. & Tempé, J. (1988). Hairy roots are more sensitive to auxin than normal roots. *Proceedings of the National Academy of Sciences, USA*, **85**, 3417–3421.

Spano, L., Mariotti, D., Pezzotti, M., Damiani, F. & Arcioni, S. (1987). Hairy root transformation in alfalfa (*Medicago sativa* L.). *Theoretical and Applied Genetics*, **73**, 523–530.

Stephens, P. A., Nickell, C. D. & Widholm, J. M.

(1991). Agronomic evaluation of tissue-cultured-derived soybean plants. *Theoretical and Applied Genetics*, **82**, 633–635.

Sukhapinda, K., Spivey R. & Shahin, E. A. (1987). Ri-plasmid as a helper for introducing vector DNA into alfalfa plants. *Plant Molecular Biology*, **8**, 209–216.

Tabaeizadeh, Z. (1989). Genetic transformation of a pasture legume, *Lotus corniculatus*, L. (Birdsfoot trefoil). *Biotechnology Letters*, **11**, 411–416.

Tepfer, D. (1990). Genetic transformation using *Agrobacterium rhizogenes*. *Physiologia Plantarum*, **79**, 140–146.

Vieira, M. L. C., Jones, B., Cocking, E. C. & Davey, M. R. (1990). Plant regeneration from protoplasts isolated from seedling cotyledons of *Stylosanthes guianensis*, *S. macrocephala* and *S. scabra*. *Plant Cell Reports*, **9**, 289–292.

Warkentin, T. D. & McHughen, A. (1991). Crown gall transformation of lentil (*Lens culinaris* Medik.) with virulent strains of *Agrobacterium tumefaciens*. *Plant Cell Reports*, **10**, 489–493.

Webb, K. J., Jones, S., Robbins M. P. & Minchin, F. R. (1990). Characterization of transgenic root cultures of *Trifolium repens*, *Trifolium pratense* and *Lotus corniculatus* and transgenic plants of *Lotus corniculatus*. *Plant Science*, **70**, 243–254.

Webb, K. J., Woodcock, S. & Chamberlain, D. A. (1987). Plant regeneration from protoplasts of *Trifolium repens* and *Lotus corniculatus*. *Plant Breeding*, **98**, 111–118.

Wei, Z. & Xu, Z. (1988). Plant regeneration from protoplasts of soybean (*Glycine max* L.). *Plant Cell Reports*, **7**, 348–351.

White, D. W. R. & Greenwood, D. (1987). Transformation of the forage legume *Trifolium repens* L. using binary *Agrobacterium* vectors. *Plant Molecular Biology*, **8**, 461–469.

Wier, A. T., Thro, A. J., Flores, H. E. & Janyes, J. M. (1988). Transformation of *Lotononis bainesii* Baker with a synthetic protein gene using the leaf disk transformation-regeneration method. *Phyton*, **48**, 123–131.

Williams, D. J. & McHughen, A. (1986). Plant regeneration of the legume *Lens culinaris* Medik. (lentil) in vitro. *Plant Cell, Tissue and Organ Culture*, **7**, 149–153.

Yang, N. S. & Christou P. (1990). Cell type specific expression of a CaMV 35S-gus gene in transgenic soybean plants. *Developmental Genetics*, **11**, 289–293.

Yu, J. P. & Shao, Q. Q. (1991). Transformation of *Lotus corniculatus* L. mediated by *Agrobacterium tumefaciens*. *Science in China B*, **34**, 932–937.

Zhou, J. H. & Atherly, A. G. (1990). In situ detection of transposition of the maize controlling element (*Ac*) in transgenic soybean tissues. *Plant Cell Reports*, **8**, 542–545.

9

Spring and Winter Rapeseed Varieties

Philippe Guerche and Catherine Primard

Introduction

Rapeseed is one of the most important oilseed crops after soybeans and cottonseed, representing 10% of world oilseed production in 1990 (McCalla & Carter, 1991). It has been domesticated for more than 3000 years, along with the other *Brassica* species having noteworthy economic value, in Asia and in Europe (where rapeseed was first mentioned as a crop in the thirteenth century). It has been introduced into Canada only very recently (1942). The production of oilseeds, meal and oil has been increasing continuously for the last 30 years for food and feed grains, mainly by expansion of the area under cultivation (Röbbelen, 1991). By the year 2000, China should be the leading producer with 9.2 Mt (26%), followed by India with 7.8 Mt (22%), European Community (12 countries), 7.6 Mt (21%), Canada 3.8 Mt (11%) and eastern Europe 2.6 Mt (7%) (Carr & MacDonald, 1991).

Oilseed rape is a cruciferous species, belonging to the genus *Brassica* (Tribe Brassiceae, Family Brassicaceae), resulting from the natural hybridization between a cabbage (*B. oleracea* L., CC, $2n = 18$, Western Europe and Northwest Africa) and a turnip rape (*B. campestris* L., AA, $2n = 20$, Europe and Asia). Rape is an amphidiploid (*B. napus*, AACC, $2n = 38$), whose center of diversity is the intersection of its parental areas. Artificial crosses (Chèvre *et al.*, 1991) or somatic fusions have been used to introgress genes from related species (for a review, see Trail, Richards & Wu, 1989). The development of a detailed genetic linkage map (isozymes, restriction fragment length polymorphism (RFLP), random amplified

polymorphic DNA (RAPD)) is of interest for analysis of genetic relationships between species.

Brassica napus is a herbaceous plant that can reach 2 m in height. The flowers and pods are produced in a mass concentrated near the top of the plant, ensuring a high harvest efficiency using modern management and harvest practices. As a result of the long history of breeding efforts focused on this crop, rapeseed is the only polyploid *Brassica* offering spring as well as winter cultivars. In western Europe, winter rapeseed completes its cycle in 250–310 days and spring rapeseed in only 120–150 days, the former having the higher yield (around 4 tons/ha) (Renard *et al.*, 1992). Rapeseed is a semi-autogamous species. Pollination is by wind and insects (bees), drawn by the nectar of its bright yellow flowers. Outcrossing is estimated to be 10%–30% in the absence of incompatibility genes. Thus, breeders could aim to create a number of varietal types, such as populations or pure lines, but synthetic varieties or hybrids (provided an allogamization system is available) are better able to exploit heterosis.

Oil and proteins are the important rapeseed products, accounting for 40%–46% and 20%–28%, respectively, of the rapeseed dry matter, followed quantitatively by carbohydrates (20%), of which starch accounts for only 2%–3%.

Rapeseed edible oil is of high quality with a high unsaturated fatty acid content (mainly oleic acid (60%) and linoleic acid (22%)). The oil also has some qualities for nonfood industrial uses (combustible, lubricating oil, paint). Rapeseed proteins have a well-balanced amino acid composition resulting in possibilities for products of high nutritive value if other antinutritional compounds present in the seeds can be avoided.

Considerable effort has been made to improve rapeseed agronomic qualities by selective breeding techniques. In this way the so-called 'double zero' cultivars were developed containing no erucic acid in the oil and a very low content of glucosinolate compounds in the meal. These types of cultivar are named canola in Canada. The future goals for improving this crop include increased yield and disease resistance, in addition to altering the oil content and composition. The protein content of the meal is negatively correlated with oil content. Conventional breeding coupled with emerging biotechnologies will continue to play a major role in this improvement.

The main objectives may be summarised as follows. To increase seed oil content without increasing erucic acid and to lower α-linolenic acid (responsible for 'room odour' at high temperature) from 9% to 5–7%. For industry, specific varieties with high erucic acid content (66%), linoleic acid (30%), oleic acid (80%), or a reduction of α-linolenic acid to 0% are desired. Biochemical and molecular analyses of biosynthesis of fatty acids from rapeseed or other species (*Arabidopsis thaliana, Borago officinalis, Cuphea ellipsis*) are developing with the aim of cloning genes implicated in fatty acid desaturation, chain elongation and transacylation reactions. The eventual goal is to modify the composition of fatty acids by gene transfer. Other important objectives are to lower the content of glucosinolate compounds, choline esters and cellulose. Modification of the protein composition to enhance the balance of essential amino acids (methionine, lysine and tryptophan) is yet another goal. All of these objectives may be achieved by gene transfer (Vandekerckhove *et al.*, 1989; De Clercq *et al.*, 1990).

More productive and better-adapted rapeseed varieties with improved tolerance to biotic or abiotic stress can be developed by breeders looking for earliness, drought and cold tolerance, lodging resistance, apetalous flowers, and insect and herbicide resistance. Some objectives may be achieved through gene transfer:

1. The introduction of a cloned foreign gene into the rapeseed genome in order to add a new character, such as a gene supplying resistance to a herbicide, a pest or a type of stress or to modify the metabolite composition of an organ, such as fatty acid or protein components.

2. The inactivation of a gene by the insertion of a marker gene in order to find mutants.
3. The study of the expression of a given gene in a homologous or heterologous system, in order to study promoters and regulation of gene expression.

Transformation techniques

A number of gene transfer methods have been used to mediate delivery of foreign genes into rapeseed. *Agrobacterium*-mediated transformation has been most commonly used but some direct gene transfer techniques have also been tried with success. Table 9.1 summarizes the reports where efforts were made to improve the efficiency of transformation and where transgenic rapeseed plants were obtained. Most of these experiments were done with genes that allowed positive selection of the transformed tissue (hairy root phenotype induced by expression of the *Agrobacterium rhizogenes* transferred DNA (T-DNA) genes or antibiotic or herbicide resistance genes). The following conclusions may be drawn from the references listed in Table 9.1.

Transformation by *Agrobacterium*

Transformation by Agrobacterium rhizogenes
Winter and spring varieties of rapeseed are very susceptible to infection by *A. rhizogenes* and most organs when inoculated with this bacteria give rise to hairy root proliferation with a very high efficiency (50%–100% of inoculated organ segments). These roots have a very high growth rate on hormone-free medium. This allows the roots to rapidly outgrow the bacteria on a medium containing standard antibiotics and thus it is easy to decontaminate the transformed tissue.

The first transgenic rapeseed plant was obtained by Ooms *et al.*, in 1985 after regeneration of a clonal hairy root explant induced by *A. rhizogenes* infection. This plant was unfortunately sterile and showed the classical phenotype of many plants from other species regenerated from hairy root (mainly wrinkled leaves, short internodes and abundant secondary root system) (Tepfer, 1984). But in later experiments, fertile plants were regenerated, although exhibiting the hairy root phenotype and with varying degrees of reduced fertility. They produced enough viable pollen to fertilize an emasculated wild-type rapeseed; in this way, backcrossed progeny could be obtained.

The 'hairy root' phenotype prevents direct use of these plants in breeding programs, but some researchers have taken advantage of this easy procedure to produce transgenic plants. Two strategies may be used: cotransformation experiments (with a wild-type (WT) *A. rhizogenes* plus a disarmed *A. tumefaciens* carrying a plasmid with a plant-selectable marker gene and/or a gene of interest) or binary vector experiments (with an *A. rhizogenes* carrying two plasmids, a wild type Ri plasmid plus a disarmed T-DNA plasmid, carrying the genes of interest). Cotransformed regenerated plants bear both Ri TL–TR DNA and the desired construct. Segregation analysis of the cotransferred *A. rhizogenes* T-DNA and the gene(s) carried by the disarmed T-DNA is performed by backcrossing progeny to the wild-type parent. When the two markers are integrated independently or at a distance that allows recombinations, the BC1 plants, which carry only the disarmed T-DNA (with the desired gene), can be selected in the progeny.

Generally, plant tissues used for transformation are segments of axis tissues, such as hypocotyl, vegetative stem or floral axis. These fragments are prepared and cultivated *in vitro*. In the case of floral or even vegetative axis tissues, the segments are set up in gelified medium upside down to avoid natural rooting. Inoculation is performed with freshly subcultured bacteria, either in liquid or on solid medium. In case of cotransformation, *A. rhizogenes* and *A. tumefaciens* are mixed just before inoculation generally in a 1:1 ratio. A small amount of bacteria is laid on the upper wounded surface of the segments. The delay before hairy root development is varied from 10 days to 3 weeks. Young roots are harvested and independently cultured on a decontamination medium. When the plasmid used allows it, cotransformed root clones can be immediately screened on the proper selective medium. The regeneration efficiency of the transformed roots has been improved by preconditioning the roots before regeneration in a 2,4-dichlorophenoxyacetic acid (2,4-D)-containing medium (Guerche *et al.*, 1987*a*). All tested cultivars are able to regenerate plants, but with varying regeneration frequency, depending on whether the cultivar used was a spring (very efficient) or winter (less efficient) variety. Each of the resulting root clones may then

be regenerated into several plants. Cotransformation rate (percentage of cotransformed hairy root clones/total number of hairy root clones) seems to be higher when the binary vector strategy is used (80%–90% versus 15%–20%) as was first shown by Hamill *et al.* (1987).

It is noteworthy that not only hairy roots but shoots can be obtained after *A. rhizogenes* inoculation. Damgaard & Rasmussen (1991) observed that when decapitated hypocotyls with an intact root system were inoculated by *A. rhizogenes* carrying either a binary vector (WT Ri plasmid and a disarmed T-DNA carrying a kanamycin resistance gene) or a *cis* vector (only the WT Ri T-DNA plasmid in which the kanamycin resistance gene was introduced), up to 58% of these hypocotyls developed shoots directly at the point of infection and that 4%–15% of these shoots show a hairy root phenotype and NPT II activity. It will be interesting to discover whether it is possible to obtain directly, by a binary vector strategy, transformed shoots containing only the kanamycin resistance gene, without the WT Ri T-DNA as has been observed with tomato (Shahin *et al.*, 1986). It is still too early to know whether the cotransformation with two bacteria leads more often to the integration of the two exogenous DNAs at independent loci than the binary vector transformation strategy.

According to the different studies and the various transformation strategies used to obtain transgenic rapeseed, the copy number of the inserted gene can vary between one and ten copies. Often, if several copies are present, they behave as a unique locus, due to tandem or inverted integrations. The copy number of the TL-DNA of *A. rhizogenes* is higher than that of the TR-DNA in the few plants studied as the TL part may be integrated alone but the TR part is always associated with the TL.

Polymerase chain reaction (PCR) techniques allow rapid assays on a large number of progeny to counter-select plants with integrated Ri T-DNA and to select those bearing the gene of interest. The ease of the procedure and the high frequency of transformation allow the possible use of this strategy for the transfer of genes without any selective means.

Transformation by Agrobacterium tumefaciens

Because of the unfavorable phenotype induced by the Ri T-DNA, transformation procedures with a

Table 9.1. *Synopsis of the major transformation experiments carried out on rapeseed*

Variety	A.t.	A.r.	DGT	Organ	Promoter	Selection	Putative transgenic tissue rate[a]	Number of transgenic plants	F	Copy number[b]	Remarks	References
Jet 9 (W)		WT		Stem		Hairy-root	50% root	1p	—	>5 (1p)	Sterility of the regenerated plant. Hairy root phenotype	Ooms et al. (1985)
Brutor (S)		WT		Stem		Hairy-root	70% root	173p	+	1–4 (5p)	High efficiency of root regeneration (up to 90%). Hairy root phenotype	Guerche et al. (1987a)
HM 81 (S)		WT		Stem		Hairy-root		1p		2–3 (1p)	Hairy root phenotype	Hrouda et al. (1988)
Bienvenue (W) Brutor (S)		WT		Cotyledonary node	NOS	Hairy-root-Kan 50		17p (90% co-transformed)	+	5–10 (1p)	Binary vector. High frequency co-transformation. Possible segregation of the two markers in the progeny	Boulter et al. (1990)
Line (S) 1046 (S)		WT		Decapitated hypocotyl	NOS	Kan (no selection on Kan-medium)	19%–58% shoots	9%–30% op 4%–15% NPTII	+ +		Cis and binary vectors. Direct regeneration of shoots at the infection point. Selection with NPTII and opine assays	Damgaard & Rasmussen (1991)
Samourai (W)	Dis nop	WT		Floral axis	19S 70	CS100 Kan 60	30% root	3p (15% co-transf root)	+	2–3 (19p)	Binary vectors. Co-transformation.	Primard et al. (unpublished data)
Samourai (W)	Dis nop	WT		Floral axis	ALS	CS100	30% root	3p (21% co-transf root)	+		Attempts at plant regeneration were performed on only a limited number of root clones. Possible segregation of the two markers in the progeny	
Samourai (W)	Dis nop	WT		Floral axis	ALS	CS100	100% root	13p (80% co-transf root)	+			
Westar (S)	Dis nop Dis oct			Stem disc	NOS 35S	Kan 100	7% shoot	200p	+	2–3 (5p)	Cis and binary vectors. More than 50% escapes on Kan 50 and 10% on Kan 100. Low efficiency of octopine strain	Fry et al. (1987)
Westar (S)	Dis nop			Floral stem epiderm	35S	Mtx 0.01	32% shoot	24p (10%)	+		Cis vector. Unsuccessful selection on Kan. One-third escapes	Pua et al. (1987)
Westar (S)	Dis nop			Floral stem epiderm	NOS NOS 35S	Kan 15 CAM7 CAM7	6.6% shoot 5% shoot 5% shoot	12p (2%) 10p (1.6%) 5p (2.5%)	+ + +	1–2 (6p)	Cis and binary vectors. Low efficiency selection on Kan and CAM. One-third escapes	Charest et al. (1988)
Westar (S)	WT nop Dis nop			Hypocotyl	35S 35S	Kan 25 Kan 25	1.5% shoot 3.2% shoot	3p (0.7%) 9p (1.9%)	+	1–3 (12p)	Binary vectors	Radke et al. (1988)
Westar (S) R8494 (W)	Dis nop Dis nop			Hypocotyl	35S 35S	Kan 50 PPT 20	40% (S,W) 30% (S,W) shoot	748p (S,W) 550p (S,W)	+	1–8 (26p)	Binary vector. 25% of the transgenic plants showed no detectable PAT or NPTII activity	De Block & Debrouwer (1989)

Variety			Explant	Promotor	Selection	Transformation rate			Copy number	Comments	Reference
Westar (S)	Dis nop		Cotyledonary petiole	35S	Kan 15	55% shoot	22p	+	1–4 (6p)	Binary vector. Few escapes. 55% is the result of 1 experiment with 40 infected explants	Moloney et al. (1989)
Topaz (S)	Dis nop		Microspore-derived embryo			14.7% 29% embryo	40% NPTII + 25% NPTII+ (of resistant embryos)			Binary vector. 1000 plantlets regenerated, NPTII test on 75. More than 60% escapes. Doubling needed. Risk of chimeras	Pechan (1989)
Topaz (S)	Dis oct		Microspore-derived embryo	NOS	Kan 50	0.6% embryo	3p (on 600 cocultivated embryos)	+		Binary vector. Low kanamycin resistance level. Doubling needed. Risk of chimeras	Swanson et al. (1989)
Westar (S) Cobra (S)	Dis oct Dis nop	WT	Floral stem		Kan 150 GUS	30% shoot	13p (7% of the shoots)	+	5–10 (1p)	Binary vectors. High frequency of escapes. Nop strain of A.t. allowed transformation of Cobra (W)	Boulter et al. (1990)
Westar (S) Westar (S) Drakkar (S) R8494	Dis oct Dis nop		Stem Hypocotyl	NOS 35S PTR	Kan 100 PPT20 Kan 50	2% shoot 34% 18% (W) shoot	6p 228p	+ +	≥1 (3p) 2–5 (18p)	Cis vectors. No escapes. Binary vectors. Cotransformation with 2 different strains carrying 2 different marker genes. 57% of the cotransformed plants have the 2 copies linked	Misra (1990) De Block & Debrouwer (1991)
Brutor (S)		EP	Leaf protoplasts	19S	Paro 20 Kan 70	5.2×10^{-6} colony	2p	+	1 (2p)	Paromomycine allowed selection of colonies at higher density than Kan	Guerche et al. (1987b)
Winter varieties		MIJ	Microspore-derived embryos	35S, SV40	Kan	28% (35S) 49% (SV40) embryo		+	>2 (8p)	Necessity of a second round of embryogenesis to avoid chimeras, no tissues selection on Kan	Neuhaus et al. (1987)
Spring varieties		PEG	Leaf protoplasts	35S	Hyg 30	$1–4 \times 10^{-3}$ colony	8p		1, ≥2 (12col)	Transformation experiment using total DNA of a Hyg resistant B. nigra line	Golz et al. (1990)
Brutor (S)		EP	Leaf protoplasts	19S NOS 35S	Par 20 Hyg 25 PPT 5	2.6×10^{-4} col 2×10^{-4} col 3.8×10^{-4} col	30p			Improvement of electroporation efficiency by GUS transient expression assays	Rouan & Guerche (1991)

Notes:

Abbreviations: (S), Spring; (W), Winter; A.t., Agrobacterium tumefaciens; A.r., A. rhizogenes; WT, wild-type; Dis nop, disarmed T-DNA (nopaline strain); Dis oct, disarmed T-DNA (octopine strain); DGT, direct gene transfer; EP, electroporation; MIJ, microinjection; PEG, polyethylene glycol treatment; NOS, nopaline synthase promotor; 19S, CaMV 19S RNA promotor; 35S, CaMV 35S RNA promotor; 70, CaMV 35S RNA promotor with duplicated enhancer; PTR, 1' and 2' gene promotors from A. tumefaciens T-DNA; SV40, simian virus 40 promotor; ALS, promotor from the mutated acetolactate synthase gene from Arabidopsis thaliana; Kan 15, kanamycin (15 mg/l); Hyg, hygromycine; GUS, β-glucuronidase; PPT, phosphinothricin; Paro, paromomycin; CS, chlorsulfuron; Mtx, methotrexate; CAM, chloramphenicol; F, fertility of the plant; p, plant; col, colony.

[a] Putative transgenic tissue rate is the fraction of presumed transgenic shoots or roots divided by the total number of treated explants × 100 for Agrobacterium transformation and the % ratio of the number of resistant colonies in the total number of colonies submitted to selection for EP and PEG experiments.

[b] Copy number: the minimal and maximal copy number found in the plants studied (in parens).

disarmed *A. tumefaciens* that have proven highly successful in tobacco have been applied by several groups to rapeseed. For large-scale transformation programs it is necessary to have an efficient method of transformation. One of the main pre-requisites for this is a high level of regeneration from the chosen explant. Leaf disc, the tissue the most commonly used for tobacco transformation, does not seem to have the highest regeneration potential in rapeseed. Stem sections, thin stem layers, cotyledon petioles, hypocotyls or microspore-derived embryos appear to have more regeneration potential. The case of microspores or young-derived embryos is of interest because of the high potential number of embryos obtained from microspore culture of *B. napus*, and because they are haploid, giving the opportunity to pro-duce (after chromosome doubling (spontaneous or with colchicine)) directly homozygous trans-genic plants. The risk of chimeras obtained when transformations take place in an already pluri-cellular embryo may be avoided by the ability of *B. napus* microspore-derived embryos to undergo an embryogenic process. These embryogenic strains give rise to secondary embryos issued from one cell, so cloning a transformation event. However, to date, diploid tissues have been more commonly used.

When different factors influencing the shoot regeneration of cotyledon petioles from various *Brassica* species and cultivars of *B. napus* are com-pared, substantial differences in the regeneration rate between genotypes are displayed (Dale & Ball, 1991). A precise study of the regeneration ability of sections and thin epidermal layers from the floral axis points to the importance of the auxin content in the culture medium. On auxin-containing medium, two winter genotypes (cvs. Bienvenu and Darmor) show very few regenerated shoots from thin layers, whereas the spring-type cv. Brutor shows a high regeneration rate (>50%). Conversely, axis sections from all three genotypes produce shoots on a medium without auxin. Moreover the winter cv. Darmor showed the high-est regeneration rate (>90%). According to the source of the explant and the composition of plant growth regulators in the medium, the cultivars tested may have regeneration potential indepen-dent of their spring or winter type (Julliard *et al.*, 1992). Some reports in Table 9.1 underline the greater difficulty in transforming the winter culti-vars, at least those presently tested. Experiments

were done mainly with the spring variety Westar with variable success, but are generally low com-pared to tobacco (Horsch *et al.*, 1983).

The other prerequisite is the ability of *A. tume-faciens* to transform the rapeseed tissues. Two fac-tors are combined here: the virulence of the bacteria strain towards *B. napus*, and the cell accessibility to the bacterium. Both octopine and nopaline strains have been used to obtain trans-genic rapeseed plants, but nopaline strains have been more widely used, partly because undis-armed nopaline strains show greater virulence, even for winter varieties (Holbrook & Miki, 1985). However, Charest *et al.* (1989) showed that adding acetosyringone to the bacteria culture medium could enhance the virulence of octopine strains towards *B. napus*. As regards accessibility, in published protocols, the bacterium will have access mainly to the surface cells. Cytological studies show that if shoots originate from cells in the inner cell layers of the treated tissue, the fre-quency of transformation is much lower than if they are on or near the surface (Moloney *et al.*, 1989).

The conditions of the plant culture, the type of organ and age of the plant are chosen primarily with regard to the regeneration capacity, and explants are prepared to be cultured *in vitro*. The agrobacteria are prepared so as to be as virulent as possible: culture conditions and subculture rhythm should provide cells in their fast growth phase and, finally, the density of the last resuspen-sion of the bacteria in the inoculation medium is important. For inoculation, explants are generally dipped entirely into the bacterial suspension. Time of inoculation is variable (from a few min-utes to several days) according to the organ type, as too long a delay may affect the survival of the tissues or their regeneration ability (Radke *et al.*, 1988).

The duration of the coculture (fragments are blotted dry and transferred on to a solid medium with or without filter paper) is also to be assessed. Little is known concerning the T-DNA integra-tion event in the cell genome except that dividing cells are said to be the best targets. Thus, the coculture time may be a crucial phase with regard to the physiological state of the tissue being trans-formed.

Most of the authors noticed that cocultivation and further decontamination of the tissues before or during regeneration have dramatic effects on the plant regeneration rate, lowering by at least one order of magnitude its efficiency. There is no precise study of the effect on this point of the *Agrobacterium* strain used, but *in vitro* culture conditions and the antibiotic used to eliminate the *Agrobacterium* have been more thoroughly investigated. For the former, two main areas were investigated in order to reduce hypersensitivity to bacteria-induced damage: preconditioning of the tissue before inoculation, and complementing the coculture medium with a feeder layer of healthy fast growing cells (references cited in Table 9.1). Carbenicillin seems to be the most commonly used for tissue decontamination, but this antibiotic also has an auxin-like effect that has to be taken into account in the medium composition for tissue regeneration.

The other limiting step is the selection of transgenic shoots. Again three main factors are to be considered: the construct characteristics, the timing and severity of the selection of transformed cells. First, the promoter strength controlling the expression level of the resistance of the transformed cells surrounded by dying tissues is an important factor. Different promotors and marker coding sequences (all of them widely used for tobacco transformation (for a review, see Klee & Rogers, 1989)) have been used with success to select transgenic rapeseed plants. However, the study of the relative strengths of some promoters (cauliflower mosaic virus 35S, 2′ and nopaline synthase (*nos*)) demonstrated that the level of gene expression in transformed calli may be less in rape calli than in tobacco calli, although the ranking of these promoters is conserved (35S constructs give the the highest level, *nos* the lowest) (Harpster *et al.*, 1988). Kanamycin has been used as selective agent in many systems. For rapeseed transformation, except for one report (Pua *et al.*, 1987), kanamycin is the most efficient selective agent. Hygromycin also works well.

Most publications show that the number of putative transgenic shoots (or embryos) increases when the selective agent concentration is lowered in the medium but in return the number of escapes increases dramatically when the concentration approaches the minimum lethal dose for untransformed tissue. The same problem is observed if application of selection is delayed for too long after cocultivation.

At the last *in vitro* culture step, some authors noticed that the transgenic shoots were often vitri-fied or difficult to root (Fry, Barnason & Horsch, 1987; De Block, Debrouwer & Tenning, 1989). Ethylene production, which inhibits shoot formation and causes senescence of the tissue, can be limited by the addition of $AgNO_3$ and good ventilation of the *in vitro* cultivated tissues. Vitrification can be avoided by decreasing the water potential of the medium, reducing the relative humidity in the culture vessels and lowering the cytokinin concentration. Rooting can be improved by adding a low concentration of auxin in a well-aerated gelling agent medium (using perlite for example).

Lastly, it seems that the variability in the different transformation experiments is extremely high and that the repeatability of the same transformation procedure between different laboratories is extremely low, so few general rules can be drawn from these investigations for the moment.

The copy number of the transgene(s) in the few transgenic plants that have been analyzed at the DNA level varies from one to more than ten copies of the foreign gene. Varying degrees of complexity in their integration patterns can be observed but it seems that multiple copies of the T-DNA integrate mainly as inverted or tandem repeats. These integrations behave as single genetic loci rather than independent copies. In some studies, it seems that up to 25% of the transgenic rapeseed shoots that were obtained on medium containing a selective agent gave rise to plants in which the transgene was present but its expression was not detectable (De Block *et al.*, 1989).

Direct gene transfer

Few transgenic plants have been obtained by direct gene transfer experiments. One paper reports the microinjection of eight-cell microspore-derived embryoids with a very high efficiency (around 40%), but no transgenic plants were shown (Neuhaus *et al.*, 1987).

To date, only direct gene transfer of protoplasts has allowed the recovery of transgenic rapeseed plants. Generally, mesophyll or hypocotyl protoplasts derived from *in vitro*-cultured plants were subjected to electroporation or polyethylene glycol (PEG) treatment. Experimental conditions used for these two techniques are approximately the same as those developed for tobacco protoplast transformation.

The frequency of transformation is always low

Table 9.2. *Synopsis of the major transformation experiments with agronomic applications carried out on rapeseed*

Variety	Trans	Promotor and selective agent	Gene of interest	Expression		Results	Reference
				RNA	Protein		
Westar	A.t.	35S Kan	2S albumin gene (napin) from *Brassica napus*	+		Seed specific expression	Radke et al. (1988)
Westar	A.t.	NOS Kan	35S pro-Metallothionein I 3' NOS		+	Constitutive expression. The transgenic plants are 10× more $CdCl_2$ tolerant than WT	Misra (1989)
Westar	A.t.	PTR Kan Hyg	Enkephalin substitution in *Arabidopsis* 2S albumin gene	+	+	Seed specific expression. 10–50 nmol enkephalin/g of seed	Vandekerckhove et al. (1990)
Westar Drakkar	A.t.	PTR Kan Hyg	*Arabidopsis*/Brazil nut 2S gene; Methionine rich substitution in 2S *Arabidopsis* gene	+; +	+; +	Seed specific expression 0.1% of salt soluble seed protein; 1–2% of salt soluble protein	De Clercq et al. (1990)
Brutor	EP	NOS Kan	Pro lectin-2S Brazil nut gene	+	+	Lectin-like expression. 0.02–0.06% of salt soluble seed protein	Guerche et al. (1990)
Drakkar	A.t.	35S PPT	Anther specific promotor TA29-barnase coding sequence	+		Anther specific expression. Male sterility of transgenic plants	Mariani et al. (1990)
Westar Profit	A.t.	35S CS, Kan	Mutated acetolactate synthase gene from *A. thaliana*	+	+	Transgenic plants are 30× more resistant to CS than controls	Miki et al. (1990)
Westar	A.t.	35S Kan	Maize oleosin coding sequence with napin promotor	+	+	Seed specific expression. 1% of total seed proteins. Well targeted in protein bodies	Lee et al. (1991)

Notes:
Abbreviations: Trans, transformation technique; 35S, CaMV 35S RNA promotor; NOS, nopaline synthase promotor; PTR, 1' and 2' gene promotors from *A. tumefaciens* T-DNA; Kan, kanamycin; CS, chlorsulfuron; PPT, phosphinothricin.

in comparison with that obtained from tobacco protoplasts (1%–5% by both methods (Shillito *et al.*, 1985)) mainly because the plating efficiency is lower for rapeseed protoplasts (30%) than for tobacco protoplasts (90%–100%) and because of the difficulties of selecting the transgenic colonies (Rouan & Guerche, 1991). Indeed, low density cultivation of rapeseed protoplast-derived micro-colonies is difficult and, for this reason, the authors use agarose-embedded culture (Shillito *et al.*, 1983) to select and regenerate the few candidate colonies into plants. The selection of the transformed colonies on phosphinothricin-containing medium seems to give better results than selection on kanamycin or hygromycin because resistant colonies grow faster, allowing a shorter selection time and a higher regeneration efficiency of the colonies. Here, too, the efficiency of transformation is largely dependent on the genotype and the protoplast culture conditions.

The electroporation efficiency can be improved by the use of GUS transient expression assays (Jefferson *et al.*, 1987) to optimize the parameters and Chapel & Glimelius (1990) noticed that a combination of PEG and electroporation treatment led to a better transformation efficiency.

All these experiments were performed on spring varieties for which protoplast culture is better controlled than for winter varieties.

Conclusion

Rapeseed is one of the species for which all the tissue culture techniques required for transformation experiments exist but all the necessary steps to achieve transformation are less efficient than for model transformation species (tobacco, etc.). Except for experiments by De Block *et al.* (1989; De Block & Debrouwer, 1991) transformation efficiency is reduced by one or two orders of magnitude in comparison with tobacco (Horsch *et al.*, 1983). It seems that the constructs with selectable marker genes having strong promotors (35S, 70) are recommended for lowering the escape / transgenic plants ratio. Considering all the literature, it is always possible to select transformants with any selective agent, except with chloramphenicol, provided selective conditions are well established (Table 9.1).

By using one of these different techniques, it is also always possible, with time and effort, to obtain transgenic rapeseed plants in laboratories where some experience with rape tissue culture exists. To ensure reliable transformation rates, several modifications of already published protocols have been necessary, either to adapt them to new cultivars or simply to experimental conditions in each laboratory. It is clear that if these described techniques are to be used, much work needs to be done to improve regeneration of tissue from cultivated varieties (mainly winter types). It will also be necessary to find tricks to decontaminate tissue from *Agrobacterium* without injuring its regeneration potential, and to select the transgenic shoots whilst minimizing escapes in the process. At present, spring cultivars, especially Westar, have higher transformation rates than winter types. Two strategies are then possible: (i) to choose to introduce the gene of interest into a model spring cultivar, easily transformed and less time-consuming for the analysis of the genetic transmission, followed by at least four backcrosses to a more recalcitrant but agronomically desired cultivar; or (ii) to choose a different transformation technique, such as cotransformation with *A. rhizogenes*/*A. tumefaciens*. The latter is more time-consuming because one more step of segregation is required to eliminate the Ri T-DNA, but is efficient enough to obtain the required number of transformed plants. Little is known on the stability of the transferred genes because the number of transgenic plants studied is low and no systematic or statistical studies have been applied to this field, but it seems that extinction of some transgenes may occur as noted in other species (Matzke *et al.*, 1989; Linn *et al.*, 1990).

Agronomic applications of rapeseed transformation

Some transformation experiments have already been carried out with different genes of agronomic or commercial interest. These experiments are detailed in Table 9.2. Most of them refer to seed protein expression, either the production of 'pharmaceutical' plants or the enrichment of the meal in some specific amino acids by increasing the quality or the quantity of a specific storage protein. In these preliminary experiments, the level of expression of these transgenes was too low to lead to a notable alteration of the amino acid composition of the meal. The future aims of the authors are now to enhance the level of gene expression.

Efforts are now being made to produce herbicide-resistant plants to allow treatment of rape fields with herbicide, both to preferentially kill weeds and also to kill the offspring of old rapeseed varieties (the seed can survive for ten years in the soil). Phosphinothricin (De Block et al., 1989), glyphosate (C. Primard et al., unpublished data) and sulfonylurea (Miki et al., 1990; C. Primard et al., unpublished data) resistant rapeseed plants have been obtained by transformation techniques. At present, there are no commercial varieties, but some of the transgenic plants are being used in risk assessment programs in Canada and France in order to measure the possible diffusion of the transgenes into wild species (Kerlan et al., 1991).

The nuclear male sterility created by Mariani et al. (1990) has also been transferred into rapeseed plants and is currently being tested to measure the efficiency of this hybrid production strategy. Rapeseed hybrids will soon be commercially very important and this artificial means to control pollination could compete with natural genetic controls (nuclear or cytoplasmic male sterility, incompatibility genes).

Some experiments are also in progress to modify the oil quality of the seeds. Recently a Calgene group obtained high stearate transgenic rapeseed plants by an antisense strategy (Kridl et al., 1991). The expression of the antisense gene blocks the synthesis of the $\delta 9$ desaturase, which is the first enzyme of the desaturation pathways of C_{18} fatty acid in rapeseed. These kinds of preliminary experiment have proved that it is certainly possible to modify the fatty acid metabolic pathway in rapeseed.

On one hand, rapeseed is a crop of considerable economic importance; on the other hand, it shows good aptitude for in vitro culture, and transgenic rapeseed plants can be obtained by several of the present-day techniques. Therefore, it will certainly be in the near future one of the first candidates for the commercialization of transformed cultivars. However, unless new transformation techniques are developed much work has to be done before rapeseed could be considered as a model species for transformation.

Acknowledgements

The authors thank Kirk Schnorr, Ian Small and Georges Pelletier for critical reading of the manuscript and helpful discussions.

References

Boulter, M., Croy, E., Simpson, P., Shields, R., Croy, R. & Shirsat, A. (1990). Transformation of Brassica napus L. (oilseed rape) using Agrobacterium tumefaciens and Agrobacterium rhizogenes: a comparaison. Plant Science, 70, 91–99.

Carr, R. & McDonald, B. (1991). Rapeseed in a changing world: processing and utilisation. Groupe Consultatif International de Recherche sur le Colza, Rapeseed Congress, Saskatoon, 9–11 July, 1, 39–56.

Chapel, M. & Glimelius, K. (1990). Temporary inhibition of cell wall synthesis improves the transient expression of the GUS gene in Brassica napus mesophyll protoplasts. Plant Cell Reports, 9, 105–108.

Charest, P. J., Holbrook, L. A., Gabard, J. Iyer, V. N. & Miki, B. L. (1988). Agrobacterium-mediated transformation of thin cell layer explants from Brassica napus L. Theoretical and Applied Genetics, 75, 438–445.

Charest, P., Iyer, V. N. & Miki, B. L. (1989). Factors affecting the use of chloramphenicol acetyltransferase as a marker for Brassica genetic transformation. Plant Cell Reports, 7, 628–631.

Chèvre, A. M., This, P., Eber, F., Deschamps, M., Renard, M., Delseny, M. & Quiros, F. (1991). Characterization of disomic addition lines Brassica napus-Brassica nigra by isozyme, fatty acid and RFLP markers. Theoretical and Applied Genetics, 81, 43–49.

Dale, P. J. & Ball, L. F. (1991). Plant regeneration from cotyledonary explants in a range of Brassica species and genotypes. Groupe Consultatif International de Recherche sur le Colza, Rapeseed Congress, Saskatoon, 9–11 July, 4, 1122–1127.

Damgaard, O. & Rasmussen, O. (1991). Direct regeneration of transformed shoots in Brassica napus from hypocotyl infections with Agrobacterium rhizogenes. Plant Molecular Biology, 17, 1–8.

De Block, M. & Debrouwer, D. (1991). Two T-DNA's co-transformed into Brassica napus by a double Agrobacterium tumefaciens infection are mainly integrated at the same locus. Theoretical and Applied Genetics, 82, 157–263.

De Block, M., Debrouwer, D. & Tenning, P. (1989). Transformation of Brassica napus and Brassica oleracea using Agrobacterium tumefaciens and the expression of the bar and neo genes in the transgenic plants. Plant Physiology, 91, 694–701.

De Clercq, A., Vandewiele, M., Van Damme, J., Guerche, P., Van Montagu, M., Vandekerckhove, J. & Krebbers, E. (1990). Stable accumulation of modified 2S albumin seed storage proteins with higher methionine contents in transgenic plants. Plant Physiology, 94, 970–979.

Fry, J., Barnason, A. & Horsch, R. B. (1987).

Transformation of *Brassica napus* with *Agrobacterium tumefaciens* based vectors. *Plant Cell Reports*, **6**, 321–325.

Golz, C., Köhler, F. & Schieder, O. (1990). Transfer of hygromycin resistance into *Brassica napus* using total DNA of a transgenic *B. nigra* line. *Plant Molecular Biology*, **15**, 475–483.

Guerche, P., Charbonnier, M., Jouanin, L., Tourneur, C., Paszkowski, J. & Pelletier, G. (1987*b*). Direct gene transfer by electroporation in *Brassica napus*. *Plant Science*, **52**, 111–116.

Guerche, P., De Almeida, E. R. P., Schwarzstein, M. A., Gander, E., Krebbers, E. & Pelletier, G. (1990). Expression of the 2S albumin from *Bertholletia excelsa* in *Brassica napus*. *Molecular and General Genetics*, **221**, 306–314.

Guerche, P., Jouanin, L., Tepfer, D. & Pelletier, G. (1987*a*). Genetic transformation of oilseed rape (*Brassica napus*) by the Ri T-DNA of *Agrobacterium rhizogenes* and analysis of inheritance of the transformed phenotype. *Molecular and General Genetics*, **206**, 382–386.

Hamill, J. D., Prescott, A. & Martin, C. (1987). Assessment of the efficiency of cotransformation of the T-DNA of disarmed binary vectors derived from *Agrobacterium tumefaciens* and the T-DNA of *A. rhizogenes*. *Plant Molecular Biology*, **9**, 573–584.

Harpster, M. H., Townsend, J. A., Jones, J. D. G., Bedbrook, J. & Dunsmuir, P. (1988). Relative strengths of the 35S cauliflower mosaic virus, 1′, 2′ and nopaline synthase promoters in transformed tobacco, sugarbeet and oilseed rape callus tissue. *Molecular and General Genetics*, **212**, 182–190.

Holbrook, L. & Miki, B. (1985). *Brassica* grown gall tumourigenesis and *in vitro* of transformed tissue. *Plant Cell Reports*, **4**, 329–333.

Horsch, R. B., Fraley, R. T., Rogers, S. G., Sanders, P. R., Lloyd, A. & Hoffmann, N. (1983). Inheritance of functional foreign genes in plants. *Science*, **223**, 496–498.

Hrouda, M., Dusbabkova, J. & Necasek, J. (1988) Detection of Ri T DNA in transformed oilseed rape regenerated from hairy roots. *Biologia Plantarum*, **30**, 234–236.

Jefferson, R. A., Kavanagh, T. A. & Bevan, M. W. (1987). GUS fusions: ß Glucuronidase as a sensitive and versatile gene marker in higher plants. *EMBO Journal*, **6**, 3901–3907.

Julliard, J., Sossountzov, L., Habricot, Y. & Pelletier, G. (1992). Hormonal requirements and time competency for shoot organogenesis in two cultivars of *Brassica napus*. *Physiologia Plantarum*, **84**, 521–530.

Kerlan, M. C., Chèvre, A. M., Eber, F., Botterman, J. & De Greef, W. (1991). Risk assessment of gene transfer from transgenic rapeseed to wild species in optimal conditions. *Groupe Consultatif International*

de Recherche sur le Colza, Rapeseed Congress, Saskatoon, 9–11 July, **4**, 1028–1033.

Klee, H. J. & Rogers, S. G. (1989). Plant gene vectors and genetic transformation: plant transformation systems based on the use of *Agrobacterium tumefaciens*. *Cell Culture and Somatic Cell Genetics of Plants*, Vol. 6: *Molecular Biology of Plant Nuclear Genes*, ed. J. Schell & I. K. Vasil, pp. 1–23. Academic Press Inc, New York.

Kridl, J. C., Knutzon, D. S., Johnson, W. B., Thompson, G. A., Radke, S. E., Turner, J. C. & Knauf, V. C. (1991). Modulation of stearoyl-ACP desaturase levels in transgenic rapeseed. *International Society for Plant Molecular Biology*, Third International Congress, Tucson, 6–11 October, Congress Abstracts, 723.

Lee, W. S., Tzen, J. T. C., Kridl, J. C., Radke, S. E. & Huang, A. H. C. (1991). Maize oleosin is correctly targeted to seed oil bodies in *Brassica napus* transformed with the maize oleosin gene. *Proceedings of the National Academy of Sciences, USA*, **88**, 6181–6185.

Linn, F., Heidmann, I., Saedler, H. & Meyer, P. (1990). Epigenetic changes in the expression of the maize A1 gene in *Petunia hybrida*: role of numbers of integrated gene copies and state of methylation. *Molecular and General Genetics*, **222**, 329–336.

MacCalla, A. & Carter, C. (1991). Rapeseed in a changing world: the impacts of global issues. *Groupe Consultatif International de Recherche sur le Colza, Rapeseed Congress*, Saskatoon, 9–11 July, **1**, 1–14.

Mariani, C., De Beuckeleer, M., Truettner, J., Leemans, J. & Goldberg, R. (1990). Induction of male sterility in plants by a chimaeric ribonuclease gene. *Nature (London)*, **347**, 737–741.

Matzke, M., Primig, M., Trnovsky, J. & Matzke, A. (1989). Reversible methylation and inactivation of marker genes in sequentially transformed tobacco plants. *EMBO Journal*, **8**, 643–649.

Miki, B. L., Labbé, H., Hattori, J., Ouellet, T., Gabard, J., Sunohara, G., Charest, P. J. & Iyer, V. N. (1990). Transformation of *Brassica napus* canola cultivars with *Arabidopsis thaliana* acetohydroxide synthase genes: analysis of herbicide resistance. *Theoretical and Applied Genetics*, **80**, 449–458.

Misra, S. (1989). Heavy metal tolerant transgenic *Brassica napus* L. and *Nicotiana tabacum* L. plants. *Theoretical and Applied Genetics*, **78**, 161–168.

Misra, S. (1990). Transformation of *Brassica napus* L. with a 'disarmed' octopine plasmid of *Agrobacterium tumefaciens*: molecular analysis and inheritance of the transformed phenotype. *Journal of Experimental Botany*, **41**, 224, 269–275.

Moloney, M. M., Walker, J. M. & Sharma, K. K. (1989). High efficiency transformation of *Brassica napus* using *Agrobacterium* vectors. *Plant Cell Reports*, **8**, 238–242.

Neuhaus, G., Spangenberg, G., Mittelsten Scheid, O. & Schweiger, H. G. (1987). Transgenic rapeseed plants obtained by the microinjection of DNA into microspore derived embryoids. *Theoretical and Applied Genetics*, 75, 30–36.

Ooms, G., Bains, A., Burrell, M., Karp, A., Twell, D. & Wilcox, E. (1985). Genetic manipulation in cultivars of oilseed rape (*Brassica napus*) using *Agrobacterium*. *Theoretical and Applied Genetics*, 71, 325–329.

Pechan, P. (1989). Successful cocultivation of *Brassica napus* microspores and proembryos with *Agrobacterium*. *Plant Cell Reports*, 8, 387–390.

Pua, E. C., Mehra-Palta, A., Nagy, F. & Chua, N. H. (1987). Transgenic plants of *Brassica napus* L. *Bio/Technology*, 5, 815–817.

Radke, S. E., Andrews, B. M., Moloney, M. M., Crouch, M. L., Kridl, J. C. & Knauf, V. C. (1988). Transformation of *Brassica napus* L. using *Agrobacterium tumefaciens*: developmentally regulated expression of the reintroduced napin gene. *Theoretical and Applied Genetics*, 75, 685–694.

Renard, M., Brun, H., Chèvre, A., Delourme, R., Guerche, P., Mesquida, J., Morice, J., Pelletier, G. & Primard, C. (1992). Colza oléagineux. In *Amélioration des Espèces Végétales Cultivées*, ed. A. Gallais & H. Bannerot, pp. 135–145. INRA Ed., Paris.

Röbbelen, G. (1991). Rapeseed in a changing world: plant production potential. *Groupe Consultatif International de Recherche sur le Colza, Rapeseed Congress*, Saskatoon, 9–11 July, 1, 29–38.

Rouan, D. & Guerche, P. (1991). Transformation and regeneration of oilseed rape protoplasts. In *Plant Tissue Culture Manual*, ed. K. Lindsey, Vol. B4, pp. 1–24. Kluwer Academic Publishers, Dordrecht.

Shahin, E., Sukhapinda, K., Simpson, R. & Spivey, R. (1986). Transformation of cultivated tomato by a binary vector in *Agrobacterium rhizogenes*: transgenic plants with normal phenotypes harbor binary vector T-DNA, but no Ri-plasmid T-DNA. *Theoretical and Applied Genetics*, 72, 770–777.

Shillito, R. D., Paszkowski, J. & Potrykus, I. (1983). Agarose plating and a bead type culture technique enable and stimulate development of protoplast-derived colonies in a number of plant species. *Plant Cell Reports*, 2, 244–247.

Shillito, R. D., Saul, M. W., Paszkowski, J., Müller, M. & Potrykus, I. (1985). High efficiency direct gene transfer to plants. *Bio/Technology*, 3, 1099–1103.

Swanson, E. & Erickson, L. (1989). Haploid transformation in *Brassica napus* using an octopine-producing strain of *Agrobacterium tumefaciens*. *Theoretical and Applied Genetics*, 78, 831–835.

Tepfer, D. (1984). Genetic transformation of several species of higher plants by *Agrobacterium rhizogenes*: phenotypic consequences and sexual transmission of the transformed genotype and phenotype. *Cell*, 37, 959–967.

Trail, F., Richards, C. & Wu, F. S. (1989). Genetic manipulation in *Brassica*. In *Biotechnology in Agriculture and Forestry*, Vol. 9, *Plant Protoplasts and Genetic Engineering* II, ed. Y. P. S. Bajaj, pp. 197–216. Springer-Verlag, Berlin.

Vandekerckhove, J., Van Damme, J., Van Lijsebettens, M., Botterman, J., De Block, M., Vandewiele, M., De Clercq, A., Leemans, J., Van Montagu, M. & Krebbers, E. (1989). Enkephalins produced in transgenic plants using modified 2S seed storage proteins. *Bio/Technology*, 7, 929–932.

10
Sunflower

Günther Hahne

Introduction

The cultivated sunflower (*Helianthus annuus* L.) is one of the four most important species grown worldwide as oil crops. Sunflower oil is appreciated for its high nutritional value, due to a balanced content of fatty acids. This position as an important, and in some countries almost exclusive, source of one of the basic food components, has been made possible by a very successful application of conventional breeding techniques. The economic importance of sunflower has recently stimulated interest in biotechnological approaches, which are expected to further extend the possibilities for improvement of this species. A major issue in this context is, of course, the transfer of isolated genes that will confer novel characters on this important crop.

Although first reports of sunflower tissues cultured *in vitro* date back to the early times of plant tissue culture (e.g. De Ropp, 1946; Hildebrandt, Riker & Duggar, 1946; Henderson, Durrell & Bonner, 1952; Kandler, 1952), serious interest in the *in vitro* culture of sunflower tissues and cells is quite recent, a situation that is reflected in the comparatively small number of publications dealing with this issue. Moreover, this species has proven quite recalcitrant to regeneration, and technological advances as well as the comprehension of the underlying problems still flow at a slow rate. In the absence of a significant number of publications on transgenic sunflowers, the purpose of this review cannot be a critical discussion of the stability of foreign genes in such plants, nor an evaluation of the methods by which these genes could have been introduced. This review rather attempts to compile the existing tissue culture systems that might be, or have already been, used to approach the problem of sunflower transformation, and to highlight the difficulties that have been encountered at various levels.

Biological background information

The genus *Helianthus* belongs to the family Asteraceae (formerly Compositae) and is composed of 67 species (Heiser, 1976) which are annual or persistent herbaceous plants, growing in climates ranging from arid to more temperate conditions (Heiser *et al.*, 1969). While some of these species are known as noxious weeds in certain areas, and others are grown as ornamental plants, the only species with some economic value are *Helianthus tuberosus* (Jerusalem artichoke) and *Helianthus annuus* (the cultivated sunflower). Most sunflower lines cultivated today are F_1 hybrids, the production of which necessitates an intricate system of cytoplasmic male sterility comprising sterile and fertile inbred lines as well as restorer lines. This fact has important consequences for the strategies to be chosen for a practically useful transformation protocol. A truly useful protocol must be applicable to inbred lines and must not be limited to a small range of genotypes.

Any approach to transformation has to take the particular biological properties of the studied plant into account. A brief description of the growth cycle of sunflower will facilitate the understanding of the specific problems encountered during transformation and regeneration, and will make the choice of certain explants more obvious.

Sunflower seeds are quite large (8–10 mm) and enclosed in a hard shell (pericarp) composed of two halves containing not only the embryo, but also an air cavity that generally harbors fungal spores often difficult to eliminate. Elimination of bacterial and fungal contaminants from sunflower tissues is perceived as a major obstacle in establishing tissue cultures. Upon germination, sunflower kernels (achenes) quickly produce a long and sturdy hypocotyl bearing two large cotyledons. Adult sunflowers are tall plants with large leaves. The size of cultivated lines ranges from 1.5 m to more than 2.5 m. Their size, their requirement for high light intensities, and their attractiveness to a number of insect and fungal pests can present problems for the greenhouse culture of sunflower (Jeannin & Hahne, 1991). Male sterile lines used as female parents, and F_1 hybrids used for seed production, usually have one single stem terminating in one head of impressive size. Pollinator lines are in general branched and bear multiple, smaller inflorescences. The transition from the vegetative to the reproductive phase is in most cultivars relatively independent of environmental factors such as day length, and appears to be developmentally regulated (Steeves et al., 1969). Depending on genotype and growth conditions, anthesis is approximately 8–12 weeks after sowing. As is characteristic for Asteraceae, the inflorescences are radially symmetrical and composed of a multitude of small florets. The sterile ray flowers with their fused petals are responsible for the attractive yellow outer whorl, whereas the center contains the inconspicuous but fertile disc flowers. The opening of these flowers is progressive and proceeds from the circumference to the center of the inflorescence. Depending on the size of the flower head, this process may take 5–10 days. Embryo development is consequently asynchronous within one inflorescence. The resulting embryos develop quickly and reach their final length approximately 12–15 days after fertilization, but full maturity requires that the seeds remain some two additional months on the plant. For experimental purposes, however, this time can be shortened and the dehusked immature seeds can be germinated in vitro when harvested only 10–20 days after pollination. More detailed information on sunflower anatomy and development can be found in the authoritative publication edited by Carter (1978).

Approaches to sunflower transformation

Because of the convenient experimental properties of sunflower seedlings (size of hypocotyl and cotyledons), sunflower has been a favorite subject for physiological studies. Its susceptibility to *Agrobacterium tumefaciens* has been recognized already quite early, and crown gall tumors have been the subject of numerous studies (e.g. De Ropp, 1946; Matzke et al., 1984; Ursic, 1985; Yao, Jingfen & Kuochang, 1988). An adaptation of the early protocols for the production of transgenic callus rather than tumors, however, has only been taken quite recently (Everett, Robinson & Mascarenhas, 1987; Nutter et al., 1987; Escandón & Hahne, 1991). These experiments made use of disarmed strains of *A. tumefaciens*. The leaf disc approach, which has been so successful for tobacco (Horsch et al., 1985), is not applicable to sunflower because no reliable regeneration system from leaves is available. Long-term clonal propagation of sunflower shoot cultures as a convenient and axenic source of leaves for such experiments is difficult or even impossible, due to the tendency of precocious flowering in vitro that puts an end to the vegetative multiplication phase after only a few weeks (e.g. Henrickson, 1954; Paterson, 1984). Regeneration from cultured tissues, in particular from transgenic cultures such as callus or cell suspensions, remains difficult. The cases where regeneration from (nontransformed) callus has been successful either made use of immature embryos as donor material (Wilcox McCann, Cooley & van Dreser, 1988; Espinasse & Lay, 1989; Espinasse, Lay & Volin, 1989) or must today be considered as an exceptional event, restricted to specific conditions (Greco et al., 1984; Paterson & Everett, 1985; Lupi et al., 1987).

Although gene transfer to sunflower tissues has been routine practice for a long while, and although transformed plants have been described, the production of transgenic sunflowers is far from being a routine procedure. The approaches published to date suffer from very low efficiencies and poor reliability (Everett et al., 1987; Schrammeijer et al., 1990). One procedure that is reported not to suffer from these shortcomings has recently been communicated by Malone-Schoneberg et al. (1991; Bidney, Malone-Schoneberg & Scelonge, 1992a). Their protocol consists of a combination of *A. tumefaciens* and the particle gun. Details of this approach have been published for tobacco (*Nicotiana* species; Bidney *et al.*, 1992*b*). Information concerning sunflower is available only in conference abstracts (Malone-Schoneberg *et al.*, 1991; Bidney *et al.*, 1992*a*), and details on the production of transgenic sunflowers are not available.

Regeneration: a fundamental necessity

The production of transgenic plants involves two distinct processes: (i) the transfer of the foreign gene into a cell, and (ii) the regeneration of a whole plant from the progeny of this cell. Since regeneration is the limiting step for sunflower transformation, I will devote a rather large section to the review of the different regeneration systems available.

Adventitious regeneration from tissues

Adventitious regeneration from an explant can occur either directly, i.e. some cells of the tissue are stimulated to develop into a somatic embryo or a shoot without a prolonged dedifferentiated phase, or indirectly, where cells of the explant first form a callus, which can later be induced to regenerate plants. Both pathways are compatible with transformation. Direct regeneration systems are faster and usually show less somaclonal variation than indirect ones, but selection for transformed tissues is much easier on the callus stage.

Direct regeneration
Both direct somatic embryogenesis and direct shoot morphogenesis have been described for sunflower. Immature zygotic embryos have a high morphogenic capacity for somatic embryogenesis (Finer, 1987; Freyssinet & Freyssinet, 1988) as well as shoot morphogenesis (Jeannin & Hahne, 1991). To induce direct regeneration from immature embryos, these are generally isolated from the inflorescence when they have reached a length of 1–5 mm. They are then cultured on a medium supplemented with either auxin (Finer, 1987) or cytokinin (Freyssinet & Freyssinet, 1988; Jeannin & Hahne, 1991). A high osmotic pressure appears to favor the embryogenic response. Fertile plants have been obtained from a range of genotypes, although with different efficiencies. For regeneration, this approach is fast and relatively reliable,

but overall efficiency is limited (0.1–0.5 plants regenerated/explant). The constant availability of the explant material requires a considerable effort in the greenhouse, and the long exposure of the donor plant to varying environmental conditions results in an elevated experimental variability (Jeannin & Hahne, 1991). Direct morphogenesis from immature embryos has not yet been used for transgenic plant production.

The cotyledons of mature seeds or germinating seedlings give rise to shoots directly formed on their surface (Power, 1987; Nataraja & Ganapathi, 1989; Chraibi *et al.*, 1991, 1992; Knittel, Escandón & Hahne, 1991). Experimentally, this system is convenient because the period required for the production of the donor material is less than a week, and this process can easily be performed in a strictly controlled environment. The number of initially induced shoots can be rather high, although a large proportion of them may be abortive and, of the remainder, not all shoots may develop to full plants owing to competition. The morphogenic response is critically dependent on the developmental stage of the explant (Knittel *et al.*, 1991) and shows a strong genotype dependence. This latter limitation, however, can be overcome by culturing the explants in liquid rather than solid medium for 2 weeks (Chraibi *et al.*, 1992), or in the presence of inhibitors of ethylene production or action (Chraibi *et al.*, 1991). Fertile, regenerated plants are obtained quickly: the time between sowing of the donor plant and harvesting seeds from the regenerated plant may be as short as 4 months. This regeneration system should be sufficiently efficient to be used for transformation, but the cotyledon is not easily infected with *A. tumefaciens*.

Indirect regeneration
Callus induction is possible on virtually any tissue of young or adult sunflower plants. Because of its convenience, hypocotyl is often used for callus induction (Greco *et al.*, 1984; Paterson & Everett, 1985; Piubello & Caso, 1986; Robinson & Adams, 1987). These calli may be vigorously growing and can be maintained for long periods. They also represent a convenient source for cell suspension cultures. Mature tissues that are suitable as explants for plant regeneration from callus, however, are limited to the hypocotyl of certain genotypes (Greco *et al.*, 1984; Paterson & Everett, 1985; Lupi *et al.*, 1987). Unfortunately, such a

genotype with high regeneration capacity is not available to the public. Indirect embryogenesis has been demonstrated on epidermal thin layer explants from hypocotyl of an inbred line (Pélissier *et al.*, 1990). In this system, a high number of embryos could be induced, but failed to germinate. Secondary embryogenesis has permitted the regeneration of fertile plants, but frequencies are not indicated for this step.

As is true for direct regeneration, immature embryos are a good source also for morphogenic callus (Wilcox McCann *et al.*, 1988; Witrzens *et al.*, 1988; Espinasse & Lay, 1989; Espinasse *et al.*, 1989). This material shows a pronounced genotype × medium interaction (Espinasse & Lay, 1989). For callus induction, embryos must be isolated early, at approximately heart-to-torpedo stage (0.2–1.5 mm), corresponding to approximately 4 days after pollination. In most cases the calli are not cultured for long periods, because their regeneration potential is highest shortly after callus induction and then decreases rapidly. Although induction of embryogenesis in suspension cultures of sunflower has been described (Prado & Bervillé, 1990), I am not aware of any callus or cell suspension culture that has been maintained for a long period and still shows a reasonably high morphogenic activity.

Regeneration from callus may occur via shoot morphogenesis (Cavallini & Lupi, 1987; Espinasse *et al.*, 1989) or somatic embryogenesis (Paterson & Everett, 1985), irrespective of the type of donor material used for callus induction. The efficiency of the infection of sunflower hypocotyl with *A. tumefaciens* is well documented (Matzke *et al.*, 1984; Everett *et al.*, 1987), and the production of transformed callus presents no problem (Escandón & Hahne, 1991). The feasibility of this approach for plant transformation has been demonstrated, albeit with low efficiency and reliability (Everett *et al.*, 1987). No information is available concerning the use of immature embryos for the production of transgenic callus and the possibility of subsequent plant regeneration.

Regeneration from existing meristems

Because of the difficulties encountered with regeneration from adventitious shoots and somatic embryos, alternative strategies have been explored. The apical meristem of young plantlets can be induced to proliferate and allows the regeneration of a relatively high number of shoots that can be transferred to the greenhouse (Paterson, 1984). However, the meristem is not easily accessible for *A. tumefaciens*, and thus the use of this approach for the production of transgenic plants is limited by its low efficiency (Schrammeijer *et al.*, 1990). This limitation seems to be eliminated by the combination of *A. tumefaciens* and the particle gun already mentioned above (Malone-Schoneberg *et al.*, 1991; Bidney *et al.*, 1992*b*), but full experimental details concerning sunflower explants are not yet published.

Regeneration from single cells

Protoplasts

Protoplasts have been isolated and cultured from a variety of sunflower tissues, including hypocotyl (Lenée & Chupeau, 1986; Moyne *et al.*, 1988; Chanabé, Burrus & Alibert, 1989; Schmitz & Schnabl, 1989; Dupuis, Péan & Chagvardieff, 1990), cotyledons (Bohorova, Cocking & Power, 1986), petioles (Schmitz & Schnabl, 1989), leaves (Guilley & Hahne, 1989; Schmitz & Schnabl, 1989), and suspension cultures (G. Hahne, unpublished data). The isolation of protoplasts presents no particular difficulties, and the obtained yields are comparatively high. Yields average 2–5 million protoplasts/g tissue for most donor materials. Media have been defined that are suitable for the regeneration of callus from all these materials. The genotype appears to be of subordinate importance for the ability of protoplasts to develop into callus, and most of the described media can be used for all protoplast types. Sunflower protoplasts are in general quite stable, with the exception of mesophyll protoplasts, which tend to be fragile under some conditions. The plating efficiencies for the first divisions are high, a property that makes them useful for transient expression studies. In order to obtain reasonable plating efficiencies for callus production, it is advisable to embed the protoplasts in a semi-solid matrix. Both agarose (Moyne *et al.*, 1988; Dupuis *et al.*, 1990) and alginate (Fischer & Hahne, 1992) are convenient and have been used. While it is not difficult to produce vigorously growing callus from protoplasts, successful shoot induction has rarely been reported for *H. annuus*, and fertile plants have been obtained in only two cases. Binding *et al.* (1981) reported shoot regeneration from shoot-tip-derived protoplasts, but unfortunately the experimental details and the genotype were insufficiently described for the results to be reproduced. Burrus *et al.* (1991) obtained a small number of fertile plants from hypocotyl protoplasts of a genotype that had been specially selected for its regeneration capacity. The feasibility of plant regeneration from sunflower protoplasts has thus been demonstrated. The regeneration protocol itself is quite classic; it can therefore be assumed that the genotypic effect is the decisive factor. This genotype, derived from an interspecific hybrid of *H. annuus* and *H. petiolaris*, is unfortunately privately owned and not available to the public. We have recently established a protocol for the regeneration of fertile plants from protoplasts of an inbred sunflower line (Fischer, Klethi & Hahne, 1992). This protocol differs from that described by Burrus *et al.* (1991) in many aspects, and both appear to be specific for their respective genotypes. Given the availability of a suitable genotype, there is no fundamental obstacle to the use of protoplasts for the production of transgenic plants other than the overall efficiency, which is still rather low even in the specially selected genotype (Burrus *et al.*, 1991).

Pollen and oocytes
Although pollen and oocytes are not usually considered as objects for plant transformation, they do have some potential for genetic manipulation (e.g. Neuhaus *et al.*, 1987). Since none of the more traditional approaches forms the basis of a sound transformation system for sunflower, I include these reproductive cells in this brief overview of the state of the *in vitro* techniques for sunflower.

Plants have been regenerated from cultured anthers of interspecific hybrids in the genus *Helianthus* (Bohorova, Atanassov & Georgieva-Todorova, 1985; Gürel, Nichterlein & Friedt, 1991), and haploid plants have been obtained through androgenesis in sunflower (Mezzaroba & Jonard, 1986). However, frequencies and reproducibility are still too low to permit routine application of this technique.

Although the isolation of oocytes is more difficult and yields are much lower than is the case for microspores, gynogenesis is an alternative approach for haploid plant production in sunflower (Hongyuan *et al.*, 1986; Gélébart & San, 1987). This technique, although successful in

some cases, is not widely applied because of technical difficulties and low efficiency.

Possible strategies to use pollen, microspores, or oocytes for the production of transgenic plants include microinjection and the use of the particle gun (see below).

Properties of regenerated plants

While shoots obtained from preexisting meristems present no major difficulties for the transfer into the greenhouse, except premature flowering *in vitro*, most adventitious shoots and even somatic embryos are vitrified to a more or less serious extent. Vitrification is a phenomenon encountered in many tissue culture systems (e.g. see Böttcher, Zoglauer & Göring, 1988), but its causes are poorly understood. It must, at present, be resolved empirically for each case. No general recipe exists to prevent vitrification in the case of sunflower, but factors that might be considered include the gas phase, relative humidity, water potential of the medium, and hormonal composition of the media used throughout the regeneration procedure. Trivial factors such as the type of container and its position in the culture room may be decisive for the production of shoots with a quality sufficient for a transfer into the greenhouse. With few exceptions, adventitious shoots and even somatic embryos present enormous difficulties in root induction, and this step is often responsible for a very low efficiency of the transfer of shoots to the greenhouse (Freyssinet & Freyssinet, 1988; Burrus *et al.*, 1991). Once roots have been obtained, plants can usually be acclimated without major difficulties, and most of them will proceed to the flowering stage. Alternatively, shoots that fail to produce roots can still be transferred to the greenhouse when grafted on to an established rootstock. We have obtained vigorously growing, protoplast-derived shoots by this technique, and seed set was unproblematic (Fischer *et al.*, 1992).

Due to the generally very low regeneration efficiencies, not many studies are available concerning tissue-culture induced variability. The regenerated plants themselves (R0) are generally stunted (e.g. Finer, 1987; Wilcox McCann *et al.*, 1988; Jeannin & Hahne, 1991) and are often not taller than 20–50 cm, although they may reach the size of normal sunflower in exceptional cases. Even in lines that are characterized by only one stem bearing a single head, it is not unusual for R0 plants to be highly branched and to produce multiple inflorescences. In short, the overall appearance of regenerated sunflowers hardly resembles that of normal plants. However, the reasons for this are physiological rather than genetic, and this phenotype is not transmitted to ensuing generations (Freyssinet & Freyssinet, 1988; Jeannin & Hahne, 1991).

From the few published studies on genetic stability of tissue culture-derived sunflower plants it emerges that plants derived from immature embryos by direct organogenesis or somatic embryogenesis appear to be genetically quite stable. Significant deviations from the parent phenotype have not been detected in the offspring of regenerated plants during several generations (Freyssinet & Freyssinet, 1988; Jeannin & Hahne, 1991). Karyotypic variations have not been detected in three generations after plant regeneration from immature embryos (Jeannin, Poirot & Hahne, 1990; Jeannin & Hahne, 1991).

Somaclonal variants with altered coumarin content have been obtained after mercuric chloride-treatment of immature embryo-derived callus (Roseland, Espinasse & Grosz, 1991). No detailed analysis is available concerning plants derived by any of the other regeneration routes. Only Witrzens et al. (1988) have indicated the occurrence of possible somaclonal variants after regeneration from immature embryo-derived callus of one genotype.

The basic problems in culture of sunflower *in vitro*

We can conclude that a variety of regeneration systems is available for sunflower. Regeneration has been demonstrated to be possible from almost any conceivable multi- or unicellular explant, but only a few of them can be considered as actually established, universally applicable protocols (these are based mainly on the use of immature embryos). The recurring problems are (i) low efficiency, (ii) low reliability, and (iii) low universality, i.e. the frequently encountered limitation to one or few genotypes. The approaches to regeneration that are more generally applicable and have a reasonable efficiency for plant regeneration, i.e. direct regeneration systems and callus induced on immature embryos, cannot easily be used in conjunction with *A. tumefaciens*. In contrast, those

systems that are quite amenable to *Agrobacterium*-mediated transformation have only low regeneration efficiencies or very limited applicability. The efficiency of virtually any regeneration system is critically dependent on the genotype used, with the more efficient lines often not being available to the public. Of the public lines, the inbred line HA 300 in both its male sterile (A) and male fertile (B) forms presents perhaps the highest regeneration potential, although its performance is in many respects not satisfactory. A good regeneration potential from immature embryos is found in HA 89B. Moreover, the performance of a public line such as HA 300 may vary in function because of the supplier from whom the seeds have been obtained.

Transformation

Genetic markers

Any method used for the transformation of plant tissues or cells occurs with an efficiency far below 100%. If long and tedious screening on the level of regenerated plants is to be avoided, a powerful and reliable selection for transformed cells is indispensable. A number of selectable markers have been used in transformation experiments of various sunflower explants. The *npt*II gene coding for the enzyme, neomycin phosphotransferase II (NPT II), is perhaps the most universally utilized genetic marker for transformed plant tissues. This enzyme inactivates by phosphorylation a number of compounds belonging to the class of aminoglycoside antibiotics, including kanamycin, paromomycin, gentamycin, etc. Kanamycin has proven its efficiency for tobacco and many other species (e.g. see Fraley, Rogers & Horsch, 1987; Potrykus, 1991), and has been employed for the selection of transformed callus obtained from sunflower hypocotyl (Everett et al., 1987) or protoplasts (Moyne et al., 1989). Selection has been possible in these cases, but the regeneration capacity of the callus was severely diminished (Everett et al., 1987). Other authors observed a high spontaneous resistance of sunflower tissues towards kanamycin and were unable to use this compound for selection, while other aminoglycosides such as paromomycin or G418 could be used successfully (Escandón & Hahne, 1991). Phosphinothricin (PPT; 'Basta'), inactivated by the product of the *bar* gene (phosphinothricin acetyl transferase (PAT; De Block *et al.*, 1987)) also has proven to be a useful selective marker (Escandón & Hahne, 1991). The reasons for the conflicting observations concerning kanamycin are not clear, but may be due to a genotypic effect. Furthermore, factors such as the auxin : cytokinin ratio and N-source in the culture medium may be important for the efficiency of the selection (Escandón & Hahne, 1991).

Quantitative and semi-quantitative markers that have been used successfully with sunflower tissues include β-glucuronidase (GUS), NPT II, and chloramphenicol acetyl transferase (CAT). The availability of several quantifiable markers allows the use of an internal standard for transient expression assays. The experimental variability can thus be decreased, and individual experiments rendered more comparable (as shown for tobacco by Lepetit *et al.* (1991)). Quantitative assays are usually employed for transient expression studies. The histological GUS reaction, which is semi-quantitative when interpreted with caution, has also been utilized for the detection of stable transformation events in sunflower. No background problems are encountered with the GUS stain when bacterial expression is excluded (e.g. by using the GUS-intron construction (Vancanneyt *et al.*, 1990)), but sunflower tissue has a tendency to turn brown during the fixation and staining process, rendering the blue color difficult to detect.

Agrobacterium-mediated gene transfer

All experiments resulting in transformed sunflower tissues have made use of *A. tumefaciens*, with one exception (microinjection; Espinasse-Gellner, 1992). All transformed sunflower plants published to date with a detailed experimental description have been obtained using this vector (Everett *et al.*, 1987; Schrammeijer *et al.*, 1990).

Crown gall tumors of sunflower were used early on to study the integration (Ursic, 1985) and expression (Matzke *et al.*, 1984) of foreign genes. The *Agrobacterium* strains used in these studies were of wild-type, and although the resulting tumors could be established *in vitro* (Ursic, 1985), no plants have been regenerated from these cultures for obvious reasons. This aspect will not be considered in more detail here.

Two publications exist to date describing transformed sunflower plants, and two additional approaches have been described in conference abstracts. First, Everett *et al.* (1987) used a disarmed strain of *A. tumefaciens* to transform hypocotyl explants of a genotype that regenerated efficiently from callus induced on this explant. Upon selection on kanamycin, however, the regeneration potential was lost to a great extent. It was necessary to subculture the selected callus on nonselective medium for several months before shoots could be regenerated. Seventeen plants were obtained and were shown to be transgenic by kanamycin resistance and Southern analysis. No further publication followed this initial report, but a short note by Hartmann (1991) indicated that some difficulties were encountered and that the regeneration and selection system needed further work. Second, Schrammeijer *et al.* (1990) used longitudinally cut meristems of mature embryos for their transformation experiments; they treated the specimens with a suspension of *A. tumefaciens*. The bacteria were indeed able to penetrate into the meristematic tissue in a few cases, which then gave rise to plants containing transformed cell lines. This approach has the benefit of being very straightforward, and at least theoretically being independent of genotypic restrictions. However, the efficiency is extremely low. Of the 1500 meristems treated, two transgenic shoots were obtained with chimeric transgene expression. Progeny could be obtained from one shoot, but no expression of the *uid*A gene was detectable in the offspring.

The approach of Schrammeijer *et al.* (1990) has been used in a modified version by Malone-Schoneberg *et al.* (1991; Bidney *et al.*, 1992*a*) to produce transgenic sunflower plants. Their original technique combines the particle gun approach and *Agrobacterium* as a very efficient vector for DNA delivery. Some experimental details are given in the related paper (Bidney *et al.*, 1992*b*) using tobacco as a model, and also describing transformed sectors on sunflower shoots. The small, dense projectiles of the particle gun are used in this approach not for DNA delivery (as in the technique that has become standard today; for a review, see Christou, 1992), but to create microlesions that allow access of *Agrobacterium* to the tissue in a very efficient way (Bidney *et al.*, 1992*b*). This combination increases the transformation efficiency for tobacco leaves by 100-fold, and it can be expected that it is equally efficient for sunflower meristems. A number of transgenic sunflowers have been produced using this approach (Bidney et al., 1992a). The description of technical details and characterization of the obtained plants (Bidney et al. 1992c) is not very detailed, but it appears that most of Bidney et al.'s primary transformants produced with this technique are chimeric for the expression of the introduced gene(s). Our own experiments with embryonic axes indicate that an appropriate selection scheme can limit the occurrence of chimeras (N. Knittel et al., unpublished data).

Direct gene transfer

Particle gun

The particle gun in its many versions has been used successfully for the transformation of a number of plant species that had hitherto defied all efforts towards transformation (e.g. see Klein et al., 1988; McCabe, Martinell & Christou, 1988), and for transient expression studies in tissues that are inaccessible for protoplast isolation (e.g. see Hamilton et al., 1992; van der Leede-Plegt et al., 1992). It would therefore appear logical that this powerful tool had been applied to sunflower. However, apart from the novel approach used by Bidney et al. (1992a,b), no successful application of the particle gun to any sunflower tissues or organs has been reported so far.

Microinjection

Transgenic plants of Brassica napus have been produced by microinjection in young, microspore-derived somatic embryos at the four to eight cell stage (Neuhaus et al., 1987). Homogeneous transgenic plants could be obtained from the primary chimeric regenerants by secondary embryogenesis. Microinjection thus appears to be a feasible approach to the transformation of a recalcitrant species, provided the regeneration has been established for an organ that can be microinjected.

A similar approach has been taken for sunflower, where embryos cultured in ovulo (Espinasse et al., 1991) are injected with DNA. The technique has been presented only as a conference abstract (Espinasse-Gellner, 1992); experimental details and a characterization of the transgenic plants are thus not available. The authors have, however, reported on their success to regenerate plants from 3-day-old embryos of interspecific hybrids cultured in ovulo (Espinasse

et al., 1991). An evaluation of efficiency and universality of this approach must await further details.

Protoplasts

Direct gene transfer into protoplasts is a well-established technique that is used extensively in many species in one of two ways, namely electroporation and polyethylene glycol (PEG)-mediated DNA uptake. Analysis of the treated protoplasts after only a short time (24–48 h) allows the detection of a fairly strong signal in a significant number of protoplasts (Negrutiu et al., 1988; Jung et al., 1992), a phenomenon that has been termed 'transient expression'. When protoplasts are cultured for longer periods, the burst of transgene expression due to nonintegrated DNA disappears, and the expression of DNA stably integrated in the genome can be detected in a considerably lower percentage of protoplasts. This technique has yielded transgenic plants in a steadily increasing number of species (for a review, see Potrykus, 1991).

The feasibility of transient expression has been demonstrated for sunflower using both electroporation and PEG to stimulate DNA uptake (Kirches, Frey & Schnabl, 1991). Kirches et al. have defined the experimental parameters for mesophyll protoplasts of the hybrid line 'Primasol'. The optimal conditions are comparable to those optimized for protoplasts from other species.

Stably transformed transgenic callus has been obtained after PEG-mediated gene transfer into sunflower protoplasts (Moyne et al., 1989). The frequency of callus resisting kanamycin selection and expressing the nptII gene was relatively low (4×10^{-6}) in this study. Since no regeneration protocol is available for the genotype used, transgenic plants have not been obtained from this callus.

From these experiments it is clear that sunflower protoplasts do not have properties that prevent their use in DNA transfer experiments. However, this approach is likely to develop into a viable alternative for the production of transgenic sunflower plants only if the frequencies can be considerably improved for both stable transformation events and, even more important, for plant regeneration.

Conclusion and perspectives

A certain number of gene transfer techniques have been established for sunflower, and, likewise, a diverse range of regeneration systems is available. It is the combination of these two components into one efficient protocol that is the source of problems for the production of transgenic sunflower plants. The tissues for which relatively reliable and universal regeneration protocols are available, i.e. immature embryos and cotyledons, are not easily transformed by *A. tumefaciens*. In contrast, hypocotyl is relatively efficiently transformed, but its regeneration is difficult and seems possible with a very limited choice of genotypes and in only few laboratories. The cases are rare where both gene transfer and subsequent plant regeneration have been successful. They are mostly characterized by extremely low efficiencies and a poor reproducibility. A possible exception is the transformation of meristems by a combination of *A. tumefaciens* and the particle gun (Bidney *et al.*, 1992*a,b*).

The most important obstacle in the way towards the transgenic sunflower is beyond doubt the still very poorly developed *in vitro* culture of sunflower cells and tissues. With few notable exceptions, sunflower cell and tissue culture is characterized by a very pronounced genotype dependence, poor reproducibility in different experimental environments, and insufficient efficiency. It is at present unclear what the reasons are for this recalcitrance. Wild relatives of sunflower (Bohorova *et al.*, 1986; Chanabé *et al.*, 1991; Krasnyanski *et al.*, 1992) and certain interspecific hybrids (Chandler & Beard, 1983; Burrus *et al.*, 1991) are in many respects less problematic, although less effort has been devoted to them. Perhaps the major handicap for workers in sunflower *in vitro* culture is the strong genotype dependence of the regeneration process. Although genotypes suitable for different aspects have been identified, lines that regenerate from the undifferentiated state have in general been private property (e.g. see Paterson & Everett, 1985; Everett *et al.*, 1987; Burrus *et al.*, 1991) that is accessible for other laboratories only under restrictive conditions, if at all. One or several model genotypes that may have only a mediocre regeneration potential or agronomic value, but could serve as a positive control, have not yet been identified among the public lines. The establishment of such a reference line could considerably facilitate and accelerate research towards the transformed sunflower.

In view of the low frequencies obtained in the few published experiments that resulted in transgenic sunflower plants, the absence of information on the stability and inheritance of the foreign genes is hardly surprising.

Interest in sunflower tissue culture has been rising during the last years, and this is reflected by the increasing number of publications on this subject. Original approaches have been developed and considerable advances have been made concerning both gene transfer and plant regeneration, and there is reason to hope that it is only a question of time until both aspects are combined into a protocol for routine sunflower transformation.

The list of some potentially useful genes to be transferred into sunflower, once a reliable transformation system has been established, will not be much different from a list applicable to any other species. Apart from the study of genes and gene constructions interesting for fundamental studies, but without immediate interest for application, a number of biological and technological problems might be addressed by using transgenic sunflowers. Sunflowers are grown virtually all over the world, and the needs are evidently different for each region. Immediate application could improve resistance against herbicides and insect attack, and in the more distant future against fungal pathogens. The other domain where it is hoped that biotechnology will contribute to a new development, in due term, is the change of product quality. Novel utilizations of vegetable oil lead to a changing demand for oils of different quality, concerning not only the degree of desaturation but also characters such as chain length. Furthermore, sunflower is used in some areas as a fodder crop, and improved amino acid composition of its storage proteins could render this application more interesting.

Today, all commercially successful hybrids, i.e. the overwhelming majority of sunflower seeds grown worldwide, are based on the same cyctoplasmic male sterile (cms) cytoplasm. This potentially hazardous situation could be overcome by producing novel cms lines and their restorer systems by interspecific hybridization. The utilization of genetic engineering approaches that are very promising in other crops might provide for more flexibility and less complicated breeding.

Other breeding goals identified today concern more complex characters, such as drought resistance, early flowering for culture in more northern latitudes, and height reduction. In the absence of a clear understanding of the underlying physiology, these characters may at present be more successfully addressed by conventional breeding and *in vitro* culture techniques, such as exploiting somaclonal variation or somatic hybridization.

References

Bidney, D. L., Malone-Schoneberg, J. & Scelonge, C. J. (1992*a*). Stable transformation of sunflower and field evaluation of transgenic progeny. In *Sunflower Research Workshop*, pp. 57–58. National Sunflower Association, Fargo.

Bidney, D. L., Scelonge, C. J. & Malone-Schoneberg, J. B. (1992*c*). Transformed progeny can be recovered from chimeric plants regenerated from *Agrobacterium tumefaciens* treated embryonic axes of sunflower. In *Proceedings of the 13th International Sunflower Conference*, pp. 1408–1412. International Sunflower Association, Pisa.

Bidney, D., Scelonge, C., Martich, J., Burrus, M., Sims, L. & Huffman, G. (1992*b*). Microprojectile bombardment of plant tissues increases transformation frequency by *Agrobacterium tumefaciens*. *Plant Molecular Biology*, **18**, 301–313.

Binding, H., Nehls, R., Kock, R., Finger, J. & Mordhorst, G. (1981). Comparative study on protoplast regeneration in herbaceous species of the Dicotyledoneae class. *Zeitschrift für Pflanzenphysiologie*, **101**, 119–130.

Bohorova, N., Atanassov, A. & Georgieva-Todorova, J. (1985). In vitro organogenesis, endrogenesis and embryo culture in the genus *Helianthus* L. *Zeitschrift für Pflanzenzüchtung*, **95**, 35–44.

Bohorova, N. E., Cocking, E. C. & Power, J. B. (1986). Isolation, culture and callus regeneration of protoplasts of wild and cultivated *Helianthus* species. *Plant Cell Reports*, **5**, 256–258.

Böttcher, I., Zoglauer, K. & Göring, H. (1988). Induction and reversion of vitrification of plants cultured in vitro. *Physiologia Plantarum*, **72**, 560–564.

Burrus, M., Chanabé, C., Alibert, G. & Bidney, D. (1991). Regeneration of fertile plants from protoplasts of sunflower (*Helianthus annuus* L.). *Plant Cell Reports*, **10**, 161–166.

Carter, J. F. (1978). *Sunflower Science and Technology*. American Society of Agronomy, Crop Science Society of America, Soil Science Society of America, Inc., Publishers, Madison.

Cavallini, A. & Lupi, M. C. (1987). Cytological study of callus and regenerated plants of sunflower (*Helianthus annuus* L.). *Plant Breeding*, **99**, 203–208.

Chanabé, C., Burrus, M. & Alibert, G. (1989).

Factors affecting the improvement of colony formation from sunflower protoplasts. *Plant Science*, **64**, 125–132.

Chanabé, C., Burrus, M., Bidney, D. & Alibert, G. (1991). Studies on plant regeneration from protoplasts in the genus *Helianthus*. *Plant Cell Reports*, **9**, 635–638.

Chandler, J. M. & Beard, B. H. (1983). Embryo culture of *Helianthus* hybrids. *Crop Science*, **23**, 1004–1007.

Chraibi, K., Castelle, J. C., Latche, A., Roustan, J. P. & Fallot, J. (1992). Enhancement of shoot regeneration potential by liquid medium culture from mature cotyledons of sunflower (*Helianthus annuus* L.). *Plant Cell Reports*, **10**, 617–620.

Chraibi, K. M., Latche, A., Roustan, J. P. & Fallot, J. (1991). Stimulation of shoot regeneration from cotyledons of *Helianthus annuus* by the ethylene inhibitors, silver and cobalt. *Plant Cell Reports*, **10**, 204–207.

Christou, P. (1992). Genetic transformation of crop plants using microprojectile bombardment. *Plant Journal*, **2**, 275–281.

De Block, M., Botterman, J., Vanderwiele, M., Dockx, J., Thoen, C., Goessele, V., Movva, N. R., Thompson, C., Van Montagu, M. & Leemans, J. (1987). Engineering herbicide resistance in plants by expression of a detoxifying enzyme. *EMBO Journal*, **6**, 2513–2518.

De Ropp, R. S. (1946). The isolation and behavior of bacteria-free crown-gall tissue from primary galls of *Helianthus annuus*. *Phytopathology*, **37**, 201–206.

Dupuis, J. M., Péan, M. & Chagvardieff, P. (1990). Plant donor tissue and isolation procedure effect on early formation of embryoids from protoplasts of *Helianthus annuus* L. *Plant Cell, Tissue and Organ Culture*, **22**, 183–189.

Escandón, A. J. S. & Hahne, G. (1991). Genotype and composition of culture medium: factors important in the selection for transformed sunflower (*Helianthus annuus*) callus. *Physiologia Plantarum*, **91**, 367–376.

Espinasse, A. & Lay, C. (1989). Shoot regeneration of callus derived from globular to torpedo embryos from 59 sunflower genotypes. *Crop Science*, **29**, 201–205.

Espinasse, A., Lay, C. & Volin, J. (1989). Effects of growth regulator concentrations and explant size on shoot organogenesis from callus derived from zygotic embryos of sunflower (*Helianthus annuus* L.). *Plant Cell, Tissue and Organ Culture*, **17**, 171–181.

Espinasse, A., Volin, J., Dybing, C. D. & Lay, C. (1991). Embryo rescue through in ovulo culture in *Helianthus*. *Crop Science*, **31**, 102–108.

Espinasse-Gellner, A. (1992). A simple and direct technique of transformation in sunflower. In *Sunflower Research Workshop*, p. 50. National Sunflower Association, Fargo.

Everett, N. P., Robinson, K. E. P. & Mascarenhas, D. (1987). Genetic engineering of sunflower (*Helianthus annuus* L.). *Bio/Technology*, **5**, 1201–1204.

Finer, J. (1987). Direct somatic embryogenesis and plant regeneration from immature embryos of hybrid sunflower (*Helianthus annuus* L.) on a high sucrose-containing medium. *Plant Cell Reports*, **6**, 372–374.

Fischer, C. & Hahne, G. (1992). Structural analysis of colonies derived from sunflower (*Helianthus annuus* L.) protoplasts cultured in liquid and semisolid media. *Protoplasma*, **169**, 130–138.

Fischer, C., Klethi, P. & Hahne, G. (1992). Protoplasts from cotyledon and hypocotyl of sunflower (*Helianthus annuus* L.): shoot regeneration and seed production. *Plant Cell Reports*, **11**, 632–636.

Fraley, R. T., Rogers, S. G. & Horsch, R. B. (1987). Genetic transformation in higher plants. *CRC Critical Reviews in Plant Sciences*, **4**, 1–46.

Freyssinet, M. & Freyssinet, G. (1988). Fertile plant regeneration from sunflower (*Helianthus annuus* L.) immature embryos. *Plant Science*, **56**, 177–181.

Gélébart, P. & San, L. H. (1987). Obtention de plantes haploïdes par culture in vitro d'ovaires et d'ovules non fécondés de tournesol (*Helianthus annuus* L.). *Agronomie*, **7**, 81–86.

Greco, B., Tanzarella, O. A., Carrozo, G. & Blanco, A. (1984). Callus induction and shoot regeneration in sunflower (*Helianthus annuus* L.). *Plant Science Letters*, **36**, 73–77.

Guilley, E. & Hahne, G. (1989). Callus formation from isolated sunflower (*Helianthus annuus*) mesophyll protoplasts. *Plant Cell Reports*, **8**, 226–229.

Gürel, A., Nichterlein, K. & Friedt, W. (1991). Shoot regeneration from anther culture of sunflower (*Helianthus annuus*) and some interspecific hybrids as affected by genotype and culture procedure. *Plant Breeding*, **106**, 68–76.

Hamilton, D. A., Roy, M., Rueda, J., Sindhu, R. K., Sanford, J. & Mascarenhas, J. P. (1992). Dissection of a pollen-specific promoter from maize by transient transformation assays. *Plant Molecular Biology*, **18**, 211–218.

Hartmann, C. L. (1991). *Agrobacterium* transformation in sunflower. In *Sunflower Research Workshop*, pp. 95–99. National Sunflower Association, Fargo.

Heiser, C. B. (1976). *The Sunflower*. University of Oklahoma Press, Norman.

Heiser, C. B., Smith, D. M., Clevenger, S. B. & Martin, W. C. J. (1969). *The North American Sunflowers*. In T. Delevoryas (Ed.), *Memoirs of The Torrey Botanical Club*. The Seeman Printery, Durham, NC.

Henderson, J. H. M., Durrell, M. E. & Bonner, J. (1952). The culture of normal sunflower stem callus. *American Journal of Botany*, **39**, 467–473.

Henrickson, C. E. (1954). The flowering of sunflower explants in aseptic culture. *Plant Physiology*, **29**, 536–538.

Hildebrandt, A. C., Riker, A. J. & Duggar, B. M. (1946). The influence of the composition of the medium on growth in vitro of excised tobacco and sunflower tissue cultures. *American Journal of Botany*, **33**, 591–597.

Hongyuan, Y., Chang, Z., C., D., Hua, Y., Yan, W. & Xiaming, C. (1986). *In vitro* culture of unfertilized ovules in *Helianthus annuus* L. In *Haploids of Higher Plants in Vitro* ed. H. Han & Y. Hongyuan, pp. 182–190. Springer Verlag, Berlin.

Horsch, R. B., Fry, J. E., Hoffman, N. L., Eichholz, D., Rogers, S. G. & Fraley, R. T. (1985). A simple and general method for transferring genes into plants. *Science*, **227**, 1229–1231.

Jeannin, G. & Hahne, G. (1991). Donor plant growth conditions and regeneration of fertile plants from somatic embryos induced on immature zygotic embryos of sunflower (*Helianthus annuus* L.). *Plant Breeding*, **107**, 280–287.

Jeannin, G., Poirot, M. & Hahne, G. (1990). Régénération de plantes fertiles à partir d'embryons zygotiques immatures de tournesol. In *Cinquantenaire de la culture in vitro Versailles* (France), 24–25 October 1989, ed. C. Doré, pp. 275–276. Paris: Ed. INRA.

Jung, J. L., Bouzoubaa, S., Gilmer, D. & Hahne, G. (1992). Visualization of transgene expression on the single protoplast level. *Plant Cell Reports*, **11**, 346–350.

Kandler, O. (1952). Über eine physiologische Umstimmung von Sonnenblumenstengelgewebe durch Dauereinwirkung von ß-Indolylessigsäure. *Planta*, **40**, 346–349.

Kirches, E., Frey, N. & Schnabl, H. (1991). Transient gene expression in sunflower mesophyll protoplasts. *Botanica Acta*, **104**, 212–216.

Klein, T. M., Fromm, M. E., Weissinger, A., Tomes, D., Schaaf, S., Sletten, M. & Sanford, J. C. (1988). Transfer of foreign genes into intact maize cells using high velocity microprojectiles. *Proceedings of the National Academy of Sciences, USA*, **85**, 4304–4309.

Knittel, N., Escandón, A. S. & Hahne, G. (1991). Plant regeneration at high frequency from mature sunflower cotyledons. *Plant Science*, **73**, 219–226.

Krasnyanski, S., Polgar, Z., Németh, G. & Menczel, L. (1992). Plant regeneration from callus and protoplast cultures of *Helianthus giganteus* L. *Plant Cell Reports*, **11**, 7–10.

Lenée, P. & Chupeau, Y. (1986). Isolation and culture of sunflower protoplasts (*Helianthus annuus* L.): Factors influencing the viability of cell colonies derived from protoplasts. *Plant Science*, **43**, 69–75.

Lepetit, M., Ehling, M., Gigot, C. & Hahne, G.

(1991). An internal standard improves the reliability of transient expression studies in plant protoplasts. *Plant Cell Reports*, **10**, 401–405.

Lupi, M. C., Bennici, A., Locci, F. & Gennai, D. (1987). Plantlet formation from callus and shoot-tip culture of *Helianthus annuus* (L.). *Plant Cell, Tissue and Organ Culture*, **11**, 47–55.

Malone-Schoneberg, J. B., Bidney, D., Scelonge, C., Burrus, M. & Martich, J. (1991). Recovery of stable transformants from *Agrobacterium tumefaciens* treated split shoot axes of *Helianthus annuus*. *In Vitro Cellular and Developmental Biology*, **27**, 152A.

Matzke, M. A., Susani, M., Binns, A. N., Lewis, E. D., Rubenstein, I. & Matzke, A. J. M. (1984). Transcription of a zein gene introduced into sunflower using a Ti plasmid vector. *EMBO Journal*, **3**, 1525–1531.

McCabe, D. E., Martinell, B. J. & Christou, P. (1988). Stable transformation of soybean (*Glycine max*) plants. *Bio/Technology*, **87**, 923–926.

Mezzaroba, A. & Jonard, R. (1986). Effets du stade de prélèvement et des prétraitements sur le développement in vitro d'anthères prélevées sur le tournesol cultivé (*Helianthus annuus* L.). *Comptes Rendus de l'Académie des Sciences*, Paris, Série III, tôme **303**, 181–186.

Moyne, A. L., Tagu, D., Thor, V., Bergounioux, C., Freyssinet, G. & Gadal, P. (1989). Transformed calli obtained by direct gene transfer into sunflower protoplasts. *Plant Cell Reports*, **8**, 97–100.

Moyne, A. L., Thor, V., Pelissier, B., Bergounioux, C., Freyssinet, G. & Gadal, P. (1988). Callus and embryoid formation from protoplasts of *Helianthus annuus*. *Plant Cell Reports*, **7**, 437–440.

Nataraja, K. & Ganapathi, T. R. (1989). In vitro plantlet regeneration from cotyledons of *Helianthus annuus* cv. Morden (sunflower). *Indian Journal of Experimental Botany*, **27**, 777–779.

Negrutiu, I., Shillito, R., Potrykus, J., Biasini, G. & Sala, F. (1988). Hybrid genes in the analysis of transformation conditions. *Plant Molecular Biology*, **8**, 363–373.

Neuhaus, G., Spangenberg, G., Mittelsten-Scheid, O. & Schweiger, H. G. (1987). Transgenic rapeseed plants obtained by the microinjection of DNA into microspore-derived embryoids. *Theoretical and Applied Genetics*, **75**, 30–36.

Nutter, R., Everett, N., Pierce, D., Panganiban, L., Okubara, P., Lachmansingh, R., Mascarenhas, D., Welch, H., Mettler, I., Pomeroy, L., Johnson, J. & Howard, J. (1987). Factors affecting the level of kanamycin resistance in transformed sunflower cells. *Plant Physiology*, **84**, 1185–1192.

Paterson, K. E. (1984). Shoot tip culture of *Helianthus annuus* – flowering and development of adventitious and multiple shoots. *American Journal of Botany*, **71**, 925–931.

Paterson, K. E. & Everett, N. P. (1985). Regeneration of *Helianthus annuus* inbred plants from callus. *Plant Science*, **42**, 125–132.

Pélissier, B., Bouchefra, O., Pepin, R. & Freyssinet, G. (1990). Production of isolated somatic embryos from sunflower thin cell layers. *Plant Cell Reports*, **9**, 47–50.

Piubello, S. M. & Caso, O. H. (1986). In vitro culture of sunflower (*Helianthus annuus* L.) tissues. *Phyton*, **46**, 131–137.

Potrykus, I. (1991). Gene transfer to plants: assessment of published approaches and results. *Annual Reviews of Plant Physiology and Plant Molecular Biology*, **42**, 205–225.

Power, C. J. (1987). Organogenesis from *Helianthus annuus* inbreds and hybrids from the cotyledons of zygotic embryos. *American Journal of Botany*, **74**, 497–503.

Prado, E. & Bervillé, A. (1990). Induction of somatic embryo development by liquid culture in sunflower (*Helianthus annuus* L.). *Plant Science*, **67**, 73–82.

Robinson, K. E. P. & Adams, D. O. (1987). The role of ethylene in the regeneration of *Helianthus annuus* (sunflower) plants from callus. *Physiologia Plantarum*, **71**, 151–156.

Roseland, C., Espinasse, A. & Grosz, T. J. (1991). Somaclonal variants of sunflower with modified coumarin expression under stress. *Euphytica*, **54**, 183–190.

Schmitz, P. & Schnabl, H. (1989). Regeneration and evacuolation of protoplasts from mesophyll, hypocotyl, and petioles from *Helianthus annuus* L. *Journal of Plant Physiology*, **135**, 223–227.

Schrammeijer, B., Sijmons, P. C., van den Elzen, P. J. M. & Hoekema, A. (1990). Meristem transformation of sunflower via *Agrobacterium*. *Plant Cell Reports*, **9**, 55–60.

Steeves, T. A., Hicks, M. A., Naylor, J. M. & Rennie, P. (1969). Analytical studies on the shoot apex of *Helianthus annuus*. *Canadian Journal of Botany*, **47**, 1367–1375.

Ursic, D. (1985). Eight DNA insertion events of *Agrobacterium tumefaciens* Ti-plasmids in isogenic sunflower genomes are all distinct. *Biochemical and Biophysical Research Communications*, **131**, 152–159.

van der Leede-Plegt, L. M., van der Ven, B. C. E., Bino, R. J., van der Salm, T. P. M. & van Tunen, A. J. (1992). Introduction and differential use of various promoters in pollen grains of *Nicotiana glutinosa* and *Lilium longiflorum*. *Plant Cell Reports*, **11**, 20–24.

Vancanneyt, G., Schmidt, R., O'Connor-Sanchez, A., Willmitzer, L. & Rocha-Sosa, M. (1990). Construction of an intron-containing marker gene: splicing of the intron in transgenic plants and its use in monitoring early events in *Agrobacterium*-mediated plant transformation. *Molecular and*

General Genetics, **220**, 245–250.

Wilcox McCann, A., Cooley, G. & van Dreser, J. (1988). A system for routine plantlet regeneration of sunflower (*Helianthus annuus* L.) from immature embryo-derived callus. *Plant Cell, Tissue and Organ Culture*, **14**, 103–110.

Witrzens, B., Scowcroft, W. R., Downes, R. W. & Larkin, P. J. (1988). Tissue culture and plant regeneration from sunflower (*Helianthus annuus*) and interspecific hybrids (*H. tuberosus* × *H. annuus*). *Plant Cell, Tissue and Organ Culture*, **13**, 61–76.

Yao, X., Jingfen, J. & Kuochang, C. (1988). Transfer and expression of the T-DNA harboured by *Agrobacterium tumefaciens* in cultured explants of *Helianthus annuus*. *Acta Botanica Yunnanica*, **10**, 159–166.

11

Forest Trees

Ronald R. Sederoff

Introduction

Forest trees dominate the temperate and tropical ecosystems and the wood produced by trees is the most abundant biological material on the earth's surface (Gammie, 1981). Wood provides fuel for much of the population of the world, and wood is a leading industrial raw material. In the United States, wood accounts for 25% of the value of all industrial materials and exceeds them in amount (National Research Council, USA, 1990). Increased environmental concern requires that more forest is preserved and that less cutting be carried out in natural forests. To meet the increased demand for wood, more wood needs to be produced on less land through genetic improvement and intensive management (Buonjiorno & Grosenick, 1977; USDA Forest Service, 1982).

Genetic improvement of forest trees is a slow process. The most advanced tree-breeding programs have carried out only a few generations of selection. Forest trees are largely undomesticated plants and differ from many agronomic crops that have been cultivated and selected for thousands of years. Furthermore, trees are not annual crops, and one generation for many species can take 20 years or more.

Genetic engineering has great promise for agriculture because it can accelerate traditional breeding, bypass long generation times and cross reproductive barriers through gene transfer technology. If this is true for agricultural crops, then genetic engineering should be a greater advantage to forestry because traditional methods have been slower and more difficult. The application of

genetic engineering technology to tree improvement has been slow also because the necessary techniques and information are not available. Woody plants have been neglected as objects of interest in molecular biology and little is known about the molecular basis of biological processes that are unique to perennial woody plants, such as dormancy, wood formation, or the transition from juvenility to maturity (Hutchison & Greenwood, 1991). There is a need to acquire more basic biological information about forest trees to address many practical goals through genetic engineering.

DNA transfer and tree improvement

There are three general types of genetic modifications for accelerated tree improvement through genetic engineering. These are (i) improved resistance to pests and pathogens (biotic stress), (ii) modified metabolism and development, and (iii) resistance to different kinds of abiotic stress.

Woody plants are characterized by high levels of genetic diversity. High levels of variation are maintained within species and within populations. High diversity is associated with large geographic ranges, outcrossing breeding systems, and broad seed dispersal, but much of the variation remains unexplained (Hamrick, Godt & Sherman-Broyles, 1992). As perennial woody plants, forest trees are subjected to long-term stress from pests and pathogens. Woody perennials differ from annual crops because resistance must be maintained during different seasons over many years. Such stress is most extreme for long-lived forest trees, where selection occurs over tens or even hundreds of years. Transformation may aid in solving major problems of epidemic diseases in trees, such as those caused by white pine blister rust *(Cronartium ribicola* Fischer), Dutch elm disease *(Ceratocystis ulmi* (Buism.) C. Mor.), or chestnut blight *(Endothia parasitica)*. Insect pests, whether generally endemic, such as Southern pine bark beetle *(Dendroctonus frontalis* Zimmermann), or epidemic, such as gypsy moth *(Lymantria dispar* L.), are a serious concern in forest ecology and management. Strategies for genetic engineering using *Bacillus thuringiensis* or proteinase inhibitors could lead to reduced damage or to reduced use of chemicals in the natural environment (Strauss, Howe & Goldfarb, 1991). Damage from pests and pathogens are major factors reducing forest productivity around the world.

The second main category of traits for improved trees affect the development or metabolism of forest trees. Those of most immediate interest affect growth and yield. Growth and morphology are under moderate to strong genetic control, and accelerated improvement for yield and quality would be valuable (Zobel & Talbert, 1984). Control of fertility is an important trait because it could lead to controlled breeding, biological containment, and increased vegetative growth. Control of growth by photoperiod and the transition from juvenile to mature growth are inherited traits that have major effects on growth and quality of forest trees.

The third general category of traits for genetic engineering of forest trees are those affecting abiotic stress. Trees must resist a variety of physical and chemical stresses over long periods of time. The ability of trees to respond to temperature, flooding, drought, atmospheric pollution and global climate change are all factors related to traits that may be targets for genetic engineering. Recently, trees have been considered in combination with directed genetic modifications for bioremediation (Stomp *et al.*, 1992).

Forest trees are different from domesticated crops

The transformation systems for forest trees have some special requirements, compared to domesticated crops. Due to high levels of genetic diversity, different plants within a species vary greatly in their ability to regenerate in culture and in their response to pathogens such as *Agrobacterium* (Clapham *et al.*, 1990; Riemenschneider, 1990;

Bergmann, 1992; A.-M. Stomp, personal communication). Considerable effort is often expended in screening for the specific genotype that will perform appropriately, particularly because there are no cultivars or inbred lines. In some species, clones are available, or genotype is controlled in part, using open pollinated progeny from a known seed tree. An ideal transformation system would be relatively insensitive to variation in genotype.

Transformation studies in most forest trees are made difficult by the long reproductive cycles. It is not possible to go through several reproductive generations of segregation to sort out mosaics or genetic instabilities after transformation. Verification of transformation depends on cell culture systems and vegetative propagation. This approach has been used successfully in walnut *(Juglans regia* L.) transformation, where transformed plants were produced after repetitive somatic embryogenesis (McGranahan *et al.*, 1988, 1990). Embryos, cocultivated with *Agrobacterium tumefaciens,* were multiplied by repetitive embryogenesis, and selected for kanamycin resistance in subsequent embryonic generations. Selected embryo subclones were germinated and grown as plants or micropropagated.

A third limitation is the problem of juvenile mature correlation. Many traits cannot be readily predicted for a mature tree based on the performance or phenotype of the juvenile tree. Wood properties, for example, are significantly different in juvenile and mature trees (Zobel & van Buijtenen, 1989). Genetic engineering of mature traits would be greatly facilitated by good juvenile mature correlations.

There are two major barriers to the application of genetic engineering to forest trees. The first is the lack of information about the biological mechanisms and the genetic basis of processes that are targets for genetic engineering. The second is the absence of adequate methods for DNA transfer in virtually any forest tree species, although notable progress has been made recently with a few species.

The purpose of this chapter is to review the current status of DNA transfer technology in forest trees. We also address DNA transfer in the context of specific problems of interest in woody plants and some specific objectives related to genetic engineering of forest trees. Some aspects of genetic engineering of forest trees have recently been reviewed (Neale & Kinlaw, 1992). Many aspects of DNA transfer and genetic engineering of trees have also been discussed in other articles (see von Arnold, Clapham & Ekberg, 1990; Charest & Michel, 1991; Ahuja & Libby, 1993; Sederoff & Stomp, 1993).

Transformation in conifers

The most widely planted forest trees throughout the temperate latitudes are conifers. Among the predominant species are loblolly pine *(Pinus taeda* L.) in the southeastern United States, Douglas fir *(Pseudotsuga menziesii* (Mirb) Franco) in the northwestern United States, and Monterey pine *(Pinus radiata* D. Don) in the southern hemisphere, grown as an exotic. In the more northern latitudes, spruces (such as Norway spruce; *Picea abies* (L.) Karst), are important. Most conifers are grown for structural wood or for pulp and paper. In spite of their significant commercial value, conifers have only recently become material for experiments in DNA transformation. In the last 7 years, transfer and expression of foreign genes in cells and tissues of coniferous species have been clearly established. The majority of the evidence has come from two different technical approaches, one using *Agrobacterium* as a biological vector for DNA transfer and the other using microprojectile bombardment for physical transfer of DNA.

Agrobacterium-mediated DNA transfer in conifers

Before 1985, the only evidence for DNA transfer in conifers were morphological criteria for infection of conifers by the crown gall bacterium (De Cleene & De Ley, 1976, 1981). It is widely believed that *Agrobacterium*-induced crown gall disease is limited to angiosperms, particularly dicotyledons (dicots). Reports in the literature of infection of conifers by *Agrobacterium* are found from the 1930s (Smith, 1935*a,b*). A wide survey shows 56% of tested species of gymnosperms are susceptible, whereas 58% of angiosperms are susceptible (De Cleene & De Ley, 1976). At least 66 different conifers are susceptible to *Agrobacterium* (for a review, see Sederoff & Stomp, 1993).

Subsequently, transformation in conifers was established at a cellular or subcellular level. Evidence for transformation was obtained from expression of genes transferred by *Agrobacterium*, including those coding for opine synthesis, phytohormone synthesis, and expression of kanamycin resistance. By 1990, such evidence had been obtained in nine pine species and Douglas fir (Sederoff *et al.*, 1986; Dandekar *et al.*, 1987; Gupta, Dandekar & Durzan, 1988; Morris, Castle & Morris, 1988, 1989; Stomp *et al.*, 1988, 1990; Ellis *et al.*, 1989; Loopstra, Stomp & Sederoff, 1990). Similarly, transformation had been reported for four spruce species (Clapham & Ekberg, 1988; Ellis *et al.*, 1989; Morris *et al.*, 1989), two true firs (Clapham & Ekberg, 1986, 1988; Morris *et al.*, 1988) and western hemlock (*Tsuga heterophylla (Raf.) Sarg.*) (Morris *et al.*, 1988). Studies showing expression of marker genes for β-glucuronidase, chloramphenicol acetyltransferase, luciferase, and neomycin phosphotransferase have been successful for loblolly pine, Monterey pine, Scots pine *(Pinus sylvestris* L.), Douglas fir, jack pine *(Pinus banksiana* Lamb.), white spruce *(Picea glauca* (Moench) Voss), black spruce *(Picea mariana* (Mill.) B.S.P.) and hybrid larch (Bekkaoui *et al.*, 1988; Gupta *et al.*, 1988; Dunstan, 1989; Tautorus *et al.*, 1989; Wilson, Thorpe & Moloney, 1989; Duchesne & Charest, 1991; Stomp, Weissinger & Sederoff, 1991).

In sugar pine (*P. lambertiana* Dougl.), stable transformation has been obtained in cultured cells. Sugar pine shoots derived from cytokinin-treated cotyledons were inoculated with an hypervirulent strain of *Agrobacterium tumefaciens* containing a binary plasmid system (Stomp *et al.*, 1988; Loopstra *et al.*, 1990). Galls produced on shoots proliferated as callus in culture in the absence of phytohormone, and in the presence of kanamycin. Stable transformation was verified after more than a year in culture by Southern blots and by NPT II activity. Similar results have been obtained for Douglas fir. Phytohormone-independent callus has been grown from tumors induced in culture on shoots established *in vitro* (Dandekar *et al.*, 1987). Ellis *et al.* (1989) and Morris *et al.* (1989) have confirmed and extended evidence for physical transfer and expression of introduced genes in this species.

In the genus *Picea* several species of spruce have also been transformed by *Agrobacterium* including Norway spruce (*P. abies*), Sitka spruce (*P. sitchensis (Bong.)* Carr.), white spruce, Englemann spruce (*P. englemanni* Parry ex

Engelm.) (Clapham & Ekberg, 1986, 1988; Ellis *et al.*, 1989, 1991*a*; Hood *et al.*, 1989, 1990; Clapham *et al.*, 1990). Stably transformed callus cultures established following *Agrobacterium* infection have been obtained and characterized for Norway spruce (Hood *et al.*, 1990) and for white spruce (Ellis *et al.*, 1989). Hood *et al.* (1990) inoculated shoots of Norway spruce and then cultured the resulting tumors *in vitro*. Transformation was verified by Southern blotting and by opine synthesis.

Induction of the *Agrobacterium* virulence cascade in Douglas fir

Many strains of *Agrobacterium* that have high levels of virulence in dicotyledonous plants do not work well in conifers (Morris *et al.*, 1989; Stomp *et al.*, 1990). The native inducers of the *Agrobacterium* virulence cascade induced by dicots are acetosyringone and hydroxyacetosyringone, although many other phenolic and hydroxyphenylpropanoid compounds can act as inducers. In Douglas fir, Morris & Morris (1990) identified the phenylpropanoid glucoside coniferin as the major native phenolic virulence gene inducer. Coniferin is known to be a major soluble phenolic component of conifer xylem. It is considered to be a storage form of coniferyl alcohol, the major precursor of guaiacyl lignin (Sederoff & Chang, 1991). In developing xylem of white spruce, scraped from wood after peeling the cambium and phloem, coniferin was found at levels of more than 4% of the wet weight (Savidge, 1989). It is presumed that coniferin is located in the vacuoles because it is present at such high levels.

Transformation by bombardment

Microprojectile bombardment provides a method for studying transient expression in forest trees, particularly in systems that are either recalcitrant or have not been explored. Bombardments result in localized staining usually in single cells, and can penetrate tissue below the epidermis (Stomp *et al.*, 1991). Many forest tree species have been studied with this technique, e.g. Douglas fir (Goldfarb *et al.*, 1991); *Populus* (McCown *et al.*, 1991); black spruce (Duchesne & Charest, 1991).

An abscisic acid (ABA)-inducible promoter from wheat has been found to be particularly active in conifer cells and tissues. The promoter from the early methionine (*Em*) gene fused to *gus*A (Marcotte, Russell & Quatrano, 1989) has been used in bombardment experiments on a variety of conifer cells and tissues. These include Norway spruce somatic embryos (Robertson *et al.*, 1992), differentiating xylem (Loopstra, Weissinger & Sederoff, 1992), loblolly pine cotyledons (Stomp *et al.*, 1990) and black spruce callus (Duchesne & Charest, 1991). ABA is known to be important in the regulation of embryonic development of conifers and has been shown to affect development of somatic embryos of Norway spruce and other spruce species (Hakman & von Arnold, 1985; Hakman & Fowke, 1987; von Arnold & Woodward, 1988). Several promoters from dicots have been tested in conifers, but none is as active in conifers as the ABA-inducible *Em* promoter. Ellis *et al.* (1991*a*, 1992) observed activity for cauliflower mosaic virus (CaMV) 35S, nopaline synthase (*nos*), *Em* from wheat, a ubiquitin promoter from *Arabidopsis*, Rubisco (ribulose-1,5-bisphosphate carboxylase oxygenase) (*rbc*) promoters from *Arabidopsis*, soybean and larch (*Larix*), a phosphoenol pyruvate (PEP)-carboxylase promoter, an alcohol dehydrogenase promoter from maize, and an auxin-inducible promoter from soybean. All showed activity after bombardment, which decreased within days. A heat shock promoter from soybean could be induced and expressed for up to 4 weeks after bombardment.

Transient expression of a promoter from the *Dc*8 gene from carrot (*Daucus*) has been reported to occur following bombardment of Norway spruce cell and embryo cultures (Newton *et al.*, 1992). *Dc*8 is an ABA-responsive gene expressed abundantly and late during carrot embryogenesis (Hatzopoulos, Fong & Sung, 1990). Expression was observed in suspension cells, but less expression was observed in embryogenic callus or somatic embryos. Expression in embryos or suspension cells was enhanced if cultures were pretreated with ABA. In black spruce, *gus*A was expressed by both the CaMV 35S promoter and the *Em* promoter from wheat. The *Em* promoter appeared far more active (Duchesne & Charest, 1991). In a subsequent study, Duchesne & Charest (1992) found a similar result with embryogenic lines of hybrid larch (*Larix* sp.) and showed that expression increased following addition of ABA to the culture medium.

Microprojectile bombardment has been used in Norway spruce somatic embryos to generate stable transformation in cell lines (Robertson *et al.*, 1992). Mature somatic embryos were bombarded with pRT99gus, a plasmid that contains the *npt*II and *gus*A genes both fused to the CaMV 35S promoter. Assays of cell lines selected on antibiotic show that all of the eight lines tested show coexpression of both marker genes, but at variable levels. None of the transformed lines remained embryogenic.

Of particular interest in forest trees is the use of microprojectile bombardment to observe expression of promoters in differentiating xylem. Trees are a good system in which to study the differentiation of wood, and wood formation is an important target for genetic engineering (Whetten & Sederoff, 1991) because of the potential for modification of wood properties. Loopstra *et al.* (1992) have studied the expression of three promoters in loblolly pine stem sections (*Em* from wheat, CaMV 35S, and *nos*). Expression of the *gus*A reporter gene could be detected in different cell types, including tracheids, ray parenchyma, and axial parenchyma associated with resin canals. Again, the *Em* promoter was the most efficient.

Transgenic conifers

Perhaps the most significant recent advance in transformation of forest trees has been the production of the first transgenic conifers. Transgenic plants have been produced in white spruce and larch. Ellis *et al.* (1991*b*, 1992, 1993) were able to regenerate stably transformed plants of white spruce following microprojectile bombardment. Late stages of somatic embryos were bombarded and selected using a sublethal level of kanamycin. Cell lines were obtained that were able to form differentiated embryos. Transformed seedlings expressed genes for *gus*A and *Bacillus thuringiensis* endotoxin (Ellis *et al.*, 1993). European larch (*Larix decidua*) was transformed by *Agrobacterium rhizogenes* (Huang, Diner & Karnosky, 1991; Huang & Karnosky, 1991). *Agrobacterium rhizogenes* induced hairy roots and/or adventitious shoot buds from transformed cells. Adventitious buds were excised, cultured, rooted and transferred to the greenhouse as plantlets. Transformation was verified by opine synthesis, and Southern blot hybridization. The next steps will be the development of routine protocols for transformation in these species and other commercially important conifers.

Transformation in hardwoods

Hardwoods, the general term for dicot tree species, include a much larger and more diverse number of species than is found among the conifers (softwoods). Wood properties are more variable in hardwoods, and different hardwoods are used for a larger diversity of products (Zobel & van Buijtenen, 1989). Hardwoods are used for pulp, paper and structural material. In addition, hardwoods are a major source of fuel around the world, and are used for a great variety of specialty products. The major cultivated hardwood forest trees are in the genus *Eucalyptus*. Many other trees are cultivated to a lesser extent but have features that make them promising candidates for genetic engineering. Our discussion here is focused on trees grown for wood and wood products. Many important tree species are being investigated for other purposes, such as trees being genetically modified for improved rubber production (Kitayama *et al.*, 1990) or for horticultural crops such as apple (James *et al.*, 1989) or walnut (McGranahan *et al.*, 1990).

Populus as a model system

The genus *Populus*, which includes the aspens and the poplars, has been increasing in importance as a commercial species. Poplars, particularly hybrid poplars, have fast growth, good form, short rotation times, and are easily propagated vegetatively. They are grown in Europe and several northern regions of the USA. Poplars have become a model forest tree for molecular genetics owing to the small size of their genome (Arumuganathan & Earle, 1991), efficient regeneration in culture (Ahuja, 1986; McCown *et al.*, 1988) and ability to be transformed by *Agrobacterium* (Parsons *et al.*, 1986; Fillatti *et al.*, 1987; Pythoud *et al.*, 1987; De Block, 1990). Poplars are well suited for fundamental studies of tree physiology, and as a model system for genetic engineering of woody plants. Wood from poplars and aspens is used for a variety of processed wood products such as pulp, paper, plywood, hardboard, and packing materials, in addition to furniture (Burns & Honkala, 1990).

A major problem in *Populus* transformation is the variation within the genus for transformation and regeneration. Some sections of the genus are readily transformed and regenerated, whereas others remain recalcitrant. Often the species of interest for practical applications are poorly transformed, particularly some of the fast growing commercial hybrids. There are many advantages to the *Populus* species for basic research and it remains one of the most useful species for molecular genetics of forest trees.

Populus trichocarpa × *P. deltoides* (clone H11) a clone with significant economic potential, has been transformed with a genetically modified strain of *A. rhizogenes* (Parsons *et al.*, 1986; Pythoud *et al.*, 1987). A plasmid containing the *vir* region of the supervirulent plasmid pTiBo542 dramatically increased the infectivity. Transformation was verified by Southern blots and regeneration of shoots containing the gene for NPT II was also reported. Poplar cells transformed with these strains grow in the absence of added growth regulators, and synthesize strain-specific opines. The highest frequency of transformation was obtained with small greenwood stem sections that were cocultivated with bacteria and transferred to medium lacking hormones. Transformed cells gave rise to rapidly growing tumors that proliferated indefinitely.

Wang *et al.* (1990) reported transformation and regeneration from leaf explants of *Populus tomentosa* Carr. DNA transfer was indicated by expression of the transferred DNA (T-DNA) gene 4 and expression of chloramphenicol acetyltransferase. An alternative method for transformation of poplar *(P. tremula* L. × *P. alba* L.) by has been reported where stem internodes were inoculated with a suspension of two strains of *Agrobacterium*. A wild-type strain that induces shoot differentiation in tumors and a disarmed strain carrying a binary vector were mixed together and used to inoculate stems. Shoots were obtained from tumors that expressed both *npt*II *and gus*A derived from the binary strain (Brasiliero *et al.*, 1991). Subsequent efforts have produced transgenic plants with additional constructs in as little as 3 months (Leple *et al.*, 1992).

Herbicide resistant *Populus*

The first transgenic forest tree was a poplar carrying glyphosate tolerance. Fillatti *et al.* (1987) used an oncogenic binary system of *A. tumefaciens* (C58/587/85) containing a wild-type C58 plasmid, and an engineered plasmid (pPMG587/85), to infect leaf discs of a hybrid aspen clone NC5339 *(P. alba* L. × *P. grandidentata* Michx.).

Transformation was obtained by cocultivating leaf segments with *Agrobacterium* and a feeder layer of tobacco suspension cells (Horsch *et al.*, 1985). The regenerated transformed plants carried an active kanamycin resistance gene and an altered *aro*A gene for resistance to the herbicide glyphosate (Comai *et al.*, 1985). The *aro*A gene codes for 5-enolpyruvylshikimate synthase, an enzyme active in the synthesis of aromatic amino acids. Transformation was also confirmed by Southern and Western blotting.

The transformed poplar plants were resistant to glyphosate at levels of 0.07 kg/ha, providing further evidence that the gene for resistance is expressed and functional (Fillatti *et al.*, 1988). The level of tolerance, although clearly different from that of untransformed controls, is not at the level that would be needed for commercial application (Riemenschneider *et al.*, 1988). It may be expected that higher levels of activity will be obtained with new constructs. These experiments are a landmark in forest biotechnology because they represent the first example of a tree modified by genetic engineering.

De Block (1990) examined factors affecting transformation in a hybrid aspen (clone 357; *Populus alba* × *P. tremula*) and a hybrid poplar (clone 064; *Populus trichocarpa* Torr. & Gray × *P. deltoides* Bartr. ex Marsh). Shoot-tip necrosis effects produced in culture were overcome by modification of the nitrate/ammonia ratio of the medium (De Block, 1990). Efficient transformation was obtained using a disarmed *Agrobacterium tumefaciens* strain carrying chimeric genes for resistance to phosphotricine (glyphosinate) resistance (De Block *et al.*, 1987). The enzyme phosphotricin acetyltransferase (PAT) acetylates phosphotricine and inactivates it. The gene *bar*, codes for PAT. Glyphosinate inhibits glutamine synthetase resulting in lethal accumulation of ammonium (Murakami *et al.*, 1986). Trees transformed with the *bar* gene were normal in appearance, and were completely resistant to field level preparations of the herbicide (Botterman *et al.*, 1991). A similar study has been carried out to confer resistance to Basta where A. *rhizogenes* was used to introduce the gene into a hybrid clone *(P. tremula* × *P. alba)*. Regenerated plants were tolerant to the herbicide (Devillard, 1992).

Transformation of *Populus* for insect resistance

Populus is a preferred target of the gypsy moth, the major insect pest of hardwoods in North America (Montgomery & Wallner, 1988). Insect resistance is, therefore, an important trait for genetic improvement in *Populus*. Most studies of transformation in *Populus* use *Agrobacterium* to transfer DNA to plant cells; however, McGown *et al.* (1991) used direct gene transfer with a particle gun (electric discharge). In their system, three different target tissues (protoplast-derived cells, organized cultured nodules, and stem explants were bombarded. Two unrelated hybrid *Populus* genotypes were tested (*P. alba* × *P. grandidentata* 'Crandon' and *P. nigra* 'Beautifolia' × *P. trichocarpa*). Four plants were recovered from the more easily manipulated genotype *(P. alba* × *P. grandidentata)* that were transformed for *nos*-NPT II, CaMV 35S-GUS, and 35S-BT (modified endotoxin gene from *Bacillus thuringiensis*). One of the plants was highly resistant to feeding of two lepidopteran pests, the forest tent caterpillar (*Malacosoma disstria* Hubner) and the gypsy moth (*Lymantria dispar* L.). Transformation by bombardment has an advantage over *Agrobacterium*, where transformation is limited by host pathogen incompatibility. Recovery of transformed plants depended upon pretreatment of target tissues, optimization of bombardment parameters, and the use of a selection technique employing flooding of the target tissues.

The gene coding for potato proteinase inhibitor II (*pin2*) is one of the best-characterized plant defense genes (Ryan, 1990), and acts through production of a small proteinaceous inhibitor of animal digestive enzymes. The inhibitors accumulate in foliage after insect attack or mechanical wounding. The *pin2* gene is inducible in potato and retains inducibility when transferred to tobacco (Thornburg *et al.*, 1987). The wound-inducible *pin2* promoter from potato, in a gene fusion with *cat* (gene for chloramphenicol aminotransferase), was transferred into hybrid poplar where it was still wound inducible (Klopfenstein *et al.*, 1991). The poplar hybrid was *P. alba* L. × *P. grandidentata* Michx. clone Hansen, which is a natural hybrid originating in southeast Iowa. *Agrobacterium tumefaciens* transformation was carried out using a binary miniplasmid pRT45, in armed A281 containing the *pin2-cat* gene fusion.

Northern and Southern blot analysis verified that transformation of single gene copies had occurred and that expression of *pin2* was wound inducible.

Transformation in *Eucalyptus*

One of the world's most important genera of forest trees is *Eucalyptus*. Eucalypts are grown for pulp, paper and wood on several continents and are the most important tree for wood products in tropical forestry. Some hybrids of *Eucalyptus* (e.g. *E. grandis* Hile ex Maiden × *E. camaldulensis* Dehn.) are used in Brazil for charcoal (Zobel & van Buijtenen, 1989). In intensively managed plantations, hybrids of *Eucalyptus* (e.g. *grandis* × *urophylla* S.T. Blake) are grown on a 6 year rotation (Zobel & Talbert, 1984). Transformation in this group of forest trees could have a significant effect on the practice of plantation forestry in the tropical and subtropical regions of the world.

At least one species of *Eucalyptus* (*E. tereticornis* Sm.) has been reported to be susceptible to *Agrobacterium* (Jindal & Bhardwaj, 1986). Another report of transformation of *E. gunnii* H. by *A. rhizogenes*, was based on phenotypic effects in culture (Adam, 1986). Transient gene expression has been reported in *E. citrodora* Hook, using electroporation of protoplasts (Manders *et al.*, 1992) and in *E. gunnii* (Teulières *et al.*, 1991; Teulières, Leborgne & Boudet, 1991). In *E. gunnii*, transient expression of *gus*A was observed in protoplasts derived from callus or suspension culture after electroporation or treatment with PEG. Transient expression was also observed after electroporation of intact cells.

Transformation in energy and biomass species

Several forest tree species have been a focus of attention for biomass production for fuel. A program in the US has focused on five plants for short rotation woody crops (SRWC) for energy. In *Populus* and sweetgum (*Liquidambar styraciflua* L.), efficient transformation and regeneration of transformed plants have been achieved. In the other species of interest for SRWC, black locust *(Robinia pseudoacacia* L.), sycamore *(Platanus occidentalis* L.), and silver maple *(Acer saccarinium* L.), less work has been done (Harry & Sederoff, 1989).

In the southeastern USA, *Liquidambar styraci-*

flua (American sweetgum) is a rapidly growing high quality hardwood that is used as an ornamental species and grown for high quality paper (Kormanik, 1990). It has the potential to be improved by genetic engineering for insect resistance, lignin content or cold hardiness (Sullivan & Lagrimini, 1992). Two groups have independently used *Agrobacterium tumefaciens* to transform and regenerate sweetgum. Chen (1991) has developed a root nodule system (Z.-Z. Chen & A.-M. Stomp, personal communication) that has a high capacity for shoot regeneration. Cocultivation of leaf pieces with *Agrobacterium* gives rise to transformed cell lines from which plants can be regenerated (Chen, 1991). Sullivan & Lagrimini (1992) introduced genes for insect resistance, enhanced peroxidase activity, and reduced peroxidase activity. In both laboratories, transformation was verified by molecular hybridization or expression of reporter genes.

Black locust is susceptible to *A. tumefaciens and A. rhizogenes* (Davis & Keathley, 1989). Inoculation *in vitro* produced tumors that were phytohormone independent. All tumors had segments of T-DNA, and callus obtained from infection of A281 carrying pGA472 as a binary was kanamycin resistant and showed integration of *npt*II DNA by Southern analysis.

In Sweden, willows, *Salix* (species) have been a focus of attention for fuel production. *Salix* grows rapidly in high latitudes and is readily propagated. *Salix* clones are being tested for use as an energy-generating tree species as one component of a national alternative energy program to reduce dependence on nuclear power. Willows readily coppice and are useful for high yield cropping systems. Vahala, Stabel & Eriksson (1989) transformed *Salix* stem explants with *A. tumefaciens* strains C58 and GV3101 and observed growth on hormone-free medium containing kanamycin. Transformation was confirmed by molecular hybridization and by opine synthesis. In addition, transformation has been observed in leaf discs of willow that produced callus and roots (Rocha & Maynard, 1990). Hauth & Beiderbeck (1992) used *A. rhizogenes* to infect roots of *Salix alba* L. and established cultures of hairy roots after infection.

Transformation in yellow poplar

Yellow poplar *(Liriodendron tulipifera* L.) is native to the central eastern USA and Canada. The tree is characterized by good form, high quality wood and rapid growth. It is sometimes considered to be one of the most important hardwood species in the USA (Merkle & Sommer, 1991). Somatic embryogenesis of yellow poplar was first reported in 1986 (Merkle & Sommer, 1986). Low frequencies of conversion of embryos to plantlets limited the usefulness of this system for transformation studies. Methods for conversion of embryos at higher frequency were developed (Merkle *et al.*, 1990). Populations of synchronized mature somatic embryos were obtained from proembryonic masses cultured in liquid medium and selected for size on stainless steel sieves. Mature embryos transferred to basal medium produced plantlets.

Embryogenic suspension cultures of yellow poplar have been transformed for the *gus*A gene using either direct DNA uptake into protoplasts or microprojectile bombardment of intact cells (Wilde, Meagher, & Merkle, 1991). In subsequent studies, transgenic plants were produced from transformed suspension cells (Wilde *et al.*, 1992). Plasmid DNA containing genes for GUS and NPT II was introduced by microprojectile bombardment and mature plants were regenerated. Although, work on transformation of yellow poplar is recent, it has significant promise as a model tree species because of the excellent system for regeneration. It will be difficult to apply the system to different genotypes because somatic embryonic cultures must be produced from differentiating zygotic embryos.

Nitrogen fixation and genetic engineering in forest trees

Actinorhizal plants can establish symbiotic relationships with the actinomycete *Frankia,* and thereby fix nitrogen in root nodules (Normand & Lalonde, 1986). *Alnus* (alder) is an actinorhizal forest tree found in high latitudes that could be useful for land reclamation and reforestation. The capacity of alder *(Alnus incana; Alnus glutinosa* L. Gaertn.) for transformation by *A. tumefaciens* has been demonstrated (Mackay, Seguin & Lalonde, 1988). Tumor formation, integration of T-DNA and expression of opine synthesis were found, using standard strains of ACH5 or C58. Electroporation of *gus*A driven by the 35S promoter of CaMV into protoplasts of *Alnus incana* showed transient expression. Expression was maximal at 1 to 2 days and declined by day 4 (Seguin & Lalonde, 1988).

Betula (birch) is closely related to *Alnus* but does not have a symbiotic system for nitrogen fixation. Genes involved in symbiotic nitrogen fixation in *Alnus*, if transferred to birch, might confer the ability to establish nitrogen fixation (Normand & Lalonde, 1986). *Agrobacterium tumefaciens* was used to produce tumors in four clones of paper birch *(Betula papyrifera* Marsh.). Transformation was verified by Southern blotting, and opine synthesis (Mackay *et al.*, 1988).

Reforestation on nitrogen-deficient soils is an important problem in tropical forestry and subtropical forestry where soil degradation has occurred. Trees that fix nitrogen are also important for agroforestry. Species of Casuarinaceae that are nodulated by the actinomycete *Frankia* can also fix atmospheric nitrogen and can grow on nitrogen-deficient soils. The family includes some high yielding, fast growing trees with a high value for firewood. *Allocasuarina verticillata* Lam., a member of this family of nitrogen-fixing trees, has been transformed using *Agrobacterium rhizogenes* (Phelep *et al.*, 1991). Hairy root formation was induced, followed by induced or spontaneous shoot formation. Shoots were rooted to produce transgenic plants with extensive ageotropic root systems. Transformation has been verified by Southern blot analysis and by the detection of specific opines. Phelep *et al.* envisioned introducing genes that could contribute to pest or herbicide resistance for trees that can fix nitrogen as agronomic crops.

Some other trees of interest

Bolyard, Hajela & Sticklen (1991) have initiated transformation experiments in the hybrid pioneer elm (*Ulmus* 'Pioneer'). The purpose of these experiments is the eventual introduction of genes that could enhance resistance to the fungus that causes Dutch elm disease. Both microprojectile bombardment and *Agrobacterium* infection of stem sections resulted in transformed callus, on the basis of Southern blotting and fluorogenic detection of reporter gene activity.

Oaks are a large and diverse group of trees that have not been intensively studied at the molecular level. Initial attempts to transform *Quercus robur* L. are promising. Nodal stem explants appear to be transformed by *A. rhizogenes* 9402 or *A. tume-*

faciens A281, both containing *gus*A (Roest *et al.*, 1991). Sterile root clones were established and showed GUS activity.

Neem (*Azadirachta indica*) is an extremely versatile tropical tree that is used for veneer, firewood and as a shade tree. The leaves of neem are also used for their insecticidal properties. Neem has been transformed and regenerated (Naina, Gupta & Mascarenhas, 1989). *Agrobacterium tumefaciens* strains containing a recombinant derivative of pTiA6 were used to infect seedlings grown *in vitro*. Induced tumors gave rise to transformed callus. Plantlets were obtained from the callus and from induced shoots. Transformation was indicated by autotrophic growth of callus, synthesis of octopine and resistance to kanamycin.

Barriers to progress

Although several species will provide good model systems for the production of transgenic plants, many of the most commercially important species, for example pines, remain recalcitrant to transformation. The barriers to transformation and regeneration rest, in significant part, in limitations of the tissue culture systems. The central problem is the introduction of DNA into cells that will be capable of regeneration. In some cases, the cells that can be transformed are not the cells that can regenerate. It is essential that cells can survive the stress of the transformation protocols, whether the stress is biological, resulting from *Agrobacterium* infection, or physical, resulting from bombardment. In either case, cells that are transformed must proliferate, survive selection, and regenerate to form plantlets through organogenesis or embryogenesis. Culture systems can be very slow for many forest tree species (von Arnold & Woodward, 1988; Tautorus, Fowke & Dunstan, 1991).

Many difficulties are simply problems of efficiency, where several steps in the process each occur at low frequency. It may be that success requires only increases in frequency of rare events. Although much discussion has taken place regarding the relative merits of *Agrobacterium* transformation and bombardment, both methods have been successful in specific systems and the choice of method depends upon the biology of the system. Improvements continue in marker systems, vectors and transformation efficiency. Screening and definition of genetic variation for culture parameters could have significant effects. Development of somatic embryogenesis for conifer species, for example, has often required screening for rare genotypes that could proliferate and differentiate in culture.

Societal implications

Finally, the ecological and social implications of genetic engineering of forest trees are significant. Special interest groups in the public and private sectors are concerned about the potential threat to the value and stability of our forests posed by genetic engineering. Their concerns result from conflicting interests and the economic pressures to harvest trees from ecologically sensitive stands or from recreation areas. Opposition to genetic engineering of trees exists in part because of a presumed threat to the natural forests. However, the increasing economic pressure for wood and wood products may require use of intensive management of plantations with genetically engineered trees to produce enough wood efficiently on smaller areas of land. Another area of concern lies in the potential introduction of foreign genes into the natural populations. Methods to survey gene flow into natural populations need to be developed as well as methods to restrict potential gene flow. It may be necessary to engineer reproductive sterility into such clones to contain foreign genes within managed programs.

References

Adam, S. (1986). Obtention de racines transformées chez *Eucalyptus gunnii* H. par *Agrobacterium rhizogenes*. *Annales de Recherches Sylvicoles*, pp. 7–21.

Ahuja, M. R. (1986). Aspen. In *Handbook of Plant Cell Culture*, vol. 4, ed. D. A. Evans, W. R. Sharp & P. V. Ammirato, pp. 626–651. Macmillan, New York.

Ahuja, M. R. & Libby, W. J. (1993). *Clonal Forestry*. Vol. I *Genetics and Biotechnology*. Springer-Verlag, Berlin, Heidelberg.

Arumuganathan, K. & Earle, E. (1991). Nuclear DNA content of some important plant species. *Plant Molecular Biology Reporter*, **9**, 208–218.

Bekkaoui, F., Pilon, M., Laine, E., Raju, D. S. S., Crosby, W. L. & Dunstan, D. I. (1988). Transient gene expression in electroporated *Picea glauca* protoplasts. *Plant Cell Reports*, **7**, 481–484.

Bergmann, B. (1992). Genotype influence on adventitious organogenesis and susceptibility to *Agrobacterium tumefaciens* in *Pinus*. Ph.D. dissertation, North Carolina State University.

Bolyard, M. G., Hajela, R. K. & Sticklen, M. B. (1991). Microprojectile and *Agrobacterium*-mediated transformation of pioneer elm. *Journal of Arboriculture*, **17**, 34–37.

Botterman, J., D'Halluin, K., De Block, M., De Greef, W. & Leemans, J. (1991). Engineering of glufosinate resistance and evaluation under field conditions: phosphinothricin-acetyltransferase gene expression in transgenic tobacco, tomato, potato, poplar, alfalfa, rape and sugarbeet to confer herbicide resistance. In *Herbicide Resistance in Weeds and Crops, 11th Long Ashton International Symposium*, pp. 355–363. Butterworth-Heinemann, New York.

Brasileiro, A. C. M., Leple, J. C., Muzzin, J., Ounnoughi, D., Michel, M. F. & Jouanin, L. (1991). An alternative approach for gene transfer in trees using wild-type *Agrobacterium* strains. *Plant Molecular Biology*, **3**, 441–452.

Buonjiorno, J. & Grosenick, G. L. (1977). Impact of world economic and demographic growth on forest products consumption and wood requirements. *Canadian Journal of Forest Research*, **7**, 392–399.

Burns, R. M. & Honkala, B. H. (1990). *Silvics of North America*, Vol. 2 *Hardwoods*. Forest Service, U. S. Department of Agriculture Handbook **654**.

Charest, P. J. & Michel, M. F. (1991). Basics of plant genetic engineering and its potential applications to tree species. *Petawawa National Forestry Institute, Chalk River, Ontario. Information Report* No. PI-X–104.

Chen, Z.-Z. (1991). Nodular culture and *Agrobacterium* mediated transformation for transgenic plant production in *Liquidambar styraciflua* L.(sweetgum). Ph.D. dissertation, North Carolina State University.

Clapham, D. & Ekberg, 1. (1986). Induction of tumors by various strains of *Agrobacterium tumefaciens* on *Abies nordmanniana* and *Picea abies*. *Scandinavian Journal of Forest Research*, **1**, 435–437.

Clapham, D. H. & Ekberg, I. (1988). Induction of tumors by various strains of *Agrobacterium tumefaciens* on *Abies nordmanniana* and *Picea abies*. In *Genetic Manipulation of Woody Plants*, ed. H. W. Hanover & D. E. Keathley, p. 463. Plenum Press, New York.

Clapham, D. H., Ekberg, I., Eriksson, G., Hood, E. E. & Norell, L. (1990). Within population variation in susceptibility to *Agrobacterium tumefaciens* A281 in *Picea abies* (L.) Karst. *Theoretical and Applied Genetics*, **79**, 654–656.

Comai, L., Facciotti, D., Hiatt, W. R., Thompson, G., Rose, R. & Stalker, D. (1985). Expression in plants of a mutant *aroA* gene from *Salmonella typhimurium* confers tolerance to glyphosate. *Nature (London)*, **317**, 741–744.

Dandekar, A. M., Gupta, P. K., Durzan, D. J. & Knauf, V. (1987). Transformation and foreign gene expression in micropropagated Douglas-fir *(Pseudotsuga menziesii)*. *Bio/Technology*, **5**, 587–590.

Davis, J. M. & Keathley, D. E. (1989). Detection and analysis of T-DNA in crown gall tumors and kanamycin resistant callus of *Robinia pseudoacacia*. *Canadian Journal of Forest Research*, **19**, 1118–1123.

De Block, M. (1990). Factors influencing the tissue culture and the *Agrobacterium*-mediated transformation of hybrid aspen and poplar clones. *Plant Physiology*, **93**, 1110–1116.

De Block, M., Botterman, J., Vandewiele, M., Dockx, J., Thoen, C., Gossele, V., Rao Movva, N., Thompson, C., Van Montagu, M. & Leemans, J. (1987). Engineering herbicide resistance in plants by expression of a detoxifying enzyme. *EMBO Journal*, **6**, 2513–2518.

De Cleene, M. & De Ley, J. (1976). The host range of crown gall. *Botany Review*, **42**, 389–466.

De Cleene, M. & De Ley, J. (1981). The host range of infectious hairy-root. *Botany Review*, **47**, 147–194.

Devillard, C. (1992). Transformation *in vitro* du tremble (*Populus tremula* × *Populus alba*) par *Agrobacterium rhizogenes* et regénération de plantes tolerant au basta. *Compte Rendu de l'Académie des Sciences*. Serie 3, *Sciences de la Vie*, **314**, 291–298.

Duchesne, L. C. & Charest, P. J. (1991). Transient expression of the beta-glucuronidase gene in embryogenic callus of *Picea mariana* following microprojection-tungsten biolistic microprojectile-mediated DNA transfer. *Plant Cell Reports*, **10**, 191–194.

Duchesne, L. C. & Charest, P. J. (1992). Effect of promoter sequence on transient expression of the beta-glucuronidase gene in embryogenic calli of *Larix* × *eurolepis* and *Picea mariana* following microprojection-hybrid larch and black spruce callus culture transformation by biolistic microprojectile method. *Canadian Journal of Botany*, **70**, 175–180.

Dunstan, D. I. (1989). Transient gene expression in electroporated conifer protoplasts: electroporation of white spruce, black spruce and jack pine protoplasts. *In Vitro*, **25**, Pt.2, 63A.

Ellis, D., Roberts, D., Sutton, B., Lazaroff, W., Webb, D. & Flinn, B. (1989). Transformation of white spruce and other conifer species by *Agrobacterium tumefaciens*. *Plant Cell Reports*, **8**, 16–20.

Ellis, D. D., McCabe, D., Russell, D., Martinell, B. & McCown, B. H. (1991*a*). Expression of inducible angiosperm promoters in a gymnosperm, *Picea glauca* (white spruce). *Plant Molecular Biology*, **17**, 19–27.

Ellis, D. D., McCabe, D., McInnis, S., Martinell, B. & McCown, B. H. (1991*b*). Transformation of white spruce by electrical discharge particle acceleration. In *Application of Biotechnology to Tree Culture, Protection and Utilization*, ed. B. E. Hassig, T. K. Kirk, W. L. Olsen, K. F. Raffa & J. M. Slavicek. USDA Forest Service, Columbus, OH.

Ellis, D. D., McCabe, D., McInnis, S., Ramachandran, R. & McCown, B. (1992). Transformation of *Picea glauca*-gene expression and the regeneration of stably transformed embryogenic callus and plants. *International Conifer Biotechnology Working Group*. abstract no. 29.

Ellis, D. D., McCabe, D. E., McInnis, S., Ramachandran, R., Russell, D. R., Wallace, K. M., Martinell, B. J., Roberts, D. R., Raffa, K. F. & McCown, B. H. (1993). Stable transformation of *Picea glauca* by particle acceleration. *Bio/Technology*, **11**, 84–89.

Fillatti, J. J., Hassig, B., McCown, B., Comai, L. & Riemenschneider, D. (1988). Development of glyphosate tolerant *Populus* plants through expression of a mutant *aroA* gene from *Salmonella typhimurium*. In *Genetic Manipulation of Woody Plants*, ed. J. Hanover & D. Keathley, pp. 243–250. Plenum Press, New York.

Fillatti, J. J., Selmer, J., McCown, B., Hassig, B. & Comai, L. (1987). *Agrobacterium* mediated transformation and regeneration of *Populus*. *Molecular and General Genetics*, **206**, 192–199.

Gammie, J. (1981). World timber to the year 2000. *The Economist Intelligence Unit Special Report*, No. 98, Economist Intelligence Unit Ltd, London.

Goldfarb, B., Strauss, S. H., Howe, G. T. & Zaerr, J. B. (1991). Transient gene expression of microprojectile-introduced DNA in Douglas fir cotyledons. *Plant Cell Reports*, **10**, 517–521.

Gupta, P. K., Dandekar, A. M. & Durzan, D. J. (1988). Somatic preembryo formation and transient expression of a luciferase gene in Douglas fir and loblolly pine protoplasts. *Plant Science*, **58**, 85–92.

Hakman, I. & Fowke, L. C. (1987). Somatic embryogenesis in *Picea glauca* (white spruce) and *Picea mariana* (black spruce). *Canadian Journal of Botany*, **65**, 656–659.

Hakman, I. & von Arnold, S. (1985). Plantlet regeneration through somatic embryogenesis in *Picea abies*. *Journal of Plant Physiology*, **121**, 149–158.

Hamrick, J. L., Godt, M. J. W. & Sherman-Broyles, S. L. (1992). Factors influencing levels of genetic diversity in woody plant species. *New Forests*, **6**, 95–124.

Harry, D. E. & Sederoff, R. R. (1989). *Biotechnology in Biomass Crop Production: The Relationship of Biomass Production and Genetic Engineering*. Oak Ridge National Laboratory, Environment Sciences Division of Publications, No. 3411.

Hatzopoulos, P., Fong, F. & Sung, R. (1990). Abscisic

acid regulation of Dc8, a carrot embryonic gene. *Plant Physiology*, **94**, 690–695.

Hauth, S. & Beiderbeck, R. (1992) In vitro culture of *Agrobacterium* induced hairy roots *of Salix alba* L. *Silvae Genetica*, **41**, 46–48.

Hood, E. E., Clapham, D. H., Ekberg, I. & Johannson, T. (1990). T-DNA presence and opine production in tumors of *Picea abies* (L.) Karst induced by *Agrobacterium tumefaciens* A281. *Plant Molecular Biology*, **14**, 111–117.

Hood, E. E., Ekberg, I., Johannson, T. & Clapham, D. H. (1989). T-DNA presence and opine production in *Agrobacterium tumefaciens* A281 induced tumors on Norway spruce. *Journal of Cellular Biochemistry*, Suppl. 13 Part D, 260.

Horsch, R. B., Fry, J. E., Hoffman, N. L., Eichholz, D., Rogers, S. G. & Fraley, R. T. (1985). A simple and general method for transferring genes into plants. *Science*, **227**, 1229–1231.

Huang. Y., Diner, A. M. & Karnosky, D. F. (1991). *Agrobacterium rhizogenes*-mediated genetic transformation and regeneration of a conifer: *Larix decidua. In Vitro Cellular and Developmental Biology*, **4**, 201–207.

Huang, Y. & Kamosky, D. F. (1991). A system for gymnosperm transformation and plant regeneration: *Agrobacterium rhizogenes* and *Larix decidua*-European larch hairy root culture and propagation. *In Vitro*, **27**, 153A.

Hutchison, K. W. & Greenwood, M. S. (1991). Molecular approaches to gene expression during conifer development and maturation. *Forest Ecology and Management*, **43**, 273–286.

James, D. J., Passey, A. J., Barbara, D. J. & Bevan, M. (1989). Genetic transformation of apple *(Malus pumila* Mill.) using a disarmed Ti-binary vector. *Plant Cell Reports*, **7**, 658–661.

Jindal, K. K. & Bhardwaj, L. N. (1986). Occurrence of crown gall on *Eucalyptus tereticornis* Sm. in India. *Indian Forester*, **112**, 1121.

Kitayama, M., Takahashi, M., Surzycki, S. & Togasaki, R. (1990). Transformation of callus tissue from *Hevea brasillensis* and *Jasminium officinale*. *Plant Physiology* 93, 1, Suppl.46.

Klopfenstein, N. B., Shi, N. Q., Kernan, A., McNabb, H. S., Jr, Hall, R. B., Hart, E. R. & Thornburg, R. W. (1991). Transgenic *Populus* hybrid expresses a wound inducible potato proteinase inhibitor II-CAT gene fusion. *Canadian Journal of Forest Research*, **21**, 1321–1328.

Kormanik, P. P. (1990). *Liquidambar styraciflua, Sweetgum. In Silvics of North America*, ed. R. M. Burns & B. H. Honkala, Technical Coordinators. USDA Forest Service Publication No. 654, pp. 400–405.

Leple, J. C., Brasileiro, A. C. M., Michel, M. F., Delmotte, F. & Jouanin, L. (1992). Transgenic

poplars: expression of chimeric genes using four different constructs. *Plant Cell Reports*, **11**, 137–141.

Loopstra, C. A., Stomp, A.-M. & Sederoff, R. R. (1990). *Agrobacterium* mediated DNA transfer in sugar pine. *Plant Molecular Biology*, **15**, 1–9.

Loopstra, C. A., Weissinger, A. K. & Sederoff, R. R. (1992). Transient expression in differentiating wood. *Canadian Journal of Forest Research*. **22**, 993–996.

Mackay, J., Seguin, A. & Lalonde, M. (1988). Genetic transformation of 9 in vitro clones of *Alnus* and *Betula* by *Agrobacterium tumefaciens*. *Plant Cell Reports*, **7**, 229–232.

Manders, G., Dossantos, A. V. P., Vaz, F. B. D., Davey, M. R. & Power, J. B. (1992). Transient gene expression in electroporated protoplasts of *Eucalyptus citriodora* Hook. *Plant Cell Tissue and Organ Culture*, **30**, 69–75.

Marcotte, W. R. Jr, Russell, S. H. & Quatrano, R. S. (1989). Abscisic acid-responsive sequences from the Em gene of wheat. *Plant Cell*, **1**, 969–976.

McCown, B. H., McCabe, D. E., Russell, D. R., Robison, D. J., Barton, K. A. & Raffa, K. F. (1991). Stable transformation of *Populus* and incorporation of pest resistance by electric discharge particle acceleration. *Plant Cell Reports*, **9**, 590–594.

McCown, B. H., Zeldin, E. L., Pinkalla, H. A. & Dedolph, R. R. (1988). Nodule culture: a developmental pathway with high potential for regeneration, automated micropropagation, and plant metabolite production from woody plants. In *Genetic Manipulation of Woody Plants*, ed. J. W. Hanover & D. E. Keathley, pp. 149–166. Plenum Press, New York.

McGranahan, G. H., Leslie, C. A., Uratsu, S. L. & Dandekar, A. (1990). Improved efficiency of the walnut somatic embryo gene transfer system. *Plant Cell Reports*, **8**, 512–516.

McGranahan, G. H., Leslie, C. A., Uratsu, S. L., Martin, L. A. & Dandekar, A. M. (1988). *Agrobacterium*-mediated transformation of walnut somatic embryos and regeneration of transgenic plants. *Bio/Technology*, **6**, 800–804.

Merkle, S. A. & Sommer, H. E. (1986). Somatic embryogenesis in tissue cultures of *Liriodendron tulipifera*. *Canadian Journal of Forest Research*, **16**, 420–422.

Merkle, S. A. & Sommer, H. E. (1991). Yellow-poplar *(Liriodendron* spp.). In *Biotechnology in Agriculture and Forestry*, Vol 16 *Trees III*, ed. Y. P. S. Bajaj, pp. 94–110. Springer-Verlag, Berlin, Heidelberg.

Merkle, S. A., Wiecko, A. T., Sotak, R. J. & Sommer, H. E. (1990). Maturation and conversion of *Liriodendron tulipefera* somatic embryos. *In Vitro Cellular and Developmental Biology*, **26**, 1086–1093.

Montgomery, M. E. & Wallner, W. E. (1988). The gypsy moth: a westward migrant. In *Dynamics of*

Forest Insect Populations, ed. A. A. Berryman, pp. 353–375. Plenum Press, New York.

Morris, J. W., Castle, L. A. & Morris, R. O. (1988). Transformation of pinaceous gymnosperms by *Agrobacterium*. In *Genetic Manipulation of Woody Plants*, ed. J. W. Hanover & D. E. Keathley, p. 481. Plenum Press, New York.

Morris, J. W., Castle, L. A. & Morris, R. O. (1989). Efficacy of different *Agrobacterium tumefaciens* strains in transformation of pinaceous gymnosperms. *Physiological and Molecular Plant Pathology*, **34**, 451–462.

Morris, J. W. & Morris, R. O. (1990). Identification of an *Agrobacterium tumefaciens* virulence gene inducer from the pinaceous gymnosperm *Pseudotsuga menziesii*. *Proceedings of the National Academy of Sciences, USA*, **87**, 3614–3618.

Murakami, T., Anzai, H., Imai, S., Satoh, A., Nagaoka, K. & Thompson, C. J. (1986). The bialiphos biosynthetic genes of *Streptomyces hygroscopicus*: molecular cloning and characterization of the gene cluster. *Molecular and General Genetics*, **205**, 42–50.

Naina, N. S., Gupta, P. K. & Mascarenhas, A. F. (1989). Genetic transformation and regeneration of transgenic neem (*Azadirachta indica*) plants using *Agrobacterium tumefaciens*. *Current Science*, **58**, 184–187.

National Research Council, USA (1990). *Forestry Research: A Mandate for Change*. A report from the committee on Forestry Research, Commission on Life Sciences, Board on Biology & Board on Agriculture.

Neale, D. B. & Kinlaw, C. S. (1992). Forest biotechnology. *Forest Ecology and Management*, **43**, 179–324.

Newton, R. J., Yibrah, H. S., Dong, N., Clapham, D. H. & von Arnold, S. (1992). Expression of an abscisic acid responsive promoter in *Picea abies* (L.) Karst. following bombardment from an electrical discharge particle accelerator–microprojectile particle acceleration application to Norway spruce tissue culture and cell culture transformation. *Plant Cell Reports*, **11**, 188–191.

Normand, P. & Lalonde, M. (1986). The genetics of actinorhizal Frankia, a review. In *Third International Symposium on Nitrogen Fixation with Nonlegumes*, *Plant Soil*, **90**, 429–454.

Parsons, T. J., Sinkar, V. P., Stettler, R. F., Nester, E. W. & Gordon, M. P. (1986). Transformation of poplar by *Agrobacterium tumefaciens*. *Bio/Technology*, **4**, 533–536.

Phelep, M., Petit, A., Martin, L., Douhoux, E. & Tempé, J. (1991). Transformation and regeneration of a nitrogen fixing tree, *Allocasuarina verticillata* Lam. *Bio/Technology*, **9**, 461–466.

Pythoud, F., Sinkar, V. P., Nester, E. W. & Gordon, M. P. (1987). Increased virulence of *Agrobacterium rhizogenes* conferred by the vir region of pTiBo542: application to genetic engineering of poplar. *Bio/Technology*, **5**, 1323–1327.

Riemenschneider, D. E. (1990). Susceptibility of intra and interspecific hybrid poplars to *Agrobacterium tumefaciens* C58. *Phytopathology*, **80**, 1099–1102.

Riemenschneider, D. E., Haissig, B. E., Sellmer, J. & Fillatti, J. J. (1988). Expression of an herbicide tolerance gene in young plants of a transgenic hybrid poplar clone. In *Somatic Cell Genetics of Woody Plants*, ed. M. R. Ahuja, pp. 73–80. Kluwer Academic Publishers, Dordrecht.

Robertson, D., Weissinger, A. K., Ackley, R., Glover, S. & Sederoff, R. R. (1992). Genetic transformation of Norway spruce (*Picea abies* (L.) Karst) using somatic embryo explants by microprojectile bombardment. *Plant Molecular Biology*, **19**, 925–935.

Rocha, S. P. & Maynard, C. A. (1990). *In vitro* multiplication and *Agrobacterium* transformation of willow (*Salix lucida*). *In Vitro Cellular and Developmental Biology*, **26**, 43A (abstract 111).

Roest, S., Brueren, H. G. M. J., Evers, P. W. & Vermeer, E. (1991). *Agrobacterium* mediated transformation of oak (*Quercus robur* L.)-propagation. *Acta Horticulture*, **289**, 259–260.

Ryan, C. A. (1990). Protease inhibitors in plants: genes for improving defenses against insects and pathogens. *Annual Reviews of Phytopathology*, **28**, 425–449.

Savidge, R. A. (1989). Coniferin, a biochemical indicator of commitment to tracheid differentiation in conifers. *Canadian Journal of Botany*, **67**, 2663–2668.

Sederoff, R. & Chang, H.-M. (1991) Lignin biosynthesis. In *Wood Structure and Composition*, ed. M. Lewin & I. Goldstein, pp. 263–285. Marcel Dekker, Inc., New York.

Sederoff, R. & Stomp, A.-M. (1993). DNA transfer in conifers. In *Clonal Forestry I: Genetics and Biotechnology* ed. M. R. Ahuja & W. J. Libby, pp. 241–254. Springer-Verlag, Berlin, Heidelberg.

Sederoff, R. R., Stomp, A.-M., Chilton, W. S. & Moore, L. W. (1986). Gene transfer into loblolly pine by *Agrobacterium tumefaciens*. *Bio/Technology*, **4**, 647–649.

Seguin, A. & Lalonde, M. (1988). Gene transfer by electroporation in Betulaceae protoplasts: *Alnus incana*. *Plant Cell Reports*, 7, 367–370.

Smith, C. O. (1935*a*). Crown gall on conifers. *Phytopathology*, **25**, 894.

Smith, C. O. (1935*b*). Crown gall on the *Sequoia*. *Phytopathology*, **25**, 439–440.

Stomp, A.-M., Han, K.-H., Perkins, E. J. & Gordon, M. P. (1992). The use of genetically engineered trees and plants for environmental restoration and waste management. In *Proceedings of the Waste*

Management and Environmental Sciences Conference, San Juan, Puerto Rico, pp. 498–503. Department of Energy Publication.

Stomp, A.-M., Loopstra, C., Chilton, W. S., Sederoff, R. R. & Moore, L. W. (1990). Extended host range of *Agrobacterium tumefaciens* in the genus *Pinus. Plant Physiology,* 92, 1226–1232.

Stomp, A.-M., Loopstra, C., Sederoff, R. R., Chilton, S., Fillatti, J., Dupper, G., Tedeschi, P. & Kinlaw, C. (1988). Development of a DNA transfer system for pines. In *The Genetic Manipulation of Woody Plants,* Basic Life Sciences, vol. 44, ed. J. W. Hanover & D. E. Keathley, pp. 231–241. Plenum Press, New York.

Stomp, A.-M., Weissinger, A. K. & Sederoff, R. R. (1991). Transient expression from microprojectile-mediated DNA transfer in *Pinus taeda. Plant Cell Reports,* 10, 187–190.

Strauss, S. H., Howe, G. T. & Goldfarb, B. (1991). Prospects for genetic engineering of insect resistance in forest trees. *Forest Ecology and Management,* 43, 181–209.

Sullivan, J. & Lagrimini, L. M. (1992). Transformation of *Liquidambar styraciflua. Plant Physiology,* 99, 1, Suppl. 46.

Tautorus, T. E., Bekkaoui, F., Pilon, M., Dalta, R. S. S., Crosby, W. L., Fowke, L. C. & Dunstan, D. (1989). Factors affecting transient gene expression in electroporated black spruce (*Picea mariana*) and jack pine (*Pinus banksiana*) protoplasts. *Theoretical and Applied Genetics,* 78, 531–536.

Tautorus, T. E., Fowke, L. C. & Dunstan, D. I. (1991). Somatic embryogenesis in conifer propagation; a review. *Canadian Journal of Botany,* 69, 1873–1899.

Teulières, C., Grima-Pettenati, J., Curie, C., Teissie, J. & Boudet, A. M. (1991). Transient foreign gene expression in polyethylene glycol treated or electropulsated *Eucalyptus gunnii* protoplasts. *Plant Cell Tissue and Organ Culture,* 25, 125–132.

Teulières, C., Leborgne, N. & Boudet, A. M. (1991). Direct gene transfer procedures for *Eucalyptus* genetic transformation: *Eucalyptus* protoplast transformation using polyethylene glycol or electroporation for beta-glucuronidase reporter gene expression in culture. *In Vitro,* 28, 121A.

Thornburg, R. W., An, G., Cleveland, T. E., Johnson, R. & Ryan, C. A. (1987). Wound-inducible expression of a potato inhibitor II-chloramphenicol acetyltransferase gene fusion in transgenic tobacco plants. *Proceedings of the National Academy of Sciences, USA,* 84, 744–748.

USDA Forest Service (1982). An analysis of the timber situation in the United States, 1952–2030. *Forest Resources Report No. 23.* US Government Printing Office, Washington, DC.

Vahala, T., Stabel, P. & Eriksson, T. (1989). Genetic transformation of willows (*Salix* spp.) by *Agrobacterium tumefaciens. Plant Cell Reports,* 8, 55–58.

von Arnold, S., Clapham, D. & Ekberg, I. (1990). Has biotechnology a future in forest tree breeding? *Forest Tree Improvement,* 23, 31–48.

von Arnold, S. & Woodward, T. (1988). Organogenesis and embryogenesis in mature zygotic embryos of *Picea sitchensis. Tree Physiology,* 4, 291–300.

Wang, S. P., Xu, Z. H. & Wei, Z. M. (1990). Genetic transformation of leaf explants of *Populus tomentosa. Acta Botanica Sinica,* 3, 172–177.

Whetten, R. & Sederoff, R. (1991). Genetic engineering of wood. *Forest Ecology and Management,* 43, 301–316.

Wilde, H. D., Meagher, R. B. & Merkle, S. A. (1991). Transfer of foreign genes into yellow-poplar (*Liriodendron tulipifera*). In *Woody Plant Biotechnology,* ed. M. R. Ahuja, pp. 227–232. Plenum Press, New York.

Wilde, H. D., Meagher, R. B. & Merkle, S. A. (1992). Expression of foreign genes in transgenic yellow-poplar plants. *Plant Physiology,* 98, 114–120.

Wilson, S. M., Thorpe, T. A. & Moloney, M. M. (1989). PEG-mediated expression of GUS and CAT genes in protoplasts from embryogenic suspension cultures of *Picea glauca. Plant Cell Reports,* 7, 704–707.

Zobel, B. J. & Talbert, J. (1984). *Applied Forest Tree Improvement.* J. Wiley & Sons, New York.

Zobel, B. J. & van Buijtenen, J. P. (1989). *Wood Variation: Its Causes and Control.* Springer-Verlag, New York.

Index

Page numbers in *italics* refer to figures and tables

abscisic acid (ABA), 90
abscisic acid (ABA)-inducible promoter, 153
Ac transposase gene, 59–60
acetosyringone, 106, 153
Acinetobacter calcoaceticus transformation, 25
acquired immunodeficiency syndrome (AIDS), 10
*adh*1 promoter, 58
Agrobacterium, 23
  clone bank use, 29
  electroporation, 26
  Eucalyptus transformation, 156
  gene vector for maize, 66
  pKT230 mobilization, 27
  plasmid rearrangement, 28
  plasmid replication region cloning, 24
  Populus insect resistance, 156
  replication of recombinant molecules, 24
  transferred DNA, 27
  transformation, 24–5
  vector stability, 28
  vector system, 23
  vector transgene inheritance, 82
  virulence cascade in Douglas fir, 153
  virulence inhibition, 66
Agrobacterium rhizogenes
  alfalfa transformation, 114, 115
  birdsfoot trefoil transformation, 115, 116, 117
  black locust transformation, 157
  broad bean transformation, 109
  dry bean transformation, 111–12
  legume transformation, 119, 120
  nitrogen-fixing tree transformation, 158
  pea transformation, 107
  Populus transformation, 155
  rapeseed transformation, 126–7
  sanfoin transformation, 118
  soybean transformation, 103

  strain K599, 103
  TL-DNA, 127
Agrobacterium tumefaciens
  alfalfa transformation, 112, 113, 115
  birdsfoot trefoil transformation, 115–16, 117
  black locust transformation, 157
  broad bean transformation, 109
  cowpea transformation, 110, 111
  dry bean transformation, 111
  legume transformation, 118, 120
  lentil transformation, 110
  Lotononis bainesii transformation, 117
  nitrogen-fixing tree transformation, 157–8
  nopaline strains, 130
  octopine strains, 130
  pea transformation, 106, 107
  peanut transformation, 112
  Populus transformation, 155, 156
  rapeseed transformation, 127, 130–1
  soybean transformation, 101–3, 105–6
  sunflower infection efficiency, 140
  sunflower transformation, 138, 143–4, 145
  sweetgum transformation, 157
  transformation vector, 65
Agrobacterium-based vectors, small-grain cereals, 89
Agrobacterium-mediated transformation
  conifers, 152–3
  transgenic maize, 65–6, 74
agroinfection, 66
agropine, 114
alcohol dehydrogenase (ADH), 58
  anerobic induction of proteins, 58
alder, 157
alfalfa, 112–15
  Agrobacterium rhizogenes transformation, 115
  Agrobacterium tumefaciens transformation, 112, 113, 115

hairy root production, 114
ovalbumin gene insertion, 114
protoplast transformation technique, 115
regeneration, 112
somaclonal variation, 120
transformation, 118, *119*
alfalfa, transgenic
agropine, 114
Basta resistance, 120
GUS activity, 115
nopaline content, 114
NPT II activity, 112, 113, 115
ovalbumin expression, 120
Alnus, 157
aminoethoxyvinyl-glycine (AVG), 73
ampicillin, 11
resistance gene, 18
antibiotic resistance marker transduction, 16
antifungal genes, 120
antisense gene expression, 134
Arabidopsis thaliana, 53
Arachis hypogaea, 112
*aro*A gene, 155
Asparagus officinale, 65
Aspergillus
electroporation transformation, 36
invertase gene cloning, 41
protein secretion, 43–4
Aspergillus nidulans, 34
autonomously replicating sequence incorporation, 39
cloning complementing genes, 41
cosmid library, 41
gene replacement, 42
homologous integration, 39
introduced DNA sequence stability, 40
atrazine resistance, 120
gene, 105
autonomously replicating sequences (ARS), 38, 39
auxin, 130
regeneration induction, 139
auxin : cytokinin ratio for sunflower transformation, 143
auxotrophic markers, transduction, 16
auxotrophic mutants, 37
Azadirachta indica, 158
Azospirillum, 23
electroporation, 26
Ti plasmids, 27

Bacillus subtilis, PEG sensitivity, 13
Bacillus thuringiensis
endotoxin, 120, 154
δ-endotoxin gene, 61
forest trees, 151
insect control proteins, 61
bacteriophage, transducing, 16

bar gene, 85, 91
alfalfa transformation, 113
herbicide resistance in navy bean, 112
marker for small-grain cereal transformation, 93
Populus herbicide resistance, 155
selective agent for transgenic maize, 71
see also bialaphos
barley
α-amylase gene, 90
electroporation of mesophyll protoplasts, 90
foreign DNA stable integration, 90
foreign protein production, 91
protoplasts, 89–90
reporter gene expression, 90
stable transformation, 90–1
transformation, 93
transient gene expression, 89–90
barnase gene, 72
Basta
resistance, 120
see also phosphinothricin (PPT)
BCG, 10
persisting immune responses, 19
recombinant vaccine use, 19
Betula, 158
bialaphos, 71
see also bar gene
bialaphos resistance
gene, 55
navy bean, 111
biolistic transformation, 37
bioremediation, forest trees, 151
birch, 158
birdsfoot trefoil, 115–17
Agrobacterium rhizogenes transformation, 115, 116, 117
Agrobacterium tumefaciens transformation, 115–16, 117
CAT activity, 116
glutamine synthase gene, 120
glutamine synthetase-*uid*A gene fusion construct, 116
hairy root production, 116, 121
NPT II activity, 116
opines, 117
transformation, 118, *119*
black locust transformation, 157
Black Mexican Sweet (BMS) protoplast transformation, 69
Bradyrhizobium, 23
nitrogen fixation, 23
transformation, 25
Brassica napus, 125
microinjection technique, 67
broad bean, 109
hairy root induction, 109
NPT II activity, 109

cab gene, 59
CaCl₂/heat-shock protocol, 5
cadmium resistance plasmids, 15
CaMV 19S promoter, *Lotononis bainesii*
    transformation, 117
CaMV 35S promoter, 58, 84, 85, 102
  alfalfa transformation, 113
  birdsfoot trefoil transformation, 116
  conifers, 153, 154
  cowpea transformation, 110
  DNA containing, 104
  moth bean transformation, 109
  navy bean transformation, 112
  rapeseed, 131
  soybean transformation, 105
  Vicia narbonensis transformation, 109, 110
cat gene
  birdsfoot trefoil transformation, 115, 116
  Lotononis bainesii transformation, 117
  moth bean transformation, 108
  soybean transformation, 105
cell wall synthesis inhibitors, 11
cellobiohydrolase II, 43
cellulase
  gene transformation, 43
  production, 43
cellulose, 126
Ceratocystis ulmi, 151
cereal plants
  gene transfer techniques, 81
  protoplasts, 81
cereal plants, small-grain, 81
  artificial promoter, 84
  C residues at CpG dinucleotides, 86
  constitutive monocot promoters, 84
  CpG islands, 93
  DNA delivery methods, 85–9
  DNA macroinjection into floral tillers, 85–7
  DNA transfer via pollen tube pathway, 87–8
  electrophoretic migration of DNA, 89
  electroporation, 88, 89
  embryonic suspension cell cultures, 83
  foreign DNA, 86, 87
  gene expression requirements, 93
  illegitimate recombination, 84
  laser microbeam technique, 89
  marker genes, 82, 84–5
  mesocotyl-derived suspension cultures, 83
  microinjection, 89
  microprojectile-mediated gene transfer, 93
  particle gun technique, 83–4
  pEmu promoter, 85
  penetration markers, 82
  pollen-mediated indirect gene transfer, 88
  polyethylene glycol (PEG), 89
  promoters, 84–5
  protoplasts, 82–3, 93

  regeneration of transformed cells to mature plants,
    92
  reporter genes, 83, 86
  seed inhibition, 88
  silicon carbide fibers, 89
  stabilization of foreign genes, 82
  stable integration of DNA, 92
  stable transformant selection, 85
  stable transformation, 81
  suspension culture embryogenicity, 83
  transformation, 89–92
  transformation with *Agrobacterium*, 89
  transformation methods, 82–9, 93
  transgene inheritance, 82
  transient expression boosting, 84
cerulin, 11
chestnut blight, 151
chitinase gene, 120
Chlorsulfuron, 71
choline esters, 126
chromosome domains, 75
chymosin complementary DNA, 43
Clavibacter flaccumfaciens, 15
clone bank, 29
cloning vectors, 17, *18*
  function removal, 26–7
coat protein (CP) gene, 60–1
ColE1 plasmid, 27
  host range, 4
ColE1 replicon, 18
complementation analysis, 28
coniferin, 153
conifers
  abscisic acid-inducible promoter, 153
  adventitious shoot buds, 154
  Agrobacterium-mediated transfer, 152–3
  callus growth, 152
  crown gall bacteria, 152
  microprojectile bombardment transformation,
    153–4
  NPT II activity, 152
  pine family transformation, 152
  spruce transformation, 152, 153
  transformation, 152–4
  transgenic, 154
  xylem, 154
conjugal receptor function, 3
conjugal transfer, DNA, 7
conjugation
  coryneform bacteria, 15–16
  genes, 4
  nocardioform bacteria, 15–16
conjugative plasmids, mutagenesis, 7
Coprinus cinereus homologous integration, 39
Coprinus lagopus lithium acetate transformation, 36
Corynebacterium diphtheriae, 10
Corynebacterium glutamicum transformation efficiency, 13

Corynebacterium integration by homologous
    recombination, 17
Corynebacterium xerosis, 10
coryneform bacteria, 10–19
    cloning systems, 10
    cloning vectors, 17, *18*
    conjugation, 15–16
    cosmid-type vector, 16
    DNA introduction, 11, *12*, 13, *14*, 15–16
    electroporation, 13, *14*, 15
    illegitimate recombination integration, 17–18
    integration of nonreplicative, 17
    lysozyme action sensitivity, 11
    PEG-mediated DNA uptake in
        protoplasts/spheroplasts, 11, *12*, 13
    plasmids, 16–17
    self-transmissible extrachromosomal elements,
        15
    transduction, 16
cosmid library for *Aspergillus nidulans*, 41
cosmid rescue, 41
cotransformation
    filamentous fungi, 38
    gene introduction into rice plants, 57
cowpea, 110–11
    GUS activity, 111
    transformation, 118, *119*
cowpea mosaic virus (CPMV) mRNA, 110
CpG islands, 93
Cronartium ribicola, 151
crop improvement
    genetic engineering, 59
    genetic transformation, 53, 59
crop species, major, 53
crown gall
    bacteria, 152
    tumors, 138, 143
Cryptococcus neoformans, 36
cytokinin, regeneration induction, 139

2,4-D, 115, 127
Dactylis glomerata, 92
Dc8 promoter, 153
Dendroctonus frontalis, 151
δ9 desaturase, 134
2,4-diacetylphloroglucinol, 6
dicotyledenous plants
    gene expression regulation, 53
    transgene inheritance, 82
diphtheria, 10
divalent cations, 3
DNA
    cloning vectors, 23–4, 27–8
    conjugal recipient entry, 3
    conjugal transfer, 7
    conjugation introduction into
        nocardioform/coryneform bacteria, 15–16

    direct transformation, 8
    electroporation, 13, *14*, 15
    fate of transforming, 16–18
    imbibition into dry embryo, 88
    integron, 17
    linearized, 40
    microinjection techniques, 67
    PEG-mediated uptake in protoplasts/spheroplasts,
        11, *12*, 13
    stability of introduced sequences, 40
    transduction introduction into
        nocardioform/coryneform bacteria, 16
    transformation efficiency, 13
    transformation introduction into
        nocardioform/coryneform bacteria, 11, *12*, 13, *14*,
        15
    uptake by protoplasts, 13
    uptake by spheroplasts, 13
    see also transforming DNA
DNA introduction into bacteria, 24–7
    applications, 28–9
    conjugation, 26–7
    electroporation, 25–6, 29
    nocardioform/coryneform, 11, *12*, 13, *14*, 15, 16
    plasmid rearrangement, 28
    transduction, 27
    transformation, 24–5
T-DNA *see* transferred DNA
Douglas fir
    Agrobacterium virulence cascade, 153
    transformation, 152
dry bean, 111–12
    hairy root culture, 111
    transformation, 118, *119*
    tumor induction, 111
Ds element with *Ac* transposase gene, 59–60
Dutch elm disease, 151, 158

early methionine gene, 153
EDTA, *Pseudomonas* exposure effects, 7
electroduction, 26
electroporation, 3, 8, 29
    barley protoplasts, 90
    clone bank, 29
    DNA introduction into rice protoplasts, 55
    electric parameters, 15
    filamentous fungi, 26–7
    Gram-negative soil bacteria, 25–6
    high voltage of intact cells, 13, *14*, 15
    moth bean transformation, 108
    pea transformation, 108
    rapeseed transformation, 132
    small-grain cereals, 88, 89
    sunflower protoplasts, 144
    transformation efficiency, 15
elm, hybrid pioneer, 158
Em gene, 153, 154

embryonic suspension cell cultures
  protoplast transformation in rice, 54–5
  small-grain cereals, 83
endosperm, starch composition, 60
Endothia parasitica, 151
enhanced plasmid transformation efficiency (EPT)
    phenotype, 15
5-enolpyruvylshikimate synthase, 155
erucic acid, 126
Erwinia, 23
  electroporation, 26
Escherichia coli, 3
  DNA uptake, 25
  shuttle plasmids, 15–16
  transformation frequency, 26
ethylenediaminetetra-acetic acid *see* EDTA
Eucalyptus, 154
  transformation, 156

fasciation, 10
  induction, 19
fasciation-inducing genes, 19
Festuca arundinacea, 92
filamentous fungi, 34–44
  autonomously replicating vectors, 38–9
  cotransformation, 38
  dominant genes, 37, *38*
  electroporation transformation, 26–7
  gene disruption, 42
  genetic manipulation, 34
  genetic purification of transformants, 38
  genomic integration of circular plasmids, 39
  homologous recombination, 39
  lithium acetate transformation, 36
  marker genes, 37
  one-step disruption, 42
  particle bombardment in transformation, 27
  plasmid integration patterns, *40*
  polyethylene glycol-mediated transformation, 34–6
  protoplasts in transformation, 34
  self-cloning, 40–2
  transformant selection, 37–8
  transformation, 39
floral tillers, DNA macroinjection, 85–7
forage legumes
  alfalfa, 112–15
  birdsfoot trefoil, 115–17
  Lotononis bainesii, 117–18
  sanfoin, 118
  transformation success, 118–19
forest trees, 150–9
  abiotic stress, 151
  actinorhizal, 157
  Bacillus thuringiensis, 151
  biomass production, 156–7
  bioremediation, 151
  conifer transformation, 152–4

DNA transfer, 150–1, 152
ecological impact of genetic engineering, 159
energy production, 156
epidemic disease, 151
fertility control, 151
fuel production, 156–7
genetic engineering, 150, 151–2, 157–8
genetic improvement, 150
growth control, 151
hardwoods, 154–8
improvement, 150–1
insect pests, 151
juvenile mature correlation, 151
long-term stresses, 150
nitrogen fixation, 157–8
photoperiod, 151
plantlet formation, 158
proteinase inhibitors, 151
regeneration cells, 158
reproductive cycle length, 151
societal impact of genetic engineering, 159
transformation efficiency, 158–9
transformation systems, 151
fungi, transformation in soil species, *35*

Gaeumannomyces graminis var. *tritici*, 7
gene
  constitutive monocot promoters, 84
  disruption, 42
  introduction of new desirable, 28–9
  regulatory sequences of introduced, 75
gene cloning
  complementation, 41–2
  cosmid rescue, 41
gene expression
  cross-species, 43
  transient in wheat, 91
gene function analysis, 40–4
  gene disruption, 42
  gene replacement, 42
  titration of *trans*-acting gene products, 42
gene replacement, 42
  homologous recombination, 28
genetic engineering, forest trees, 157–8
germline transformation targets, transgenic maize,
    66–7
gibberellic acid (GA_3), 90
*gla*A gene, 43
β-glucanase gene, 120
glucoamylase, 43–4
glucosinolate, 126
β-glucuronidase (GUS)
  activity in fiber-mediated transformation of maize,
    68
  expression in transgenic rice plants, 58, 59
  gene, 86
  rice protoplast activity, 56–7

sunflower transformation, 143
glutamine synthase gene, 120
glutamine synthetase-*uid*A gene fusion construct, 116
glycerol, 13
glycine, 11, 13
Glycine max, 101–8
Glycine soya, 103
glyphosinate, 155
glyphosphate
    resistance, 120
    selective agent for transgenic maize, 71
grain legumes, 101–12
    broad bean, 109
    dry bean, 111–12
    lentil, 110
    moth bean, 108–9
    pea, 106–8
    peanut, 112
Gram-negative soil bacteria
    broad host range vectors, 23
    cloning vectors, *25*
    DNA introduction, 23
    DNA replicator region direct cloning, 29
    exogenous DNA uptake, 25
    introduction of new desirable genes, 28–9
    pGV910 replication, 24
    pSA vectors, 24
grasses, wound response absence, 81
guinea grass transformation, 92
*gus*A gene, 153, 154
    Populus expression, 155
    yellow poplar transformation, 157
gypsy moth, 151
    Populus target, 156

hardwoods
    cultivated, 154
    transformation, 154–8
Helianthus annuus, 137
Helianthus tuberosus, 137
herbicide resistance
    Populus, 155
    sunflower, 145
heterokaryons, 38
Hm^R
    calli, 55, 57
    phenotype, 59
    see also hygromycin
homologous recombination, integration by, 17
host range
    broad, 4
    narrow, 4, 5
hph gene, 55–6, 57, 92
    Agrobacterium tumefaciens transformation of pea, 106–7
    see also hygromycin
*hsp*60 expression signals, 19

hybridization, interspecific, 145
hydrogen-autotrophic growth, 15
hydrogen-auxotrophic growth, 16
hydrogenase gene, 29
hydroxyacetosyringone, 153
hygromycin
    alfalfa transformation, 113
    pea transformation, 106, 107
    phosphotransferase (HPT) gene, 89
    rapeseed transformation, 131
    resistance gene *see hph* gene
    selective agent for transgenic maize, 71
    Vicia narbonensis transformation, 109–10

ice nucleation active (Ina$^+$) epiphytic colonist, 5
immunosuppression, 10
ina gene, 5
*Inc*P (incompatibility group P), 4
    transfer, 16
*Inc*W plasmids, 4
insect resistance
    Populus, 156
    sunflower, 145
insecticidal genes, 120
integrative vectors, 19
integron, 17
introns, gene expression enhancement, 56–7
isonicotinic acid hydrazide (INH), 11
isopentyltransferase, 19

Jerusalem artichoke, 137

kanamycin
    alfalfa transformation, 112, 113, 115
    dry bean transformation, 111
    Lotononis bainesii transformation, 117–18
    moth bean transformation, 108, 109
    pea transformation, 106, 107
    rapeseed transformation, 131
    resistance gene *see npt*II gene
    sunflower regeneration, 142
    transgenic maize, 70–1

laser microbeam technique, 89
leafy gall, 10
leghemaglobin gene expression, 120
legumes, 101
    forage, 112–18
    grain, 101–12, 118
    hairy root formation, 119
    somaclonal variation, 120
    transformation success, 118–21
Lens culinaris, 110
lentil, 110
    transformation, 118, *119*
    tumor induction, 110
leprosy, 10

α-linoleic acid, 126
Lotononis bainesii, 117–18
  transformation, 118, *119*
Lotus corniculatus, 115–17
luc gene, 85
lucerne *see* alfalfa
Lymantria dispar, 151
lysostaphin, 13
lysozyme, 13

maize
  Agrobacterium-based vectors, 89
  agronomic performance, 73
  alcohol dehydrogenase (ADH), 58
  DNA introduction, 65
  embryo reception of gene transfer, 67
  embryogenic cultures, 69
  endophytic bacteria, 67
  inbred elite, 72–3
  insecticidal protein synthetic gene, 73
  marker gene delivery of egg cell, 67
  microspore-derived cultures for protoplasts, 69
  pollen grain, 66
  protoplast culture, 68
  T-DNA, 66
maize, transgenic, 65
  Agrobacterium-mediated DNA delivery, 65–6
  Agrobacterium-mediated transformation, 74
  agroinfection, 66
  applicability to commercial germplasm, 72–3
  Black Mexican Sweet (BMS) cells, 68
  culture productivity degeneration, 72
  direct gene transfer, 68–70, 73–4
  electroporation, 72
  fertile, 75
  fertile plant production, 69
  fertility, 71–2
  friable embryogenic response, 73
  germline transformation targets, 66–7
  GUS activity, 68
  inheritance of transgene expression, 74–5
  kanamycin selection, 70–1
  microinjection, 67–8
  microprojectile bombardment, 55
  molecular characterization, 73–5
  *npt*II marker, 87
  particle bombardment, 70–2
  performance enhancement, 75
  phenotypic abnormality, 71–2
  protoplast transformation, 54, 68–70
  proven approaches, 68–72
  seed production, 71
  stresses, 71
  techniques, 75
  transformation efficiency, 75
  transgene expression, 74–5
  transgene integration, 73–4

transgene introduction, 73
whisker (fiber)-mediated transformation, 68, 75
wounding for DNA uptake, 72
zygotic proembryo microinjection, 67–8
marker genes, filamentous fungi, 37
Medicago sativa, 112–18
Medicago varia, 113
meiosis, transgene transmission, 73
microinjection techniques
  Brassica napus, 67
  rapeseed, 131
  small-grain cereals, 89
  soybean, 105
  sunflower, 144
  zygotic proembryo in maize, 67–8
microprojectile bombardment
  conifers, 153–4
  DNA containing CaMV 35S constructs, 104
  gold particles, 104
  *npt*II gene, 104
  soybean, 103–5
  sunflower, 143–4
  transgenic rice plant generation, 55
  *uid*A gene, 104
  see also particle bombardment
mitochondrial plasmid replicons, 38–9
mob::Tn5 system, 27
mobilization genes, 4
mobilizer strains, 4
monocotyledenous plants
  gene expression regulation, 53
  transgenic, 53
moth bean
  protoplast transformation technique, 108–9
  transformation, 108–9, 118, *119*
Mucor, autonomously replicating sequence
    incorporation, 39
mutanolysin, 13
mycobacteria
  cloning vectors, 17, *18*
  phasmids, 16
  plasmids, 16–17
mycobacterial diseases, 10
Mycobacterium aureum transformation efficiency, 13
Mycobacterium avium, 10
Mycobacterium bovis bacille Calmette-Guérin (BCG),
    10
Mycobacterium leprae, 10
Mycobacterium smegmatis, integration by homologous
    recombination, 17
Mycobacterium tuberculosis, 10
  illegitimate recombination, 18
mycolic acid, 11

navy bean, 111–12
  GUS activity, 111
Nectria haematococca, 44

gene cloning, 41
neem, 158
Neurospora crassa, 34
  autonomously replicating sequence incorporation, 39
  electroporation transformation, 36
  gene identification method, 41
  homologous integration, 39
  inositol-requiring mutant, 34, 37
  lithium acetate transformation, 36
  particle bombardment, 37
  repeat-induced point mutation (RIP), 40
nitrogen fixation, forest trees, 157–8
nocardioform bacteria, 10–19
  cloning systems, 10
  cloning vectors, 17, *18*
  conjugation, 15–16
  DNA introduction, 11, *12*, 13, *14*, 15–16
  electroporation, 13, *14*, 15
  illegitimate recombination integration, 17–18
  integration of nonreplicative, 17
  lysozyme action sensitivity, 11
  pathobiology of infection in plants, 18–19
  PEG-mediated DNA uptake in protoplasts/spheroplasts, 11, *12*, 13
  plasmids, 16–17
  self-transmissible extrachromosomal elements, 15
  transduction, 16
nonreplicative vector integration, 17
nopaline
  rapeseed transformation, 130
  synthase activity, 66
nos gene
  birdsfoot trefoil transformation, 116
  conifers, 154
  soybean transformation, 104
Novozym-234, 34
*npt*II gene, 55, 85
  Agrobacterium tumefaciens transformation of pea, 106–7
  alfalfa transformation, 112, 113
  birdsfoot trefoil transformation, 116
  conifer transformation, 154
  cowpea transformation, 110
  macroinjection into floral tillers, 86
  marker for DNA transfer to pollen tube, 87, 88
  microprojectile bombardment, 104
  moth bean transformation, 108
  peanut transformation, 112
  Populus expression, 155
  reporter gene in barley, 90
  soybean transformation, 101, 102, 103
  sunflower regeneration, 142
nucleic acids, microprojectiles, 37

oaks, 158
oat gene transfer, 91–2

octopine
  rapeseed transformation, 130
  synthase activity, 66
oilseed production, 125
  see also rapeseed
oleic acid, 126
one-step disruption, 42
Onobrychis viciifolia, 118
oocyte regeneration, 141
opines
  birdsfoot trefoil transformation, 117
  synthesis, 120
  utilization genes, 120
orchardgrass transformation, 92
  protoplast, 54
*ori*T, 4
*ori*T$_{RK2}$ vector mobilization, 7
Oryza sativa, 53
osmotically sensitive cells (OSCs), 34
ovalbumin gene
  alfalfa transformation, 113–14
  expression in alfalfa, 120

Panicum maximum, 92
paromomycin, 142
particle bombardment, 83–4
  barley transformation, 93
  cell transformation rate, 70
  DNA coated, 70
  efficiency, 71
  illegitimate recombination, 84
  microcarriers, 83
  oat gene transfer, 91–2
  sorghum transformation, 93
  transgenic maize, 70–2
  wheat gene transfer, 91
  see also microprojectile bombardment
pathogenicity determinants, 44
pBR322, 4
  host range, 4
pBR322-*ori* plasmid, 5
pea, 106–8
  Agrobacterium rhizogenes transformation, 107
  Agrobacterium tumefaciens transformation, 106, 107
  electroporation, 108
  GUS expression, 107
  NPT II activity, 107
  protoplast transformation technique, 108
  transformation, 118, *119*
  tumor induction frequency, 106, 107
peanut, 112
  transformation, 118, *119*
pEmu promoter, 85
penicillin biosynthetic gene cluster cloning, 43
penicillin G, 11
pGV910 replication, 24
Phaseolus vulgaris, 111–12

phasmids, mycobacterial, 16
phenazine derivatives, 7
phenazine-1-carboxylic acid, 7
phleomycin resistance gene, 17
phosphinothricin (PPT), 142
    rapeseed transformation, 132
    resistance, 71
phosphotricine resistance, 155
phytopathogenicity, 3, 16
Phytophthora
    particle bombardment, 37
    vector for transformation, 39
pin2 gene, 156
pisatin demethylase, 44
Pisum sativum, 106–8
pJRD215, 24
plant defence genes, 156
plasma membrane permeability, high current-pulse
        effects, 25
plasma membrane pores, electroporation, 26
plasmids
    bacterial hosts, 4
    conjugative, 4
    extrachromosomal linear, 15
    fasciation-inducing linear, 19
    fungal, 38
    homologous integration, 42
    host range, 4
    instability, 6
    linear, 16
    self-transmissible, 27
    size and transformation frequency, 26
    transfer between bacterial strains, 26
    transfer of entire, 27
    transmissible, 4
    yeast, 38
pMON894, 102
pMON9749, 102
point mutation, repeat-induced (RIP), 40
pollen, 66
    regeneration, 141
pollen grain, DNA uptake, 66–7
pollen tube pathway, DNA transfer in small-grain
        cereals, 87–8
pollen-mediated indirect gene transfer, 88
polyethylene glycol (PEG)
    direct gene transfer to maize protoplasts, 68–9
    DNA introduction into rice protoplasts, 55
    electroporation, 13
    moth bean transformation, 108, 109
    small-grain cereals, 89
    sunflower transformation, 144
polyethylene glycol (PEG)-mediated DNA uptake in
        protoplasts/spheroplasts, 11, *12*, 13
Populus, 154–7
    DNA transfer, 155
    genome size, 154

*gus*A expression, 155
    herbicide resistant, 155
    insect resistance, 156
    *npt*II expression, 155
    transformation, 154–7
potato proteinase inhibitor gene, 156
pRJ1035 vector, 28
pRK290, 4, 23–4
    Agrobacterium transformation, 25
    cosmid, 24
    Rhizobium transformation, 25
pRK2013, 4, 24
protein secretion, 43
proteinase inhibitors, forest trees, 151
protoplast transformation
    embryogenic suspension cultures, 54–5
    transgenic rice plant generation, 54–5
protoplasts
    alfalfa transformation, 115
    barley, 89–90
    Black Mexican Sweet (BMS) origin, 69
    cereal plants, 81
    culture, 61
    direct gene transfer for maize transformation, 68–70
    DNA uptake, 13
    endogenous plasmid loss, 13
    feeder cell line interactions, 69
    filamentous fungi transformation, 34
    green plant regeneration, 91
    legumes, 119
    maize embryogenic cultures, 69
    maize endosperm origin, 69
    microspore-derived cultures, 69
    moth bean transformation, 108–9
    osmotic stabilizers, 35
    pea transformation, 108
    PEG-mediated DNA uptake, 11, *12*, 13
    production increase, 11, 13
    rapeseed transformation, 131–2
    small-grain cereals, 82–3, 93
    soybean transformation, 105
    stable transformation in barley, 90
    sunflower, 140–1, 144
    wheat, 91
pRSF1010, 16
pRT99gus plasmid, 154
pSA vectors, 24
*psb*A gene, 105
pseudomonads
    DNA introduction, 3
    genetic manipulation, 3
    phytopathogenic, 3
    rhizosphere colonizers, 3
    transformation, 5–7
Pseudomonas
    conjugal transfer, 3–5
    electroporation, 26

transformation, 3
Pseudomonas aeruginosa LEC1, 6
Pseudomonas aureofaciens strain 30–84, 7
Pseudomonas fluorescens
  strain 2–79, 7
  strain CHA0, 6–7
  strain HV37a, 6
  strain MS1650, 5–6
  strain Pf-5, 6
  transformation, 3
Pseudomonas syringae, 3
  pv. *syringae* J900, 5
  transformation, 3
pSUP suicide vectors, 4
pyoluteorin, 6
*pyr*F gene, 17
pyrrolnitrin, 6
Pythium ultimum, 6

Quercus robur transformation, 158

rapeseed, 125
  Agrobacterium rhizogenes transformation, 126–7
  Agrobacterium tumefaciens transformation, 127,
    130–1
  agronomic applications of transformation, 133, *132*,
    134
  auxin in regeneration, 130
  coculture time, 130–3
  cotransformation with *A. rhizogenes*/*A. tumefaciens*,
    127
  direct gene transfer, 131–3
  double zero cultivars, 126
  electroporation technique, 133
  gene transfer, 126
  hairy root production, 127
  herbicide-resistance, 134
  microinjection techniques, 131
  microspore-derived embryos, 130
  nuclear male sterility, 134
  oil quality, 125–6, 134
  polymerase chain reaction (PCR) techniques, 127
  protoplasts, 131–3
  regeneration tissue, 130
  spring cultivars, 125, 133
  transformation efficiency, 133
  transformation experiments, *128–9*
  transformation techniques, 126–7, *128–9*, 130–3
  transgenic shoot selection, 131
  winter cultivars, 125, 133
rapeseed, transgenic
  disarmed T-DNA, 127
  fertile, 127
  transgene copy number, 131
*rbc*S promoter, 58
*rbc*S-*uid*A fusion gene, 58–9
recombination
  excisive, 6

  homologous, 6
  illegitimate, 82, 84
  site-specific, 75
reforestation, 158
regeneration
  adventitious from tissues, 139–40
  direct, 139
  efficiency, 141, 142
  from existing meristems, 140
  indirect, 139–40
  morphogenic callus, 140
  oocytes, 141
  plant form, 141–2
  plant properties, 141
  pollen, 141
  root induction, 141
  single cell, 140–1
  speed, 139
  sunflower, 145
  systems for sunflower, 142
  transgenic plants, 139–42
regulatory sequences, *cis*-acting, 42
repeat-induced point mutation (RIP), 40
replicons, Gram-negative, 29
reporter genes
  construct insertion, 84
  small-grain cereals, 83
RH2, 4
Rhizobium, 23
  complementation analysis, 28
  electroporation, 26
  homologous recombination, 28
  introduction of new desirable genes, 28–9
  nitrogen fixation, 23
  phage genomes, 27
  plasmid rearrangement, 28
  plasmid transfer, 27
  replication of recombinant molecules, 24
  site-specific recombination, 28
  transduction, 27
  transformation, 24–5
  vector stability, 28
Rhizoctonia solani, 6
rhizosphere colonizers, 3
rhodococci
  cloning vectors, 17, *18*
  electrotransformation, 17
  metabolic activity, 10
  plasmids, 16
Rhodococcus bronchialis, 10
Rhodococcus fascians, 10, 15
  illegitimate recombination, 18
  pathobiology of infection in plants, 18–19
Ri plasmid systems, legumes, 119–20
rice
  Ac element introduction, 59
  Agrobacterium-based vectors, 89

rice (*cont.*)
  genome Hm^R plasmid integration patterns, 57
  pests and *Bacillus thuringiensis* gene, 61
  pollen tube pathway DNA transfer, 87
  transformation, 53–4, 57
  transposon tagging system, 59–60
rice actin
  gene, 59
  promoter, 84
rice plants, transgenic, 53
  CaMV 35S promoter, 58
  cotransformation, 57
  direct DNA transfer, 54
  efficiency, 55
  gene expression regulation, 58–9
  gene regulation, 61
  generation methods, 54–8
  genome size, 53
  GUS expression, 58, 59
  integration patterns of foreign DNA, 57
  intron enhancement of gene expression, 56–7
  microprojectile bombardment, 55
  plasmid DNA rearrangement, 57
  protoplast transformation, 54–5
  *rbc*S promoter, 58
  *rbc*S-*uid*A fusion gene, 58–9
  restriction fragment length polymorphism (RFLP)
    maps, 53, 61
  rice stripe virus coat protein, 60
  *rol*C of Ri plasmid, 59
  selectable markers, 55–6
  trans-acting factors, 59
  transgene integration, 57
  transposable elements, 61
  undesirable mutations, 61
  useful gene introduction/expression, 59–61
rice protoplasts, 54–5
  culture, 61
  electroporation for DNA introduction, 55
  PEG for DNA introduction, 55
  transformation, *56*
rice stripe virus, coat protein (CP), 60–1
RK2
  deletion, 7
  derivative plasmids, 27
  primase gene, 7
  tetracycline resistance gene, 4
  tra gene absence, 27
*rol*C
  promoter, 59
  of Ri plasmid, 59
root induction, 141
RSF1010, 4
rye
  DNA macroinjection into floral tillers, 85–6
  gene transfer, 92

Saccharomyces cerevisiae, 34
  gene disruption, 42
  gene replacement, 42
  homologous integration, 39
  linearized DNA, 40
  lithium acetate transformation, 36
  particle bombardment, 37
Salix, 157
sanfoin, 118
  transformation, 118, *119*
seed inhibition, small-grain cereals, 88
self-cloning
  filamentous fungi, 40–2
  transformant selection, 41
shoot morphogenesis, sunflower, 139
shuttle vectors, 4, 16
  for pseudomonads, 5
sibling selection, 41
silicon carbide whiskers (fibers), 75
  small-grain cereals, 89
silver nitrate, 73
sodium dodecylsulfate (SDS), 11
somaclonal variation, legumes, 120
somatic embryogenesis
  sunflower, 139
  yellow poplar, 157
sorghum transformation, 92, 93
Southern pine bark beetle, 151
soybean, 101–8
  Agrobacterium rhizogenes transformation, 103
  Agrobacterium tumefaciens transformation, 101–3,
    105–6
  cyst nematode propagation studies, 103
  hairy root cultures, 103
  leghemaglobin gene expression, 120
  microinjection transformation technique, 105
  microprojectile bombardment, 103–5
  protoplast transformation by DNA direct uptake,
    105
  somaclonal variation, 120
  transformation, 118, *119*
  transposable element integration, 120
  *uid*A gene expression, 104
soybean, transgenic, 102
  atrazine resistance, 120
  embryogenic suspension culture, 104
  glyphosate resistance, 120
  GUS activity, 102, 104, 105
  hygromycin resistance, 104
  NPT II activity, 102, 103
  RI plants, 104
spheroplasts
  DNA uptake, 13
  endogenous plasmid loss, 13
  PEG-mediated DNA uptake, 11, *12*, 13
  production increase, 11, 13
*spo*CIC gene, 42

starch composition of endosperm, 60
sterility, nuclear male, 134
Stylosanthes transformation, 118, *119*
suicide vectors, 4, 6
sunflower, 137–46
   Agrobacterium tumefaciens transformation, 138, 145
   Agrobacterium-mediated gene transfer, 143–4
   auxin : cytokinin ratio for transformation, 143
   callus induction, 139, 140
   crown gall tumors, 138, 143
   culture-derived, 142
   direct gene transfer, 144
   direct somatic embryogenesis, 139
   electroporation technique, 144
   embryo microinjection, 144
   F_1 hybrid cultivars, 137, 138
   flower structure, 138
   gene transfer, 138
   growth cycle, 137–8
   GUS activity, 143
   HA 89B line regeneration, 142
   HA 300 line regeneration, 142
   herbicide resistance, 145
   hypocotyl explant transformation, 143
   in vitro culture, 142, 145
   indirect embryogenesis, 140
   insect attack resistance, 145
   interspecific hybridization, 145
   male sterile lines, 138
   meristem regeneration, 140
   microinjection, 144
   morphogenic callus, 140
   NPT II activity, 143
   particle gun transformation, 143–4
   pollinator lines, 138
   polyethylene glycol technique, 144
   protoplasts, 140–1, 144
   regenerated plant properties, 141–2
   regeneration, 139–42, 145
   shoot formation, 139, 140–1
   shoot morphogenesis, 139
   single cell regeneration, 140–1
   somaclonal variants, 142
   tissue culture, 145
   transformation, 138–9, 142–4
sweetgum transformation, 157

take-all, 7
thallium resistance plasmids, 15
thaumatin gene, 86
Thielaviopsis basicola, 6
Ti plasmids, 27
tobacco
   barnase gene expression, 72
   black root rot, 6
   kanamycin-resitant transformants, 74
   mosaic virus (TMV) coat protein, 60

NT1 cell line, 71
trans-acting gene product titration, 42
transconjugant, 3
   recovery frequency, 7
transcriptional regulators, *trans*-acting, 42
transduction, 27
   coryneform bacteria, 16
   nocardioform bacteria, 16
transfer efficiency, 4
transfer vector mutagenesis, 7
transferred DNA (T-DNA), 66
transformants
   genetic transformation, 38
   selection for filamentous fungi, 37–8
transformation
   biotechnology applications, 42–4
   conifers, 152–4
   direct, 8
   efficiency, 13
   electroporation efficiency, 13, 15
   forest trees, 151–2
   Gram-negative soil bacteria, 24–5
   hardwoods, 154–8
   stable, 81–2
   sunflower, 142–4
transforming DNA
   fate, 38–40
   integration into chromosomes, 39–40
transgene
   insertion site, 75
   integration in maize, 73–4
   silencing, 75
transgene expression
   inheritance, 74–5
   inheritance in maize, 74
   transgenic maize, 73, 74–5
transposable elements
   activator (*Ac*), 59
   dissociation (*Ds*), 59
   introduction into rice, 59
transposon
   mutagenesis, 6
   tagging, 59–60
Trichoderma reesei
   heterologous gene expression, 43
   protein glycosylation system, 43
   protein secretion, 43
Tris, *Pseudomonas* exposure effects, 7
*trp*C$^+$, 42
tuberculosis, 10
turfgrass transformation, 92

*uid*A gene, 58
   alfalfa transformation, 115
   barley, 90
   macroinjection into floral tillers, 86
   marker for oat gene transfer, 92

*uid*A gene (*cont.*)
   microprojectile bombardment, 104
   pea transformation, 108
   peanut transformation, 112
   small-grain cereals, 84, 85
   soybean transformation, 102, 105
Ulmus, 158

vectors
   autonomously replicating, 38–9
   building, 39
   cosmid-type, 16
   integrative expression, 19
Vicia faba, 109
Vicia narbonensis, 109–10
   transformation, 118, *119*
vicilin, 91
Vigna aconitifolia, 108–9
Vigna unguiculata, 110–11
vir gene, 27
viral resistance, cereals, 61

walnut transformation, 151
waxy gene, 60

wheat
   embryogenic tissue cultures, 91
   protoplasts, 91
   stable transformation, 91
   transient gene expression, 91
white clover transformation, 118, *119*
white pine blister rust, 151
willow, 157
wilting diseases, 10
wood, 150
   processed products, 154
woody plants
   genetic diversity, 150
   molecular basis of biological processes, 150
   short rotation crops, 156
   see also forest trees

Xanthomonas, 23
   electroporation, 26
xylem, genetic engineering, 154

yellow poplar transformation, 157

Zea mays, 65

Printed in the United States
By Bookmasters